AF767632

Single Event Effects in Aerospace

IEEE Press
445 Hoes Lane
Piscataway, NJ 08854

IEEE Press Editorial Board
Lajos Hanzo, *Editor in Chief*

R. Abhari
J. Anderson
G. W. Arnold
F. Canavero
M. El-Hawary
B-M. Haemmerli
M. Lanzerotti
D. Jacobson
O. P. Malik
S. Nahavandi
T. Samad
G. Zobrist

Kenneth Moore, *Director of IEEE Book and Information Services (BIS)*

Single Event Effects in Aerospace

Edward Petersen

IEEE

IEEE PRESS

WILEY

A John Wiley & Sons, Inc., Publication

Copyright © 2011 by the Institute of Electrical and Electronics Engineers, Inc.

Published by John Wiley & Sons, Inc., Hoboken, New Jersey. All rights reserved.
Published simultaneously in Canada

No part of this publication may be reproduced, stored in a retrieval system, or transmitted in any form or by any means, electronic, mechanical, photocopying, recording, scanning, or otherwise, except as permitted under Section 107 or 108 of the 1976 United States Copyright Act, without either the prior written permission of the Publisher, or authorization through payment of the appropriate per-copy fee to the Copyright Clearance Center, Inc., 222 Rosewood Drive, Danvers, MA 01923, (978) 750-8400, fax (978) 750-4470, or on the web at www.copyright.com. Requests to the Publisher for permission should be addressed to the Permissions Department, John Wiley & Sons, Inc., 111 River Street, Hoboken, NJ 07030, (201) 748-6011, fax (201) 748-6008, or online at http://www.wiley.com/go/permission.

Limit of Liability/Disclaimer of Warranty: While the publisher and author have used their best efforts in preparing this book, they make no representations or warranties with respect to the accuracy or completeness of the contents of this book and specifically disclaim any implied warranties of merchantability or fitness for a particular purpose. No warranty may be created or extended by sales representatives or written sales materials. The advice and strategies contained herein may not be suitable for your situation. You should consult with a professional where appropriate. Neither the publisher nor author shall be liable for any loss of profit or any other commercial damages, including but not limited to special, incidental, consequential, or other damages.

For general information on our other products and services or for technical support, please contact our Customer Care Department within the United States at (800) 762-2974, outside the United States at (317) 572-3993 or fax (317) 572-4002.

Wiley also publishes its books in a variety of electronic formats. Some content that appears in print may not be available in electronic formats. For more information about Wiley products, visit our web site at www.wiley.com.

Library of Congress Cataloging-in-Publication Data is available.

ISBN: 978-0-470-76749-8

oBook ISBN: 978-1-118-08432-8
ePDF ISBN: 978-1-118-08430-1
ePub ISBN: 978-1-118-08431-1
10 9 8 7 6 5 4 3 2 1

Contents

CHAPTER 1

Introduction

1.1 BACKGROUND

In recent years the dominant radiation effect in space-borne electronic systems has become the family of single event effects (SEEs). SEEs arise through the action of a single ionizing particle as it penetrates sensitive nodes within electronic devices. Single events can lead to seemingly randomly appearing glitches in electronic systems—frustrating errors that may cause anything from annoying (at best) system responses to catastrophic (at worst) system failures. The problem is particularly insidious due to the combination of its random nature, the omnipresent spectrum of high energy particles in space, and the increasing sensitivity of devices to SEEs as miniaturization progresses.

A SEE is a phenomenon that follows from the continuing trend in electronic device design toward higher density devices with smaller feature sizes. This trend permits faster processing of information with smaller required quantities of electric charge. As the charge involved has decreased, it has entered the region where corresponding amounts of charge can be generated in the semiconductor by the passage of cosmic rays or alpha particles. This charge can look like a legitimate signal, temporarily changing memory contents or commands in an instruction stream.

The single event upset (SEU) phenomenon was first suggested in 1962 by Walkmark [Wallmark 1962] and first reported in an operating satellite system in 1975 by Binder, Smith, and Holman [Binder 1975]. Both of these reports were generally ignored as they suggested

Single Event Effects in Aerospace, First Edition. By Edward Petersen.
© 2011 the Institute of Electrical and Electronics Engineers, Inc.
Published 2011 by John Wiley & Sons, Inc.

responses well out of the mainstream of radiation effect studies of the time. However, in 1978, May and Wood [May 1979] reported alpha particle upsets in dynamic RAMs and Pickel and Blandford [Pickel 1978] analyzed upsets in RAM circuits in space due to heavy ion cosmic rays. It was in this time period that IBM started major on-again off-again programs with alpha emitters and terrestrial cosmic rays [Ziegler 1979, Ziegler 1981, Ziegler 1996a, and Ziegler 2004]. In 1979 Guenzer, Wolicki, and Allas [Guenzer 1979], and Wyatt, McNulty, and co-workers [Wyatt 1979] experimentally observed upsets due to high energy protons such as those present in the Earth's trapped proton belts. Gradually, as more and more upset related problems have been observed in spacecraft, SEU has come to be recognized as a very serious threat to system operations. Radiation hardening of devices and SEU tolerance approaches have alleviated the problem somewhat. It is still grave. Single event upset must be considered in all future space, missile, and avionics systems.

The early "common knowledge" of the effects was based on two papers. The paper by Binder, Smith, and Homan presented the basic information for cosmic ray induced upsets [Binder 1975]. They discussed the basic mechanisms and circuit effects, the cosmic ray environment, including the effects of shielding, and a basic approach to cosmic ray event rate. A paper by Petersen presented the proton environment with the variation in altitude and the effects of the South Atlantic anomaly and included the effects of shielding [Petersen 1981]. He then discussed the possible contributions of the various proton reactions in silicon and presented calculations of proton induced upset rates.

Much of the interest has been driven by developments such as:

- The critical errors caused by cosmic ions in the Voyager and Pioneer probes.
- The necessary retrofits, at great expense, of the Landsat D and Galileo systems due to heightened concern over single event upsets.
- The errors in the guidance system of the Hubble space telescope as its orbit carries it through the earth's radiation belts, requiring frequent scrub and reload of the guidance system.
- The loss of the Japanese satellite "Superbird" due to SEU followed by operator error [AWST 1992].

Table 1-1 lists a sampling of other space programs for which single event effects have had an impact. Some of these events were collected by Bedingfield and co-workers [Bedingfield 1996]. Ritter has also discussed some of these events [Ritter 1996].

There was a parallel set of problems for ground based systems as described by Ziegler [Ziegler 2004]. Soft errors from radiation are the primary limit on digital electronic reliability. This phenomenon is now more important than all other causes of computing reliability put together. Since chip single event rates (SERs) are viewed by many as a legal liability (selling something that you know may fail), the public literature in this field is sparse and always makes management nervous [Ziegler 2004, Preface and Chapter 1].

Table 1-1
A Few of the Spacecraft for Which Single Event Effects Have Had an Impact

For the Period 1970–1982			
DE-1	Galileo	INSAT-1	Intelsat - IV
Landsat-D	LES 8	LES 9	Pioneer Venus
SMM	Tiros-N	Voyager	
For the Period 1982–1990			
AMTE/CCE	DSCS	ERBS	Galileo Lander
GEOS-6	GEOS-7	Geosat	GPS 9521
GPS 9783	GPS 9794	HUT	IUS
MOS-1	OPEN	Shuttle	SPOT-1
TDRS-1	TDRS-4	UOSAT-2	
For the Period 1990–1997			
COBEERS-1 (SEL)	ETS-V (SEL)	ADEOS	
EUVE	HST	HST-STIS	Kitsat-1
NATO-3A	PoSAT-1	S80/T	SOHO
Spot-2	SPOT-3	STS-61	Superbird
TDRS-5	TDRS-6	TDRS-7	Topex/Poseidon
UoSAT-2	UoSAT-3	UoSAT-5	WIND
Yahkoh-BCS			
Amateur Radio Satellite Experiments			
AO-16	LO-19	I0-26	Spartan/OAST/SPRE

SEU in space originates from two sources in the natural environment. Satellites at geosynchronous orbit and corresponding regions outside Earth's radiation belts experience upsets due to heavy ions from either cosmic rays or solar flares. The natural cosmic ray heavy ion flux has approximately 100 particles/cm^2 per day. In very sensitive devices this flux can lead to daily upsets. Many devices can be upset in these environments at a rate of about 10^{-6} upsets/bit-day.

Also, upsets can occur within the proton radiation belts. Even though energy loss rates by direct ionization from protons are too low to upset most devices, proton induced nuclear reactions in silicon can result in heavy recoil nuclei capable of upsetting most memory cells. About one proton in 10^5 will undergo a nuclear reaction capable of SEU. Considering the large population of high energy protons capable of causing these reactions, proton induced upsets become a significant SEU mechanism. In the heart of the proton belts there are about 10^7 to 10^9 protons/cm^2 per day with energies above 30 MeV (approximately the minimum energy that will penetrate a spacecraft and then cause upsets). Thus, for example, a 1K memory with a proton upset cross section of 10^{-11} cm^2 per bit would have 10 upsets per day in the most intense part of the belt.

As the cosmic rays penetrate the atmosphere, there is a chain of nuclear reactions that produce high energy neutrons and protons. A nominal figure is 6000 neutrons per square centimeter per hour at 40,000 feet altitude and 45 degrees latitude. In the 1990s these were shown to produce single event upsets in complex integrated circuits in avionics equipment.

There are a variety of possible single event effects (SEEs). These are important as they can cause malfunctions in microelectronics devices operating in the space ionizing radiation environment. The principal effects are upset, transients, and latchup, but the others need also to be kept in mind. The basic effects are as follows:

SEU	UPSET	Temporary change of memory or control bit
SET	TRANSIENT	Transient introduced by single event
SEL	LATCHUP	Device latches in high current state
SES	SNAPBACK	Regenerative current mode in NMOS
SEB	BURNOUT	Device draws high current and burns out
SEGR	GATE RUPTURE	Gate destroyed in power MOSFETs
SEFI	FUNCTIONAL INTERRUPT	Control path corrupted by an upset
MBU	MULTIBIT UPSET	Several bits upset by the same event

Single events acquire that name because they depend on the interaction of a single particle. Most other radiation effects depend on the dose or damage deposited by large numbers of particles. SEEs can be caused by the passage of a single heavy ion—a cosmic ray in space, for example. As the cosmic ray passes through the silicon of the device, it deposits a track of ions. In space the cosmic rays are ordinarily energetic enough that they pass through the device. If these resulting ions are in the presence of the natural or applied field in an electronic device, they are collected at the device electrodes. See Figure 1-1. This produces an electric pulse or signal that may appear to the device as a signal to which it should respond. If the electrical characteristics of the device are such that the signal appears valid, then there may be a bit upset or the production of a signal in a logic device that triggers a latch later in the device.

High energy protons can also initiate single event effects. It is not the proton passage that produces the effect. The proton itself produces only a very small amount of ionization. Very few devices are sensitive enough to respond to the proton ionization. However, 1 proton in 10^5 will have a nuclear reaction in the silicon device. These reactions can produce heavy ions that in turn can deposit enough energy to cause upset. See Figure 1-2. Although this seems like a very small number of cases, in space the protons in the proton radiation belts are intense enough so that they can cause many more upsets than the heavy ion cosmic rays in the same environment.

The basic concepts are similar for both heavy ion and proton induced upsets. The prime emphasis in the present work is the heavy ion induced upsets. We will also discuss proton and neutron upsets for comparison later. The prediction of single event effect rates depends on

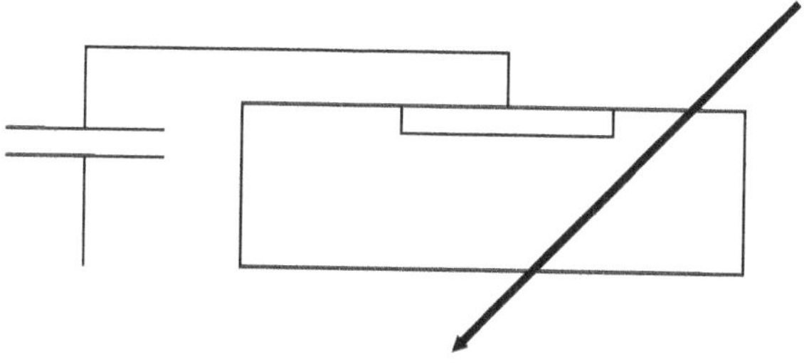

Figure 1-1 Ionization path due to direct passage of a heavy ion.

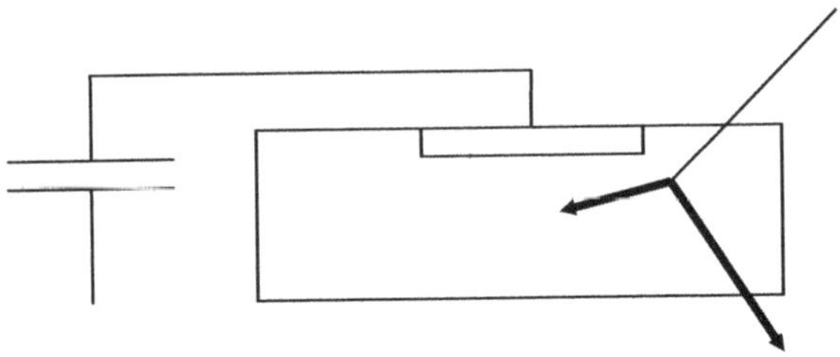

Figure 1-2 Ionization paths due to proton reaction in a device.

a number of independent models of various aspects of the phenomena involved.

Single event effects can be thought of as one of nature's ways of enforcing Murphy's Law. They can occur at any time, at any place in an electronic system. They do not depend on the cumulative exposure to the space environment and are as likely to occur during or shortly after launch as after a long time in orbit. As for location, the sorrowful words of one space system designer express it: "I know that you said I was going to have upsets, but I didn't expect an upset in *that* bit."

Because of these and other real world problems in space systems due to cosmic ions, an understanding of single particle errors in integrated circuit (IC) electronics has become an important part of the design and qualification of IC parts for space-based use.

The issue becomes even more important as device dimensions scale, and denser, more powerful integrated systems are placed in space or satellite applications. Electronics is reaching integration levels where a single bit of information is represented by an extremely small value of charge, and noise margins are very tight. For example, if a typical dynamic random access memory (DRAM) cell can tolerate approximately 100 mV of noise on the bit storage node with 100 fF (10^{-15} farads) of storage capacitance, then this value of noise corresponds to a charge of only 62,500 electrons. Any perturbation of this delicate balance by an impinging cosmic ion is intolerable. So, a recognition of, and familiarity with, the effects of space radiation on the electronics to be placed in that hostile environment is essential. Single event modeling plays a key role in the understanding of the observed-error mechanisms in existing systems, as well as the prediction of errors in newly designed systems.

There are two different aspects of interest. First is the analysis of various types of single event experiments to help understand the phenomena. Second is the modeling of the various aspects of the phenomena that allow prediction of SEE rates in space.

1.2 ANALYSIS OF SINGLE EVENT EXPERIMENTS

1.2.1 Analysis of Data Integrity and Initial Data Corrections

Single event experiments involve a mixture of electronic measurements and nuclear physics experimental techniques. This leads to a number of possible errors in the measurements, some of which are not common in standard electrical engineering experiments. Therefore, the first step in the analysis of SEEs data is the examination of the integrity of the data itself.

There are some aspects of the data that are an artifact of experimental approach. The data needs to be corrected for these before it is used for analysis and prediction.

1.2.2 Analysis of Charge Collection Experiments

The initial studies of SEEs assumed that it took a unique amount of charge to cause the effect. This corresponded to a step function as one examined device sensitivity as a function of energy deposition. The energy deposition was measured by the linear energy transfer (LET) of the ionizing particle (see Section 2.2). However, all of the experiments indicted an "S" curve of the variation of SEE rate as a function of LET. It was initially assumed that this was caused by a variation from memory cell to memory cell. Then a number of charge collection experimental and theoretical studies indicated that the variation in sensitivity corresponded to a variation across the transistor and across the memory cell.

1.2.3 Analysis of Device Characteristics from Cross-Section Data

The SEE cross-section curve is a reflection of the various charge collection processes and circuit characteristics that are involved in the

SEE. Study of the characteristics of this curve gives some insight into the processes involved in a particular device.

1.2.4 Analysis of Parametric Studies of Device Sensitivity

There are a large number of device parameters that can be adjusted when testing the device. These may be parameters that affect device design, or factors corresponding to different modes of operation in space. The variation in the experimental cross sections with parametric changes gives a direct measure of the space SEE rates dependence on these factors.

1.3 MODELING SPACE AND AVIONICS SEE RATES

1.3.1 Modeling the Radiation Environment at the Device

This involves modeling the proton and heavy ion radiation environment in space and the neutron environment in the atmosphere. (We will often refer to modeling in space, with the understanding that the discussion refers to both space and avionics environments.) The sources are cosmic rays, solar particles, and geomagnetically trapped radiation. Basic factors are intensities, elemental compositions, and energy distributions. The environment at the device is influenced by the component material shielding surrounding the device, geomagnetic shielding effects (which depend on orbit parameters), and time variations associated with the solar cycle. For single event effects by heavy ions, it is really the energy loss characteristics of the environment that are important. Therefore, the cosmic ray composition and energy variation are translated to the number of particles as a function of their energy loss per unit path length (dE/dX), normalized to the material and called linear energy transfer (LET). The LET spectrum can be expressed in either a differential form (number of particles with given energy loss) or in an integral form (number with energy loss greater than a given LET).

1.3.2 Modeling the Charge Collection at the Device

The upset depends on the amount of charge that the electronic circuit detects. This in turn depends on the amount of charge deposited by the ion. It is not sufficient for an ion to hit a device or an individual nucleus in the device, but instead it must have an appreciable track length. The amount of charge deposited depends on the product of the path length and the LET of the ion. Therefore, one must model the sensitive volume of the device. That is, you must know the area, depth, and shape of the sensitive region. With this information, the distribution of possible path lengths in the device can be modeled. These path length distributions can be expressed in either a differential or an integral form, that is, number of paths of a given length, or number of paths that equal or exceed a given length.

The charge collection may be complicated by the fact that there are several possible charge collection processes that can take place. The charge may be collected by the intrinsic and applied field along the track in the device itself. There may be additional charge collected as charge diffuses into the device. It is also possible that the fields are distorted along the ion track, leading to additional charge in the track being brought into the device.

1.3.3 Modeling the Electrical Characteristic and Circuit Sensitivity for Upset

The circuit that the device is embedded in determines what the applied pulse shape must be if it is to appear as a legitimate electrical signal. In most cases, these effects can be modeled by standard SPICE circuit modeling of the circuit. In some case, there are more complicated interactions between the charge collection and the circuit, so that the effects are better modeled if considered together.

It is not always possible to obtain enough information about the detailed device characteristics to perform electrical modeling. It has now become common to base the estimates of device sensitivity on experimental upset measurements in the laboratory.

The prediction of rate effects that occur in space then depends on models that include three basic factors: the environment, the device dimensions, and the device sensitivity. There are possible complications that involve the details of the shape of the device sensitive volumes and of the charge collection. We will analyze the various approaches to upset rate predictions, and examine how they consider both the basic factors and the complications.

One of the modeling complications that we will discuss is the treatment of charge collection in the device. The basic aspects of charge collection are included in all models. However, there are a number of possible complications that can have a significant impact on the interpretation of experimental data. These factors can also sometimes impact the rate predictions. These factors will be introduced where appropriate.

1.4 OVERVIEW OF THIS BOOK

We present an overview of the various factors and approaches that are important for single event effect measurements, rate modeling, and predictions. We then develop the concepts that are important for rate prediction, develop the history of rate predictions, outline the approaches that have developed, and present standards that are being developed. We have an extensive discussion of the interpretation of experimental results. We will also discuss some of the issues that remain.

One of the possible complications is the charge collection in the device. The basic aspects of charge collection are included in all models. We will consider the charge collection as it shows up in the cross-section measurements, but not consider the various detailed micromodeling approaches to the problem.

The book presents an overview of the methods and procedures involved in computer modeling of single event phenomena that have been proposed and utilized in recent years. The goal of these procedures is to model the interaction of a radiation environment with microelectronic circuits, and to predict the resulting influences on proper IC operation.

This book is based on published materials but does include some new material and some new examples. We hope that some of the examples will be helpful for newcomers in the field. Other examples are aimed at those experienced in the field who we believe have misunderstood some of the concepts of the integral rectangular parallelepiped (IRPP) approach to upset rate calculations.

1.5 SCOPE OF THIS BOOK

This book is intended to be a tutorial, covering the basic terminology and concepts of single event effects and rate prediction. It is based on the notes for 2008 NSREC Short Course with additions, corrections, and indexes [Petersen 2008b]. It is hoped that it presents enough information for the reader to find the relevant literature of a specific topic of interest so that he/she can study it in depth. The book will go into some depth on the issues that lay at the foundation of the subject or that are important for SEU measurements or for the interpretation of SEU measurements. Attention is paid to a number of experimental aspects from nuclear physics that may not be covered in electrical engineering courses. In some sections we may go into too much detail for the reader. Skip those sections until you need them.

There is little discussion of the actual effects in devices or circuits. It is possible to write a book or major review article on modeling and simulation of the electronic aspects of single event effects that has virtually no overlap with this book [Fleetwood 2004, Lacoe 2008, Munteanu 2008, Pease 2008]. The electronic modeling has been well presented in an IEEE NSREC (Institute of Electrical and Electronics Engineers, Nuclear and Space Radiation Effects Conference) short course by Lloyd Massengill [Massengill 1993c]. The short course presentation of Sexton is also highly recommended [Sexton 1992]. A very good update of Massengill is given by Black and Holman [Black 2006]. Reed discusses some of the basic physics underlying the electrical effects [Reed 2008].

There is a wealth of information in important IEEE NSREC short course presentations and other reviews. [Benedetto 2008, Bourdarie 2008, Hafer 2008, Munteanu 2008, Petersen 2008b, Reed 2008, Wilkinson 2008, Kastensmidt 2007, Ladbury 2007a, Black 2006, Law 2006, Santin 2006, Xapsos 2006b, Bauman 2005, Buchner 2005, Cressler 2003, Oldham 2003, Weatherford 2002, Buchner 2001, Hoffmann 2000, Dodd 1999, Dressendorfer 1998, Barth 1997, Petersen 1997b, Alexander 1996, Galloway 1996, Ritter 1996, Stapor 1995, Normand 1994a, Massengill 1993c, Sexton 1992,

McNulty 1990, Petersen 1983c, Pickel 1983]. The reader who wants a comprehensive background in single event effects should study these. At the same time, be aware that our knowledge has been changing continuously. The material in this book has been changed significantly from the corresponding material in 1997 [Petersen 1997b].

CHAPTER 2

Foundations of Single Event Analysis and Prediction

Before tackling the details of single event analysis and rate prediction, we will introduce and discuss some of the terminology used throughout this text. A portion (Section 2.1) of the summary was reprinted (with author's permission) from Massengill [Massengill 1993c, © 1993 IEEE].

2.1 OVERVIEW OF SINGLE PARTICLE EFFECTS

When an energetic nuclear particle penetrates any semiconducting material, it loses energy through Rutherford scattering (Coulombic interactions) with the semiconductor lattice structure. Through predominantly Compton interactions with the nuclei of the crystalline structure, the slowing of the particle as it transfers energy to the lattice leaves an ionization trail of free electron-hole pairs; mobile charge carriers that were electrically nonexistent before the radiation event. Within an integrated circuit structure, these excess carriers can deposit charge in unexpected and unwanted places, often leading to voltage transients on the nodes of the circuit and current transients across device junctions.

Single Event Effects in Aerospace, First Edition. By Edward Petersen.
© 2011 the Institute of Electrical and Electronics Engineers, Inc.
Published 2011 by John Wiley & Sons, Inc.

Unlike *total dose* radiation, which causes gradual global degradation of device parameters, and *dose-rate* radiation, which causes photocurrents in every junction of a circuit, a *single event* interaction is a localized effect and can lead to a seemingly spontaneous transient within a region of the circuit. If this transient influences a node that is storing information, it may lead to an *upset*; that is, the corruption of the information to an unrecognizable, unreadable, or unstable state. This upset can, in turn, lead to a *circuit error* if this corrupted state alters legitimate information stored in or propagating through the circuit. That is, an upset becomes an error when it is either latched or is misinterpreted as valid data by other circuitry. The working definition of upset in this work is a corrupted electrical state, and an error is the finalized effect of that state.

Localized information errors due to single event upsets (SEUs) can be (1) *transient*, (2) *permanent*, or (3) *static*. *Transient errors* are spurious signals that can propagate through the circuit paths during one clock cycle. These asynchronous signals can either propagate to a latch and become static, or be overwhelmed by the legitimate synchronous signals of the circuit. Timing of the radiation induced signals relative to the synchronous signals plays a key role in the possibility of errors. These types of errors are most important in combinational (nonsequential) circuitry and analog subsystems.

Permanent errors are often called *hard errors* because of their destructive, noncorrectable origins. In this case, the single event causes physical damage to the circuit, leading to a noncorrectable fault. Single event (SE) induced burnout (*SEB*) and gate rupture (*SEGR*) in power transistors are examples of hard errors. These errors are most often analyzed and modeled at the individual device level.

Single event soft errors (due to single event upsets, SEUs) and *multiple-bit soft errors* (due to multiple-bit upsets, MBUs) belong to a class of errors that are **static** (latched by the circuitry) but can be corrected by outside control. These soft errors overwrite information stored by the circuit, but a rewrite or power cycle corrects or resets the part to proper operation with no permanent damage.

A special class of single particle effects can lead to errors (2) or (3) above, depending on the severity of the circuit response. *SE induced snapback* (SES) in n-channel MOS output devices and *SE induced latchup* (SEL) in CMOS structures are regenerative current conditions which, if the current levels are benign, can be reset. However, if the regenerative current energy exceeds the thermal dissipation of the affected region, these effects can cause melting and permanent physical damage to the circuit.

2.2 PARTICLE ENERGY DEPOSITION

The effects that we are discussing arise from the energy loss of high energy ions as they pass through the devices. These are charged particles, and as they pass through any material they lose energy to the electrons of the material, leaving a track of ionization behind them. A good introduction to the processes involved is given by Reed [Reed 2008]. Figure 2-1 shows the effect of the ion tracks in an emulsion [Kolasinski 1979]. The high Z ions create much more ionization and more intense tracks. The analysis of cosmic ray upsets involves ion energy losses directly. Proton induced upsets arise from the ionization of secondary particles from nuclear scattering and reactions. The characteristics of an individual ionizing particle are not known, so in these cases we discuss the total energy available to ionization from the nuclear reaction that led to the single event effect.

There are a number of terms used to describe the ion energy loss. The nuclear physics community uses the average energy loss per unit path length (dE/dx), that is, stopping power or stopping force with units MeV/cm. The SEE community commonly uses mass stopping force, called linear energy transfer (LET), as the metric for average

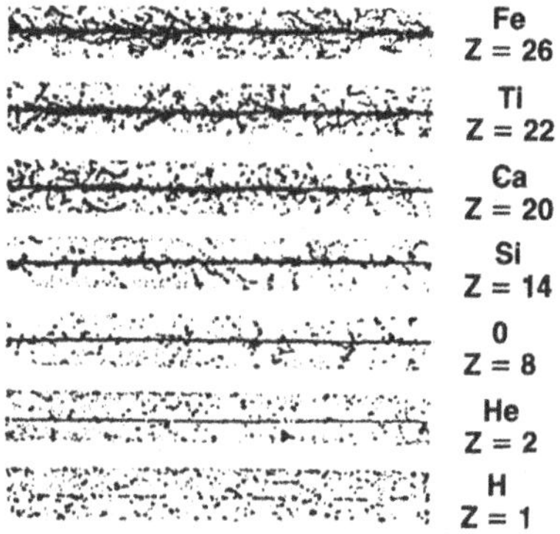

Figure 2-1 Cosmic ray ion tracks in a photographic emulsion [Kolasinski 1979]. Copyright © 1979 IEEE.

energy deposited per unit path length.

$$LET = -\frac{1}{\rho}\frac{dE}{dx'} \tag{2.1}$$

where ρ is the density of the target material (mg/cm^3) and the LET unit is typically MeV/ mg/cm^2.

The units that are commonly used in the cosmic ray community to describe the ionization of cosmic rays are MeV/g/cm^2. Note the factor of 1000 difference. We will use both sets of units. When we are discussing cosmic rays we will use MeV/g/cm^2. When we are discussing energy loss, upset rate testing, and upset rate predictions, we will use MeV/g/cm^2.

Figure 2-2 shows curves of *dE/dX* for several common ions in silicon. Tables of *dE/dX* can readily be generated using the SRIM (originally TRIM) shareware code developed by Jim Ziegler of IBM, or the CREME codes of Jim Adams [Ziegler 1996, Ziegler 2008, Adams 1986]. Reed shows values calculated by Geant4 code [Reed 2008].

As each material has a unique ionization potential, the LET can easily be converted into a charge deposition rate. Therefore, the LET can also be expressed in terms of picocoulomb/micron (pC/μm) in a given material. It takes 3.6 electron volts to produce an

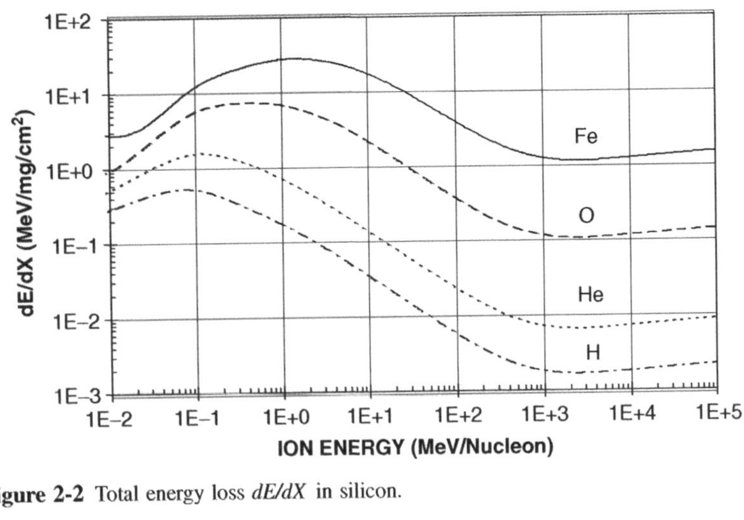

Figure 2-2 Total energy loss *dE/dX* in silicon.

electron–hole pair in silicon; 3.6 MeV energy loss will produce 10^6 electrons or 0.16 picocoulombs. [$Q = E/22.5$ (pC/MeV), 1 pC/μm = 98 MeV/(mg/cm^2).] If this charge is generated in the vicinity of a depletion region in the device, the high electric fields will cause the charge to be collected.

The ionization is often expressed in terms of specific ionization. This is the number of ion pairs produced per unit path length. It is plotted as a function of the distance traveled. For a collimated beam of particles, the relationship between the *average* specific ionization and the distance from the source is called the *Bragg curve* [Evans 1955, p. 665]. If the particles are of relatively low energy, then there is very little energy straggling, and the ion specific ionization curve and the Bragg curve are similar. This is shown in Figure 2-3. However, for high energy beams of particles, there is significant energy and range straggling. In this case the average energy loss and ionization near the end of the track are significantly smaller than the value for individual ions. Figure 2-4 shows the specific ionization and Bragg curves of 200 MeV helium in silicon. In this case there is a great deal of difference between the two curves. The Bragg curve for the average ionization was calculated assuming a monoenergetic beam. Actually, accelerator beams normally have an energy spread. This spread is magnified due

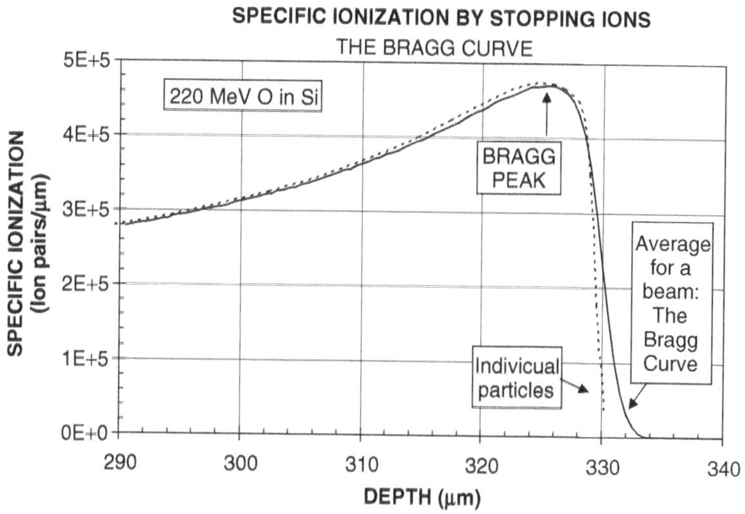

Figure 2-3 Specific ionization curves for short-range particles.

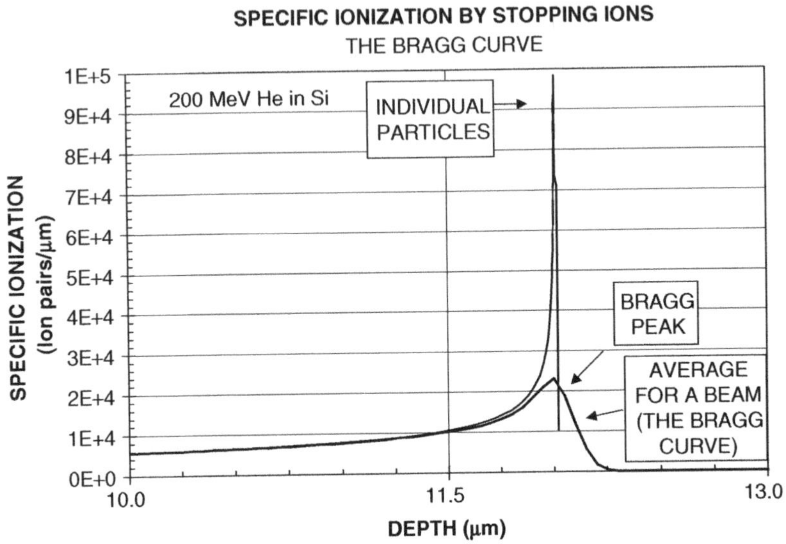

Figure 2-4 Specific ionization curves for long-range particles showing the Bragg peak.

to the energy straggling as the particles slow, so actual Bragg peaks are wider and lower than shown here.

Occasionally both the specific ionization curve and the average ionization curve are called Bragg curves [Segre 1964]. There is less confusion if the usage described by Evans is followed [Evans 1955]. It is not correct to refer to the ionization peaks shown in Figure 2-2 as Bragg peaks.

2.3 SINGLE EVENT ENVIRONMENTS

We want to consider the environment in which our electronic devices must operate. There are a number of components of the charged particle population in space. The basic group consists of galactic cosmic rays coming from outer space toward the Earth.

The trapped particles in radiation belts around the Earth and other planets can also cause upsets. Here the main single event interest is in the protons and their reactions in silicon. There are a few trapped heavy ions, but they are at low energies and will not reach into the spacecraft to cause upsets.

There are also ions that are emitted directly from the Sun in a solar energetic particle (SEP) event and which may hit the Earth from any direction. For both of these components the primary single event interest is the ionization caused by heavy ions. In some cases the protons and helium ions in the solar flares may also contribute.

Interplanetary space also contains heavy ions that have been emitted by the Sun either under normal conditions or under solar particle events. They are now floating around between the planets, their origin uncertain. This is called the anomalous component of cosmic rays and does not seem to make a major contribution to single event effects [Tylka 1994].

For avionics systems, we are primarily concerned with the neutrons produced by the high energy particles that enter the atmosphere. About 20% of the particles in the atmosphere are protons, but the effects are similar.

Note that we use two terms to describe the space environment or the particle beam used for testing: (1) flux = the rate of flow of something such as energy or particles across or onto a given area, (2) fluence = a measure of the quantity of light or other radiation falling on a surface, expressed in terms of either particles or energy per unit area.

2.3.1 The Solar Wind and the Solar Cycle

The Sun's outer atmosphere, the corona, extends several solar diameters into interplanetary space. The corona continuously emits a stream of protons, electrons, doubly charged helium ions, and small amounts of other heavy ions, collectively called the solar wind [Barth 1997]. This electrically neutral plasma streams radially outward from the Sun at a velocity of approximately 300 to 900 kilometers per second with a temperature on the order of 10^4 to 10^6 K and a density of 1 to 30 particles/cm^3. While the solar wind is millions of metric tons of matter moving at a million kilometers per hour, its density is so low that the physics is that of a vacuum. The energies of the particles range from approximately 0.5 to 2.0 keV/nuc.

The charged particle environment in near-Earth regions is dominated by the activity of the Sun, which acts as both a source and a modulator. It is a source of both protons and heavier ions via the periodic high energy solar events that accelerate large numbers of particles.

The solar wind is also a source of particles trapped in outer regions of the Earth's radiation belts. Because galactic cosmic ray (GCR) heavy ions originate outside the solar system, they must "fight" against the solar wind to reach interplanetary space. As a result, the GCR levels follow a cyclic pattern reflecting the activity level of the Sun. Atmospheric neutrons are secondary products of collisions between GCRs and oxygen or nitrogen atoms in the Earth's atmosphere; therefore, their levels are also modulated by the solar cycle. Finally, the levels of trapped particles are modulated by both long term variations in solar activity and solar storm events.

Understanding its cyclical activity is an important aspect of modeling the space radiation environment. The solar activity cycle is approximately 11 years long. During this period there are typically 7 years during solar maximum when activity levels are high and 4 years during solar minimum when activity levels are low. In reality the transition between solar maximum and solar minimum is a continuous one, but it is often considered to be abrupt for convenience. At the end of each 11-year cycle the magnetic polarity of the sun reverses and another 11-year cycle follows. Thus, strictly speaking the total activity cycle is approximately 22 years long [Bourdarie 2008]. Of the space radiations considered here, the magnetic polarity apparently only affects the galactic cosmic ray fluxes and not the trapped particle or solar particle event fluxes. Thus, things are often viewed on an approximately 11-year cyclical basis. Two common indicators of this approximately 11-year periodic solar activity are sunspot numbers and solar 10.7 cm radio flux (F10.7). The most extensive record is that of observed sunspot numbers, which dates back to the 1600s. This record is shown in Figure 2-5. The numbering of sunspot cycles began in 1749 and it is currently near the end of solar cycle 23. The record of F10.7 began part way through solar cycle 18 in the year 1947 and is shown in Figure 2-6.

Although sunspot numbers and F10.7 are commonly accepted indicators of solar activity, quantitative relations to measured radiation events and fluxes are not necessarily straightforward. Large solar particle events are known to occur with greater frequency during the declining phase of solar maximum. Trapped electron fluxes also tend to be higher during the declining phase. Trapped proton fluxes in low Earth orbit (LEO) reach their maximum during solar minimum but exactly when this peak is reached depends on the particular location.

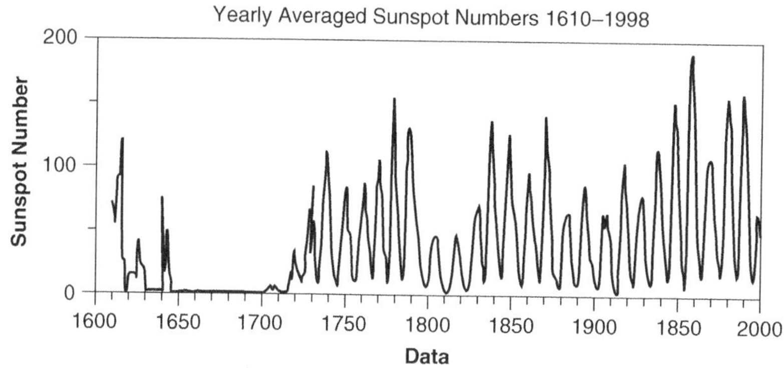

Figure 2-5 The observed record of yearly averaged sunspot numbers 1610 to 1998 [Bourdarie 2008]. Copyright © 2008 IEEE.

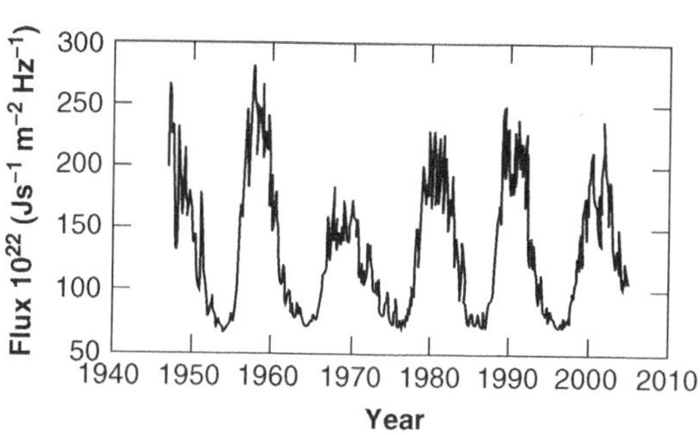

Figure 2-6 Measured values of solar 10.7 cm radio flux [Bourdarie 2008]. Copyright © 2008 IEEE.

Galactic cosmic ray fluxes are also at a maximum during solar minimum but in addition depend on the magnetic polarity of the Sun.

Figure 2-7 shows 35 years of solar sunspot activity, which defines the solar cycle, and the terrestrial flux of hadrons (mostly neutrons) in Colorado, USA. (Hadron definition: any of a class of subatomic particles, including the proton, neutron, meson, and hyperon, that consist of a combination of quarks and antiquarks.) The two data envelopes are countercyclic. Note that the terrestrial flux of hadrons has ordinate units of "% below 1954 minimum" and shows a maximum effect of the solar cycle on the particle flux of 25% [Ziegler 2004]. We should not be confused when we read that the maximum cosmic ray fluxes occur for solar minimum.

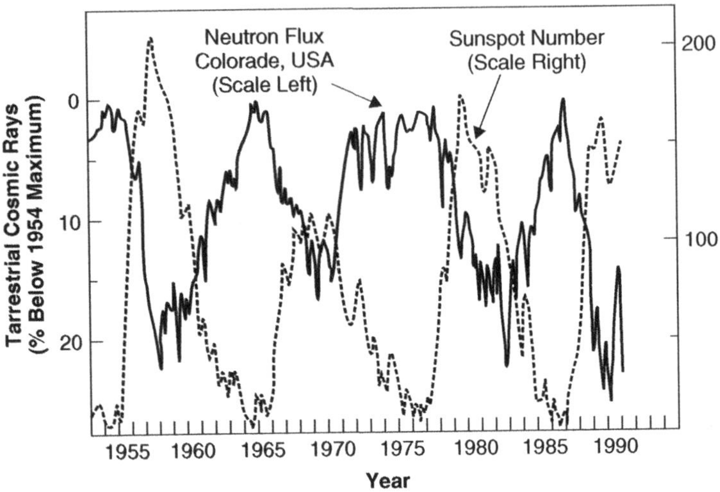

Figure 2-7 Variation in cosmic rays due to the solar cycle [Ziegler 2004]. Copyright © 2004 Cypress Semiconductor. Reproduced with permission from Cypress Semiconductor Corporation.

2.3.2 The Magnetosphere, Cosmic Ray, and Trapped Particle Motion

The Earth's magnetosphere consists of both an external and an internal magnetic field. The external field is the result of the solar wind. The internal or geomagnetic field originates primarily from within the Earth and is approximately a dipole field. Figure 2-8 shows how the solar wind and its imbedded magnetic field tend to compress the geomagnetic field. The solar wind generally flows around the geomagnetic field.

Figure 2-9 shows the geomagnetic field, which is approximately dipolar for altitudes of up to about 4 or 5 Earth radii. Charged particles become trapped because the Earth's magnetic field constrains their motion. They spiral around the field lines in a helicoidal path while bouncing back and forth between the magnetic poles. Superimposed on these spiral and bounce motions is a longitudinal drift of the particles because of the gradient of the magnetic field. Figure 2-10 illustrates the three motions.

The Earth's magnetic field must be penetrated by cosmic rays in order for them to reach a spacecraft in Earth orbit. The number of magnetic field lines a cosmic ray must cross to reach a given point within

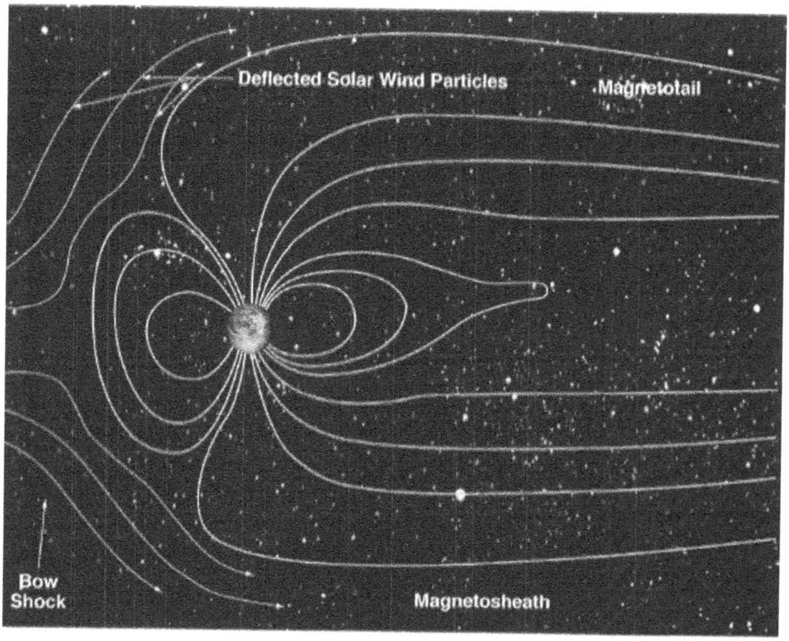

Figure 2-8 The Earth's magnetosphere [Xapsos 2006b]. Copyright © 2006 IEEE.

Figure 2-9 The internal magnetic field of the Earth is approximately a dipole field [Xapsos 2006b]. Copyright © 2006 IEEE.

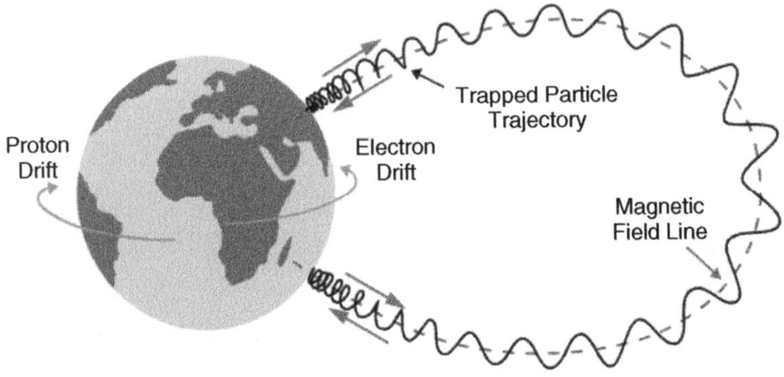

Figure 2-10 Motion of a charged trapped particle in the Earth's magnetic field [Xapsos 2006b]. Copyright © 2006 IEEE.

the magnetosphere approximately determines the minimum energy it must possess. To cross more magnetic field lines, more energy will be required. This penetrating ability is determined uniquely by the cosmic ray's momentum divided by it charge. This quantity is called magnetic rigidity. To penetrate the Earth's magnetic field, a particle must have sufficient magnetic rigidity to avoid being turned away. Regions in the outer magnetosphere and near the poles can be reached at much lower magnetic rigidities than are required to reach points near the Earth's equator [Adams 1981].

The geomagnetic cutoff can be determined by detailed ray tracing backwards from the point of interest; it determines if the trajectory leads back to outer space. Figure 2-11 shows such a computed trajectory for a cosmic ray whose magnetic rigidity is barely adequate to bring it into the atmosphere over Sinkiang Providence in western China.

Methods have been developed for determining good approximations for the rigidity in terms of the Earth's magnetic field. Figure 2-12 shows how the cutoff rigidity follows the Earth's magnetic field lines [Barth 1997]. Figure 2-13 shows a calculation of the magnetic rigidity at 800 km [Barth 1997].

2.3.3 Galactic Cosmic Rays

The cosmic rays and solar flares have a mass distribution that corresponds more or less to the natural abundance of the elements although

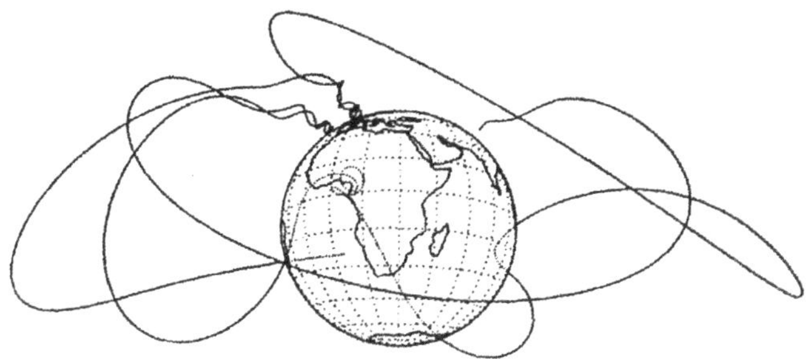

Figure 2-11 This figure shows the computed trajectory of a cosmic ray in Earth's magnetic field (after Lund) [Adams 1981].

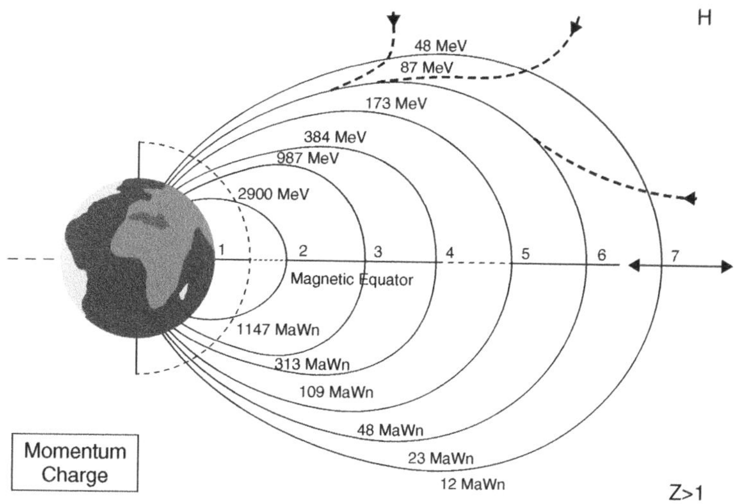

Figure 2-12 Relationship between the dipole shell parameter, L, and cutoff rigidity is used to determine the particle energy required to penetrate the magnetosphere (after Stassinopoulos) [Barth 1997]. Copyright © 1997 IEEE.

the heavy ion properties in solar flares can vary drastically from flare to flare. Most of the cosmic rays have very high energies. The large heavy ion component means that the cosmic rays contain may high LET particles with high energy loss. This contrasts with the cosmic rays that reach ground level, primarily protons, neutrons, electrons, and muons. In this situation the majority of the heavy particles have been eliminated by nuclear reactions. Few of the heavy particles with

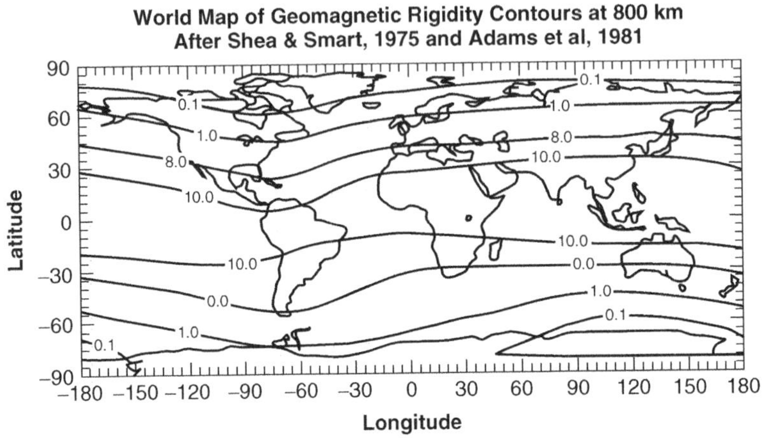

Figure 2-13 Magnetic rigidity as calculated by Shea and Smart. Note that rigidity is lower at higher latitudes, allowing increased particle exposure near the poles [Barth 1997]. Copyright © 1997 IEEE.

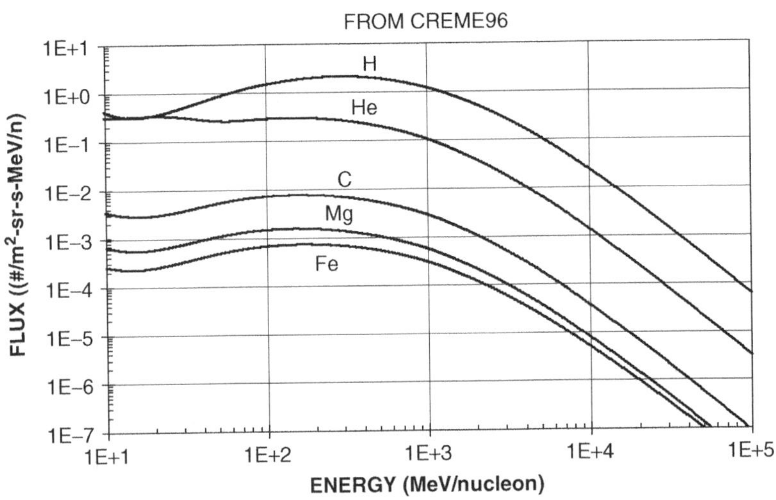

Figure 2-14 Representative cosmic ray energy spectra.

large LET reach the earth. Figure 2-14 shows representative cosmic ray energy spectra.

The cosmic ray flux is ordinarily specified by the number of particles that would hit a sphere from all directions, not necessarily passing through the center of the sphere. It is measured in terms of particles per meter squared per second, or particles per centimeter squared per second (a factor of 10^4). At other times it is useful to consider the number of particles hitting a surface. This concept allows for the directionality of cosmic rays. The flux is expressed in terms of the number of particles incident in a solid angle about a line normal to a small area on the surface of a sphere. The dimensions in this case are particles per meter squared per steradian second or particles per centimeter squared per steradian second. Remember that there are 4π steradians surrounding a point. If one has an omnidirectional flux, there is just a factor of 4π difference between the particles per meters squared per second and the particles per meter squared steradian second. Figure 2-15 shows these relationships.

The energy is expressed in MeV per nucleon rather than just MeV. This is because the equations that describe the energy spectra use the parameter of energy/nucleon, rather than just energy. It also nicely compresses the scale. Note that the energy loss in Figure 2-2 was expressed in the same way. An alpha particle (4 nucleons) at 10 MeV per nucleon has 40 MeV, whereas a proton (1 nucleon) has only 10 MeV.

Figure 2-16 shows the cosmic ray abundances for all ions. Here we have plotted the peak flux by which we mean the peak value for spectra as shown in Figure 2-14. We see two noticeable falloffs:

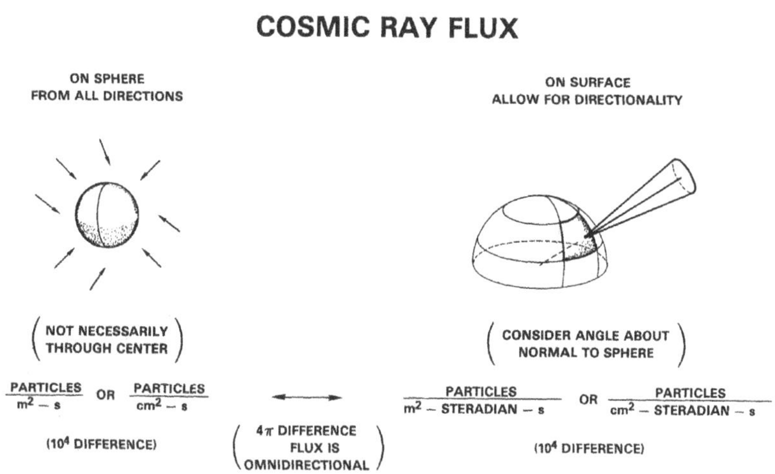

Figure 2-15 Comparison of units used for measuring cosmic ray flux.

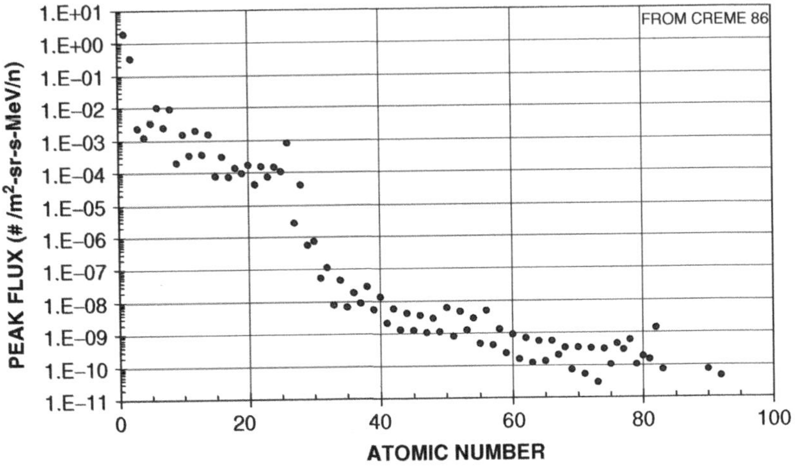

Figure 2-16 The relative abundances of all of the ions in cosmic rays. They are plotted in terms of the peak flux in their energy spectra.

one after helium and one after iron. All but seven of the 92 elements through uranium are present. Figure 2-17 expands the important range up through iron. These curves were obtained using data from the CREME codes [Adams 1986, Tylka 1996].

In order to determine the upset rates in space, we need to know how many ions deposit how much energy as they pass through the

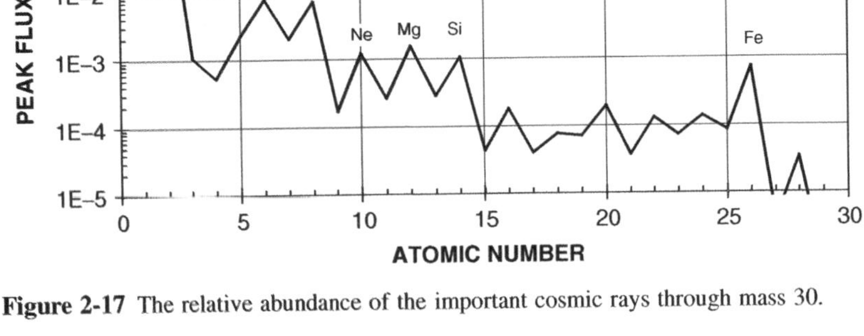

Figure 2-17 The relative abundance of the important cosmic rays through mass 30.

electronic chip. In order to do this we characterize the environment in terms of the number of ions as a function of their linear energy transfer (LET). This is called a Heinrich spectrum [Heinrich 1977]. We assume that all ions of a given LET will have the same effect. There will be discussion of this assumption later.

The most significant ion in the galactic cosmic ray (GCR) environment as far as single event effects are concerned is the iron ion. This ion has high intensity, very high energies, and high LET. It therefore penetrates all parts of a spacecraft, depositing large amounts of energy as it goes. Figure 2-18 shows the spectra of GCR iron ions. Note that the energy is expressed in terms of MeV per amu. The iron peak at 300 MeV/nucleon occurs at roughly 17 BeV iron energy. The LET of iron ions in this energy range is shown in Figure 2-19. (This is an expanded version of Figure 2-2.) Note that any energy iron ion has a LET in the range of 10^3 to 3×10^4 MeV/g/cm^2. This and the next several figures showing various aspects of cosmic rays will use units of MeV/g/cm^2. 1000 on this scale corresponds to 1 MeV/mg/cm^2 on the curves that we will use later.

The number of ions as a function of energy, shown in Figure 2-18, can be combined with the energy as a function of LET shown in Figure 2-19 to obtain the differential iron LET spectrum. Figure 2-20 shows the number of GCR iron ions as a function of their LET. All of

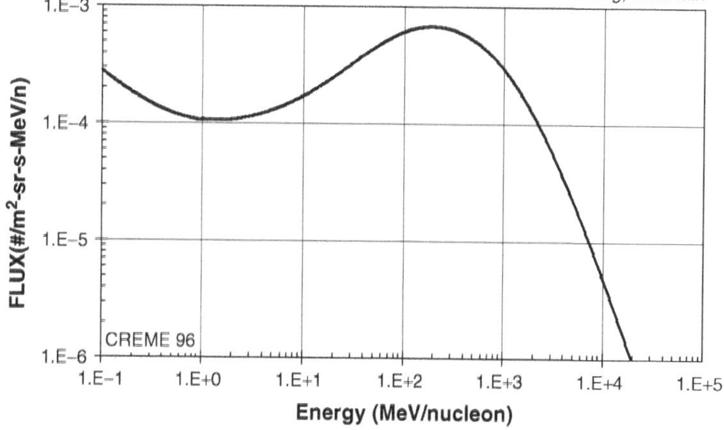

Figure 2-18 Differential galactic cosmic ray iron spectrum.

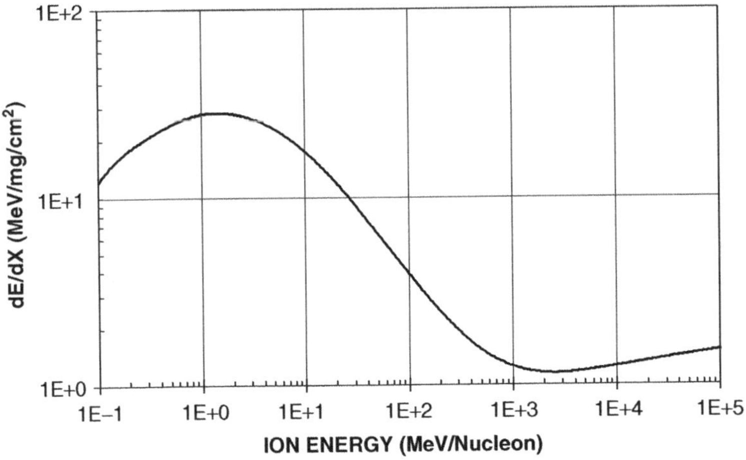

Figure 2-19 LET in silicon of the high energy iron ions present in galactic cosmic rays.

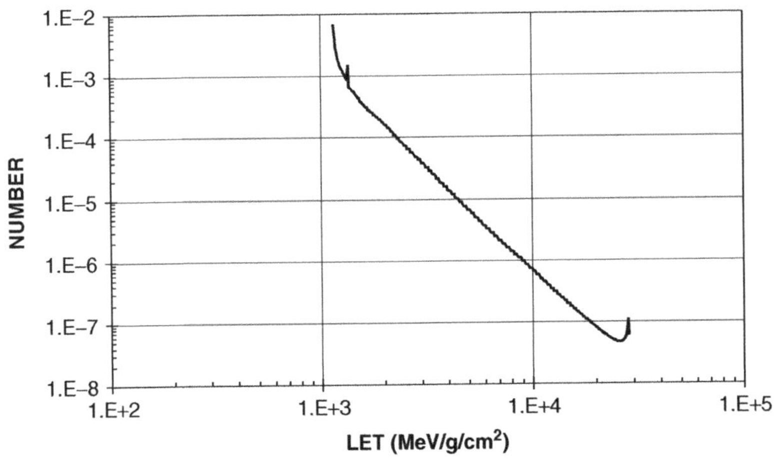

Figure 2-20 Differential GCR iron LET spectrum.

the ions contribute in a relatively narrow range of LETs. It is sometimes useful to express a spectrum in the integral form, that is, the number of ions above a given LET. The integral iron GCR LET spectrum is shown in Figure 2-21.

Figure 2-22 shows the total differential GCR LET spectrum. All of the ions heavier than hydrogen are included. This curve is obtained by summing the differential spectra of all of the individual ions.

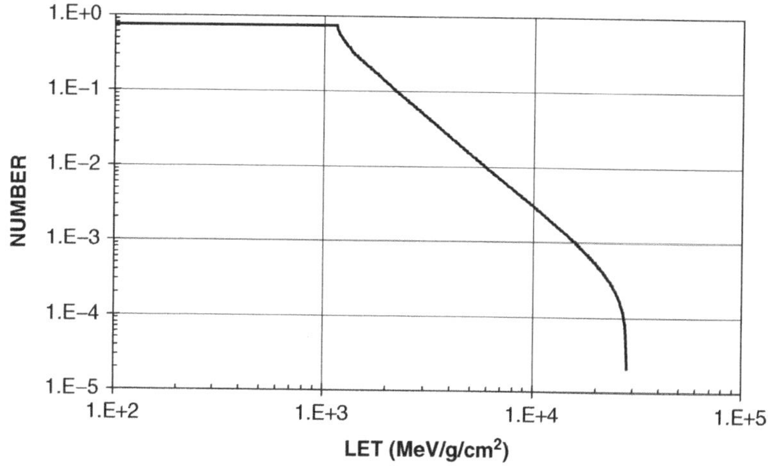

Figure 2-21 Integral iron GCR spectrum.

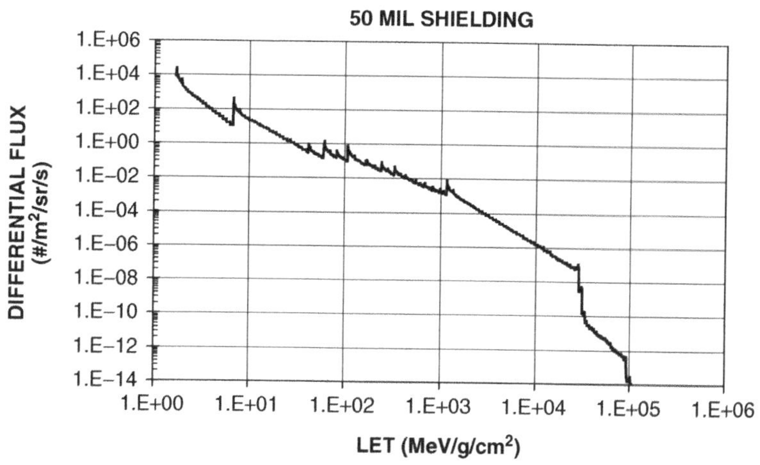

Figure 2-22 Differential spectrum from all cosmic rays.

Figure 2-23 shows more detail for the LET regime of interest for single event upsets. Figure 2-24 shows the integral LET spectrum in the upset LET region. It also shows the contribution due solely to the iron ions. They contribute approximately 50% of the intensity in this region. Figure 2-25 shows the total integral LET spectrum of all ions heavier than hydrogen.

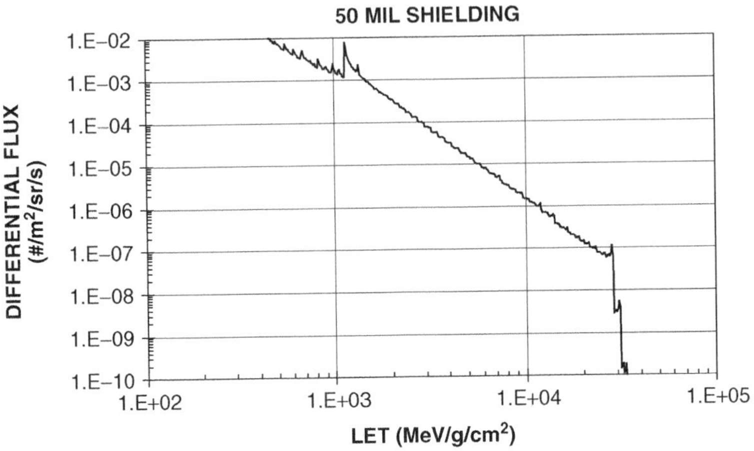

Figure 2-23 Differential GCR LET spectrum in the LET range important for single event upsets.

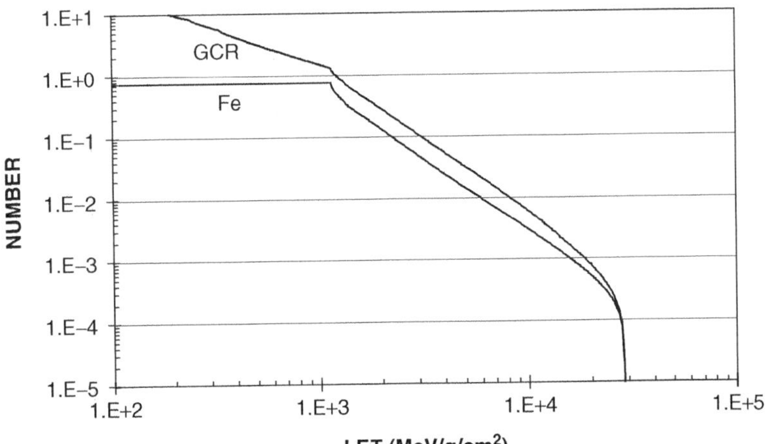

Figure 2-24 Integral iron LET spectrum and total LET spectrum.

2.3.3.1 Influence of Solar Cycle

We have presented the basic LET spectrum for galactic cosmic rays. It has been presented for solar minimum conditions. However, the intensity varies over an 11-year cycle and is rarely the maximum values that occur at solar minimum. Figure 2-26 compares the galactic cosmic ray fluences during solar maximum and solar minimum conditions. The

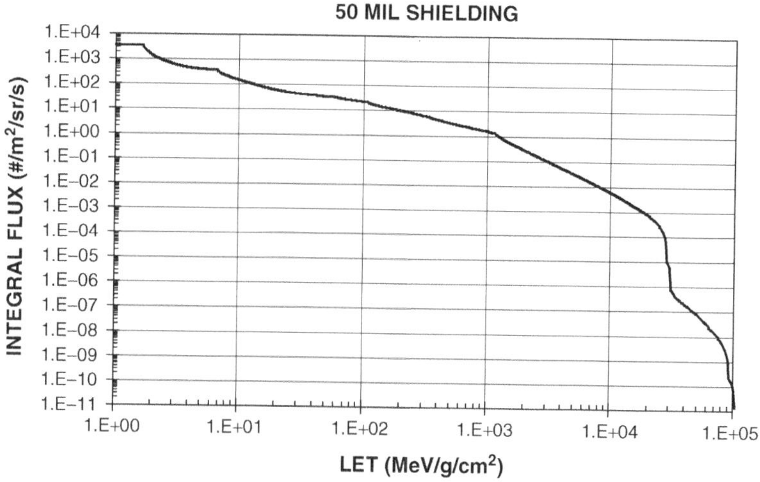

Figure 2-25 Total integral GCR LET spectrum of all ions heavier than hydrogen.

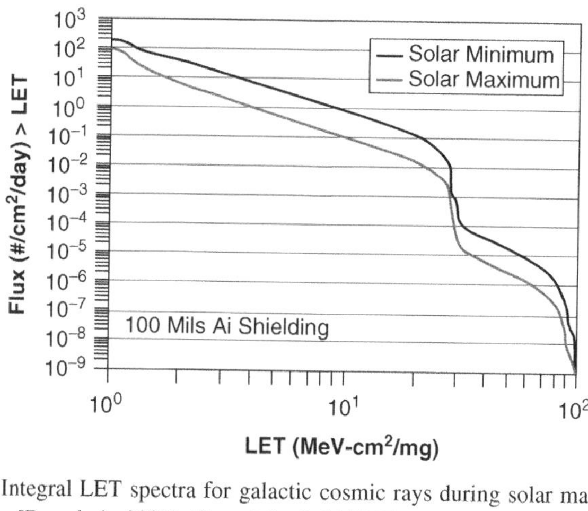

Figure 2-26 Integral LET spectra for galactic cosmic rays during solar maximum and solar minimum [Bourdarie 2008]. Copyright © 2008 IEEE.

calculations of SEE rates for most missions use solar minimum conditions for a conservative rate prediction. It is necessary to include the region above 30 MeV/mg/cm^2 (above the contributions of iron ions) for devices that have been hardened to single events or for serious errors such as latchup. This also applies to many submicron devices. Figure 2-27 shows the measured time variation of the galactic cosmic

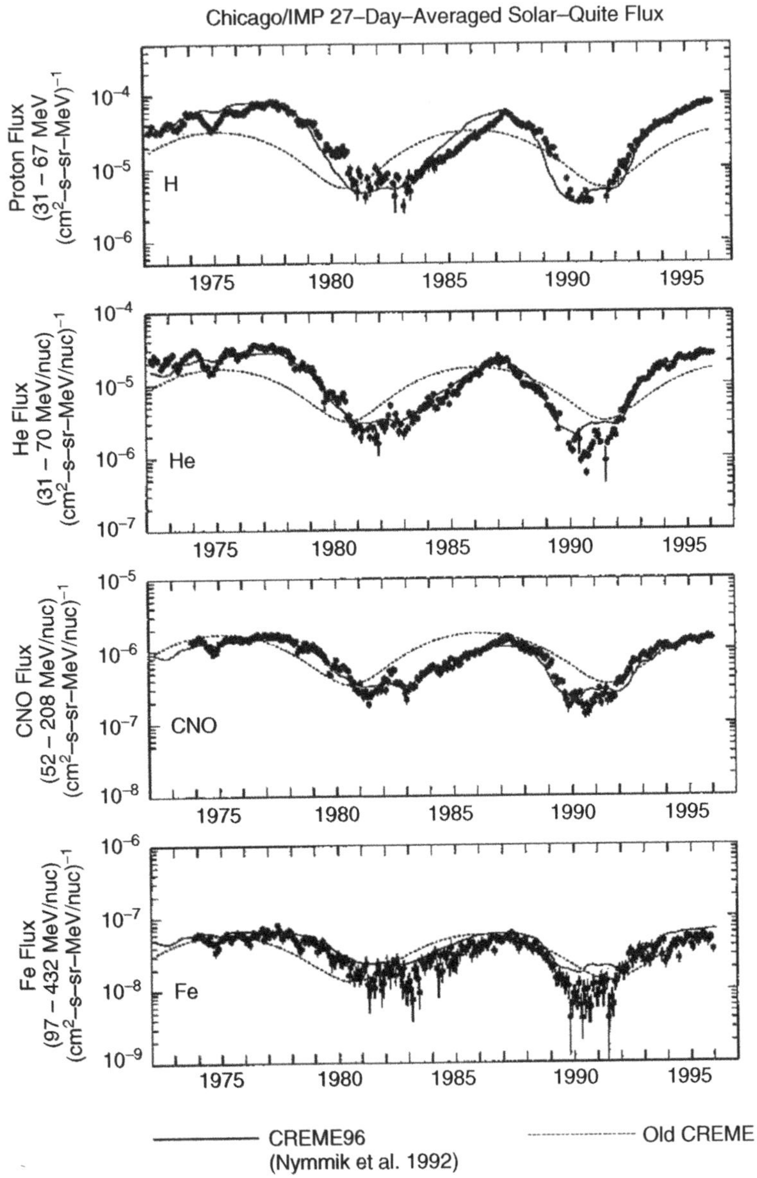

Figure 2-27 Solar cycle variation of galactic cosmic rays: Chicago IMP8/CRT data compared with CREME86 and CREME96 models of the relative isotopic intensity and time variation [Tylka 1997]. Copyright © 1997 IEEE.

rays [Tylka 1997]. Note again that the GCR flux is greatest when the solar activity shown in Figures 2-6 and 2-7 is at minimum.

2.3.3.2 Influence of Geomagnetic Shielding

We discussed the effects of the Earth's magnetic field in Section 2.3.2. As one approaches the equator, the magnetic field is more effective in rejecting incident particles. Figure 2-28 shows the low energy cutoff of cosmic rays as a function of latitude for 600 nautical mile orbits.

For single event considerations, we are more concerned about the changes in the LET spectrum. The heavy ions are attenuated as they approach Earth due to the shielding effects of the Earth's magnetic field [Barth 1997, Xapsos 2006b]. The magnetic field has little effects over the poles, but a very large effect at the equator.

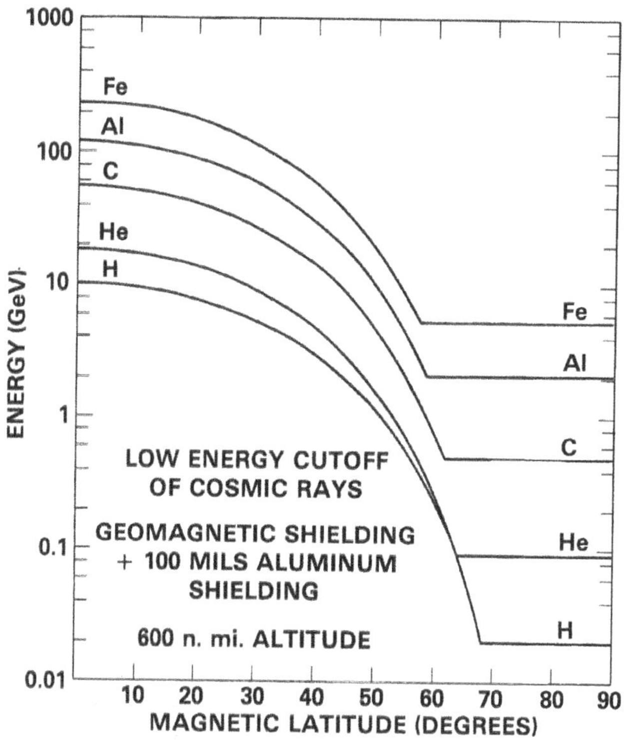

Figure 2-28 The low energy cutoff of cosmic rays as a function of latitude for 600 nautical mile orbits.

Figure 2-29 shows the change in the LET spectra with altitude for circular orbits with 60 degree inclination. The effects become much stronger toward the equator. Figure 2-30 shows the change with altitude for circular orbits at 30 degree inclination.

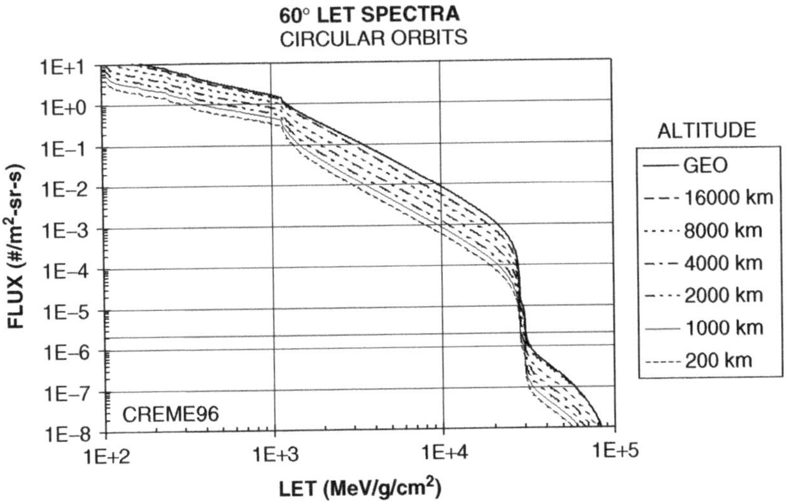

Figure 2-29 Variation of LET spectra with altitude for 60 degree inclination orbits.

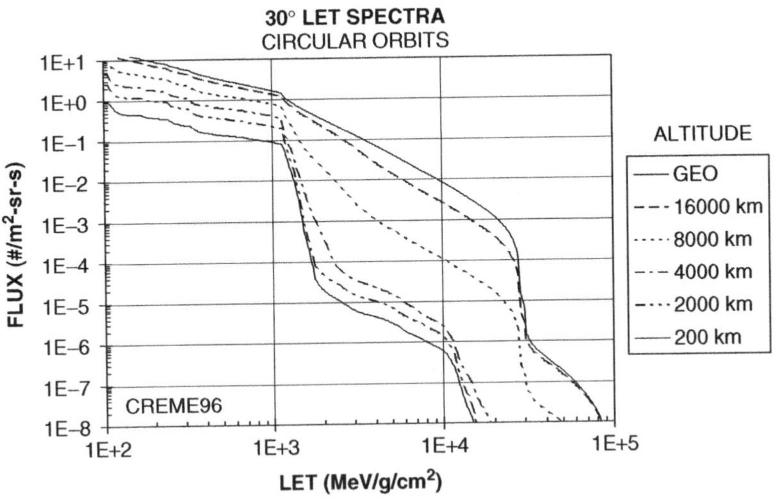

Figure 2-30 Variation of LET spectra with altitude for 30 degree inclination orbits.

Figure 2-31 shows the change in the LET spectrum as a function of inclination for circular orbits at 500 km altitude. There is a large drop at the inclination changes from 50 degrees to 40 degrees. Figure 2-32 shows the effect on the high LET portion of the heavy ion environment

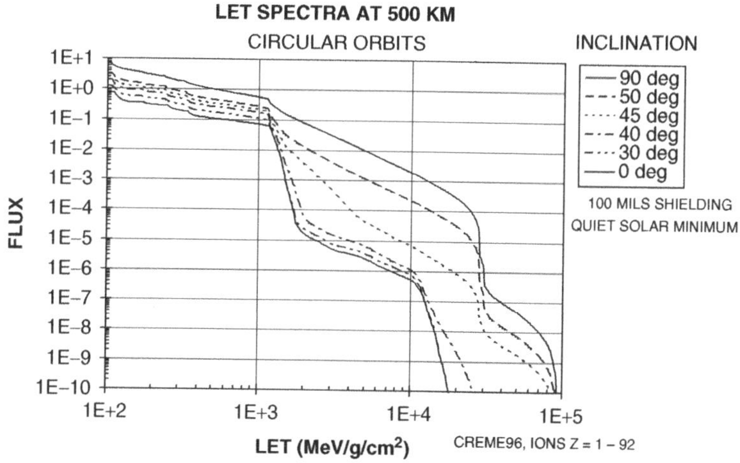

Figure 2-31 Variation of LET spectra with inclination for LEO.

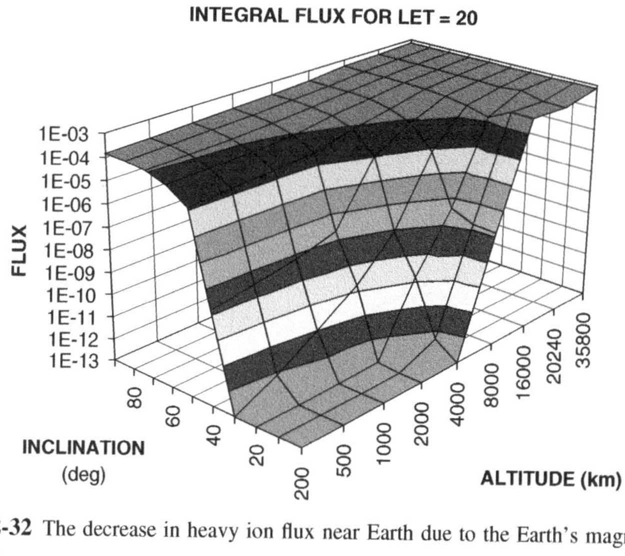

Figure 2-32 The decrease in heavy ion flux near Earth due to the Earth's magnetic shielding.

(the portion above 20 MeV/mg/cm^2). This shows the effects of the combined changes of inclination and altitude.

The region above 10^3 MeV /g/cm^2 is the region of interest for single event effects. These figures clearly indicate a reduction in cosmic ray intensities at low altitudes and low inclinations.

There is sometimes a question about the necessity of considering the cosmic ray ions heavier than the iron group. In general, the curves are basically independent of the high Z ions. This is not true for magnetic shielding at low altitude. Figure 2-33 shows a very large difference between the spectra with and without the heavy ion components. These differences will apply to the low upset rates of relatively hard devices. If you want accurate calculation of low altitude, low inclination upset rates; the complete spectrum of cosmic ray ions should be included. Figure 2-34 shows the curves for 28.5 degree, 450 km orbits. These differences will be important when calculating latchup rates, where one often wants to be assured that there will not be latchup events.

2.3.3.3 Influence of Spacecraft Shielding

The majority of cosmic rays have very high energies. Therefore, shielding has very little effect on them. However, there is a low

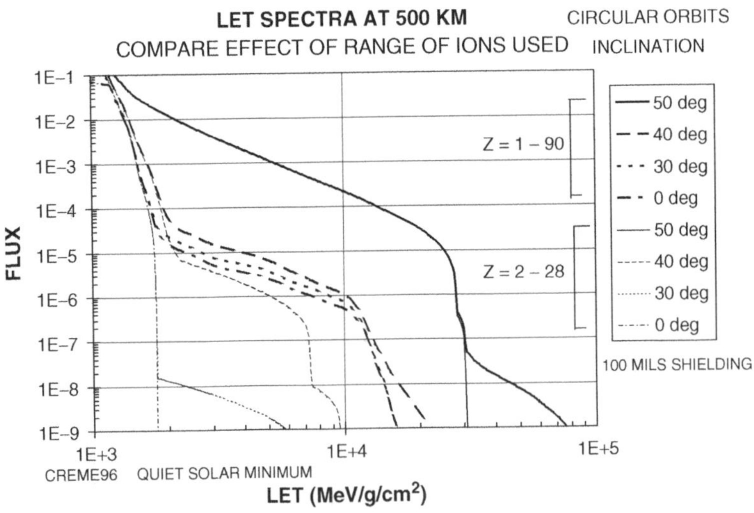

Figure 2-33 Comparison of LET for limits on Z of 2–28 and 1–90.

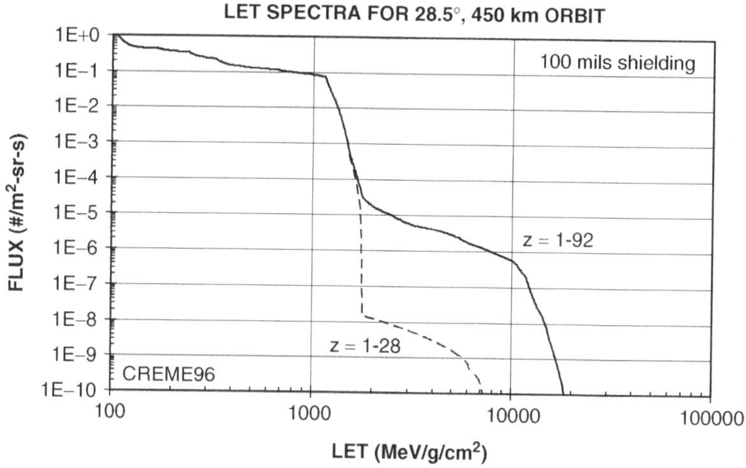

Figure 2-34 LET spectra considering two different ranges of elements for 28.5 degree LEO.

energy component that is removed by very thin layers of shielding. Figure 2-35 shows the raw iron spectrum with a large low energy component. This is removed by thin layers of shielding as shown in the figure. It is for this reason that upset rates are normally calculated using LET spectra that assume 100 mils of aluminum shielding. Although we have presented a few plots of cosmic ray flux with 50 mils of shielding, 100 mils has become the standard. Figure 2-36 shows the LET spectrum that would be used if there was no shielding. This is sometimes confusing to those who have dealt primarily with total dose effects, where the custom is to calculate the effects with no shielding. It is common to show cosmic ray LET spectra with 100 mils of shielding, often not mentioning the shielding. Satellite weight constraints make it impractical to add appreciable shielding and most devices are near some surface so detailed analysis of shielding is rarely necessary for cosmic rays. Therefore, a calculation of SEEs for 100 mils (0.1 inch) of shielding is appropriate both for the general characterization of part sensitivity and for a first order approximation for actual operational rates.

Figures 2-35 and 2-36 show the iron energy and LET spectra after various levels of shielding. The incident ions have enough energy so that it takes nearly 5 inches of aluminum to produce a major attenuation of the flux.

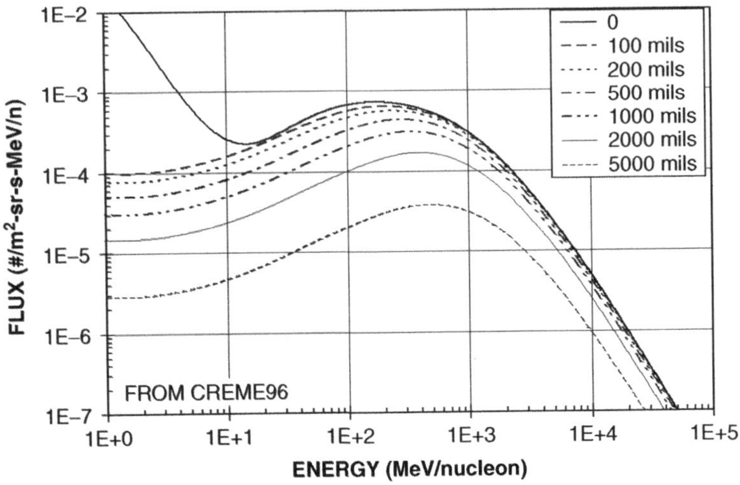

Figure 2-35 Cosmic ray iron spectra behind various levels of aluminum shielding.

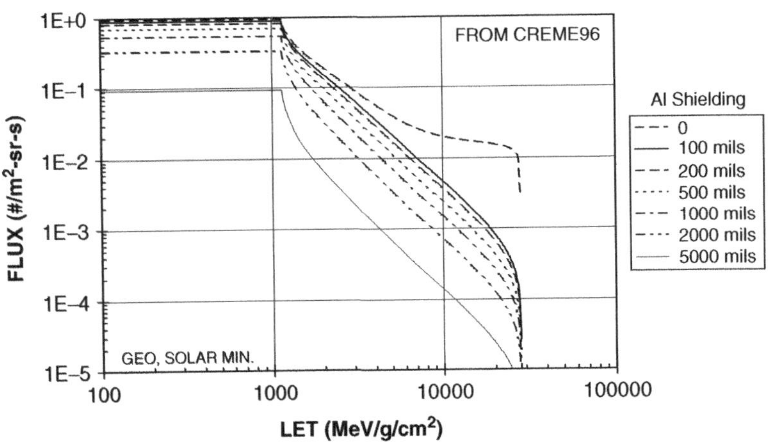

Figure 2-36 Cosmic ray iron LET spectra behind various levels of aluminum shielding.

Figure 2-37 shows the effect of shielding on the entire cosmic ray LET spectrum. We see that there is really very little effect. Even 5 inches of shielding reduces the flux by only a factor of 25. This curve is plotted for the entire cosmic ray mass spectrum with $Z = 1-92$. The portion below 30 MeV/mg/cm^2 does not change if we only plot $Z = 1-28$. The ions with $Z > 28$ become important in single event calculations for relatively hard devices. Figure 2-38 shows the effect

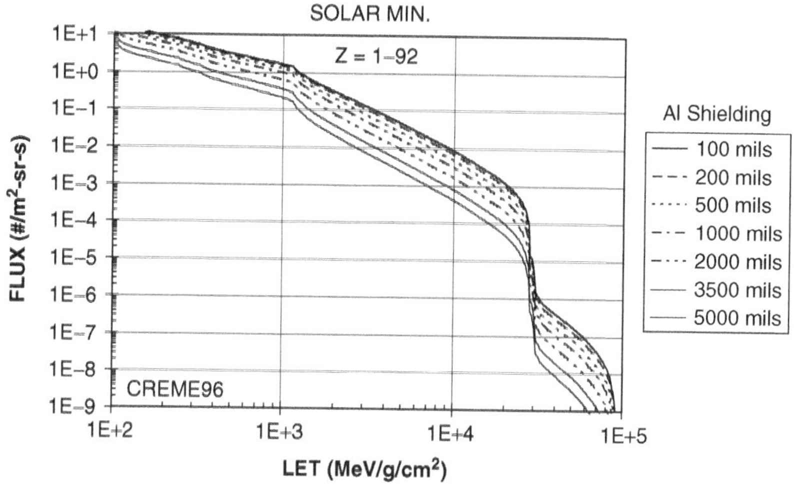

Figure 2-37 Effect of shielding on cosmic ray LET spectra.

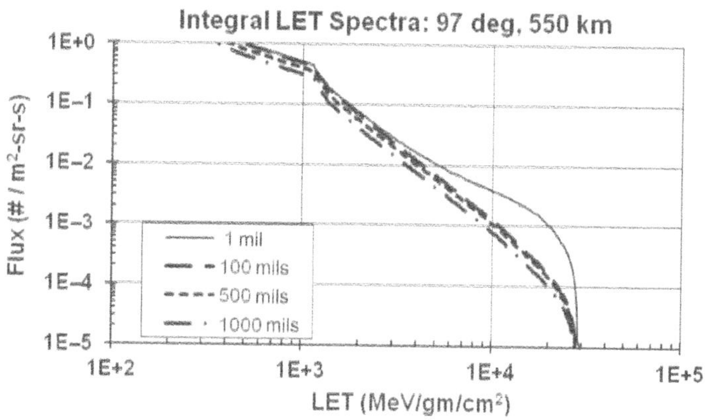

Figure 2-38 LET spectra observed in a sun-synchronous polar orbit. The 1 mil spectrum indicates the effects of a large number of very low energy particles. These are operationally unimportant and the incident spectra are usually reported for 100 mils of shielding. A shielding of 1000 mils (1 inch) has little impact on the resulting SEE rates.

of aluminum shielding in a polar orbit for which the geomagnetic shielding has little effect. This plot also demonstrates the effect of very low energy particles that do not penetrate 50 or 100 mils of shielding. There is no appreciable difference between the two. We observe that 1 inch of shielding has negligible impact on the flux of particles.

The magnetic shielding of the Earth has the effect of removing the low energy particles, while allowing the high energy particles to pass. Figure 2-39 shows the effect of additional aluminum shielding for a 51.6° orbit. A half inch of aluminum has virtually no effect. Additional shielding up to 5 inches has only slight effect. The magnetic shielding for low inclination orbits is very effective, leaving only the high energy minimum LET ions. However, this means that aluminum shielding has an effect contrary to expectations. The additional shielding slows down the very high energy ions, so that they become more ionizing, and the LET spectrum actually becomes more intense. This is shown in Figure 2-40. The effect is particularly noticeable at LETs between 10 and 30 MeV/mg/cm^2.

Section 7.10 discusses an approach to the inclusion of shielding when using the figure of merit method of rate calculation.

2.3.4 Protons Trapped by the Earth's Magnetic Fields

The previous section discussed the environments for upsets due to heavy ions. However, for many low earth orbit (LEO) satellites, the heavy ions are a minor concern. Instead it is the upsets due to protons, especially those in the South Atlantic anomaly (SAA), that are of

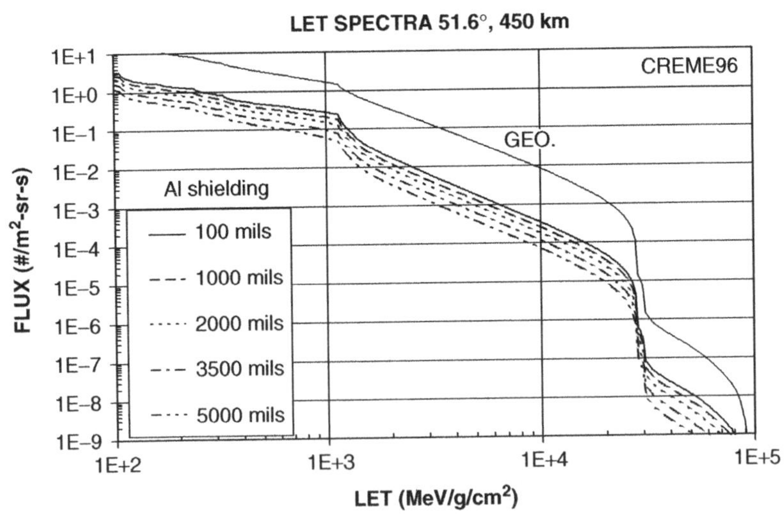

Figure 2-39 LET spectra behind shielding for 51.6 degree LEO.

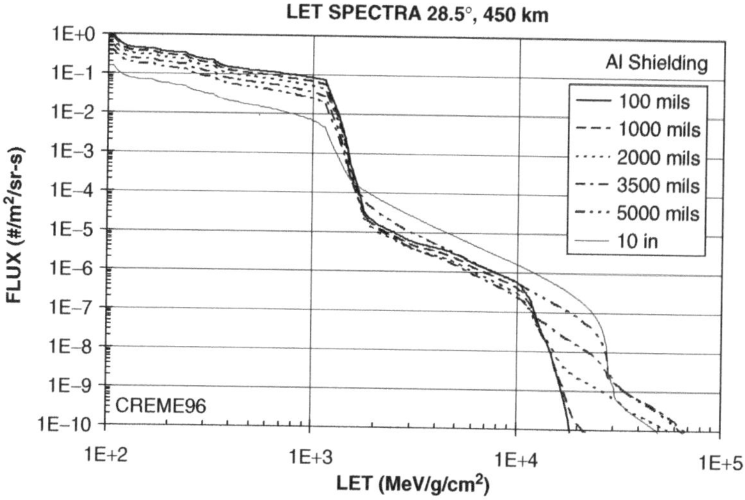

Figure 2-40 LET spectra behind shielding for 28.5 degree LEO.

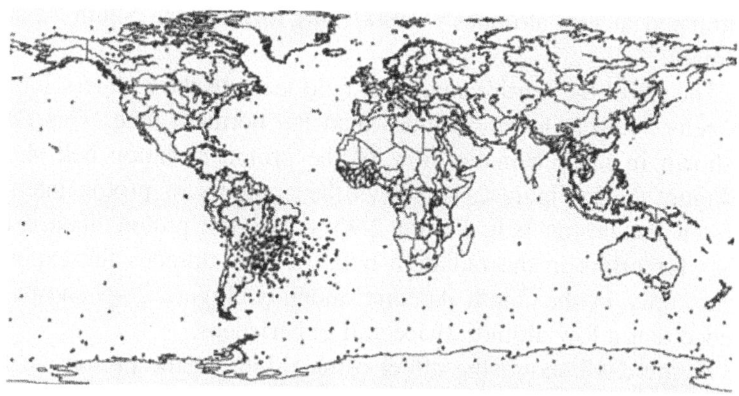

Figure 2-41 Locations of upsets observed with the TAOS satellite showing the importance of the South Atlantic anomaly. Examination of nearly 1300 single event upsets from one computer on the TAOS mission shows that nearly 50% occurred in the South Atlantic anomaly, whereas only 5% of orbital time was spent there [Aerospace 2003].

concern. This is shown by a map in Figure 2-41 of the occurrence of upsets in the TAOS mission [Aerospace 2003].

Figure 2-42 clearly shows the depression in the magnetic field in the South Atlantic caused by the 11° angle between the magnetic and geographic axes. This magnetic field sink causes charged particles to

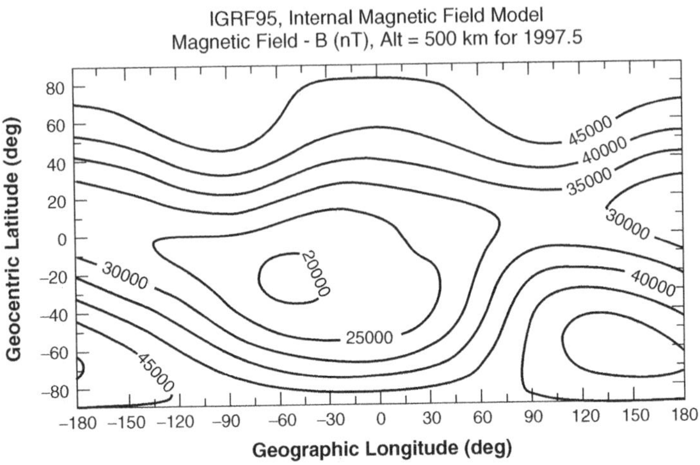

Figure 2-42 Magnetic field contours for 500 km altitude. Note the depression in the field strength in the South Atlantic region [Barth 1997]. Copyright © 1997 IEEE.

be trapped at low altitudes (~1000 km) forming the South Atlantic anomaly.

The tilt of the Earth's magnetic field leads to the southern lobe of the belts being at lower altitudes than the northern lobe. This effect is shown in the schematic view of the proton radiation belt shown in Figure 2-43. Figure 2-44 shows the contours of proton intensity through the proton belt. Figure 2-45 shows the proton fluence that exists for orbits in the radiation belt. At low altitudes the exposure occurs only in the South Atlantic anomaly. Figure 2-46 shows the fluence that a low altitude spacecraft experiences.

Figure 2-47 shows the effect of shielding on the protons in low earth orbit. Almost all of the very low energy protons are removed. However, it is virtually impossible to remove the high energy protons and eliminate proton induced SEEs by shielding. Another sample proton energy spectrum is shown later in Figure 8.7 (vide infra). That figure shows the basic proton environment and the spectrum at a part inside the spacecraft for which the shielding distribution is shown in Figure 8.6 (vide infra).

We have shown proton fluxes calculated from the AP8 model. There are other models under development. There are slight differences in the results of the various models. This has been discussed by Lauenstein and Barth [Lauenstein 2005].

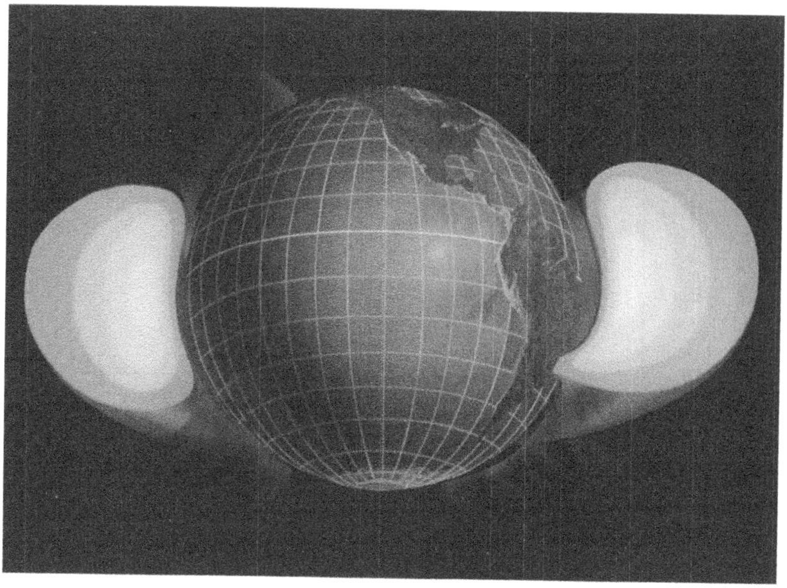

Figure 2-43 Schematic of the proton radiation belt, showing the effect of the anomalous magnetic field in the South Atlantic (from Stassinopoulos) [Ritter 1996]. Copyright © 1996 IEEE.

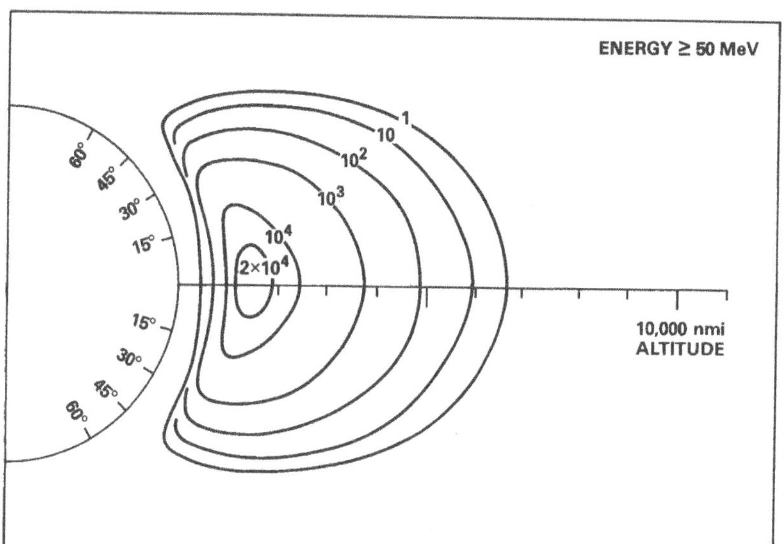

Figure 2-44 AP8Min omnidirectionnel flux (protons/cm^2-s).

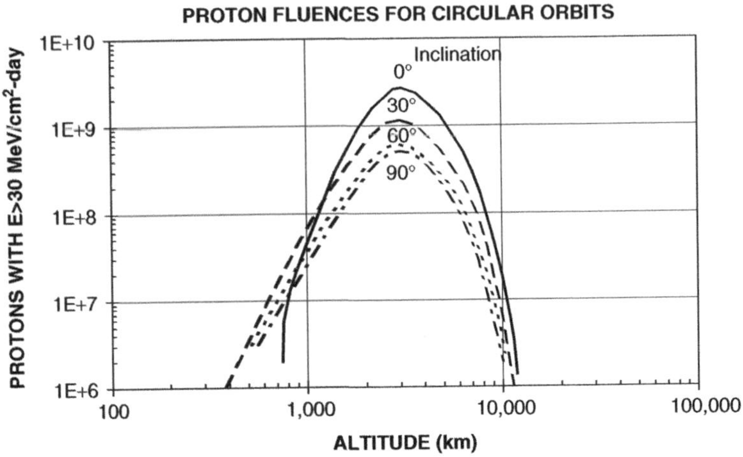

Figure 2-45 Total exposure to trapped protons as a function of altitude and inclination for circular orbits.

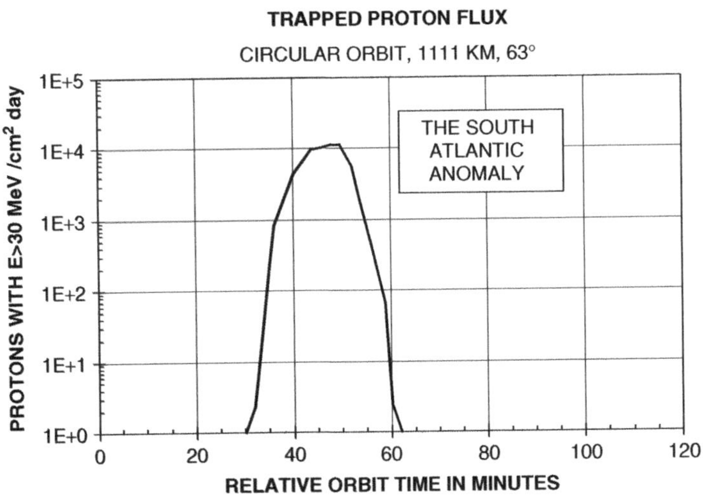

Figure 2-46 Proton exposure as a function of time in a low altitude 2 hour orbit (after Stassinopoulos) [Ritter 1996]. Copyright © 1996 IEEE.

2.3.5 Solar Events

It is believed that there are two categories of solar particle events and that each one accelerates particles in a distinct manner [Bourdarie 2008]. Solar flares result when the localized energy storage in the coronal magnetic field becomes too great and causes a burst of energy

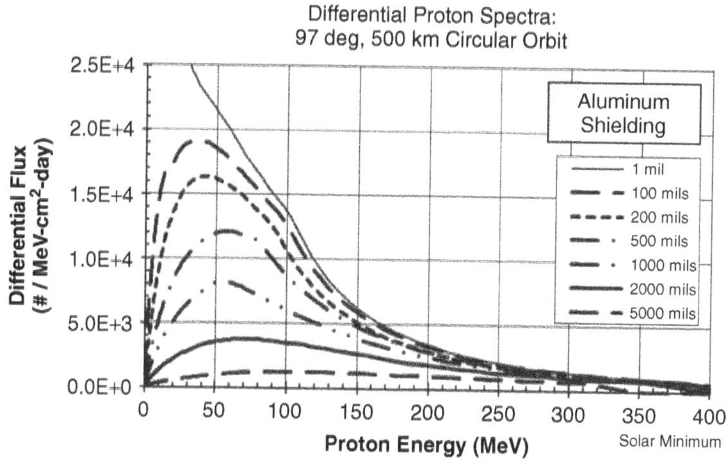

Figure 2-47 Proton spectrum in low earth orbit showing the effect of aluminum shielding.

to be released. They tend to be electron rich and last for hours. A coronal mass ejection (CME), on the other hand, is a large eruption of plasma (a gas of free ions and electrons) that drives a shockwave outward and accelerates particles. CMEs tend to be proton rich and last for days.

CMEs are the type of solar particle events that are responsible for the major disturbances in interplanetary space and the major geomagnetic disturbances at Earth when they impact the magnetosphere. The total mass of ejected plasma in a CME is generally around 10^{15} to 10^{17} grams. Its speeds can vary from about 50 to 2500 km/s. It can take anywhere from about 12 hours to a few days to reach the Earth. All naturally occurring chemical elements ranging from protons to uranium are present in solar particle events. They can cause permanent damage such as total ionizing dose (TID) and displacement damage (DD) that is due mainly to the proton and possibly the alpha particle component. Just because the heavy ion content is a small percentage does not mean it can be ignored. Heavy ions, as well as protons and alpha particles in solar particle events, can cause both transient and permanent SEEs. Figures 2-48 and 2-49 illustrate the periodic yet stochastic nature of solar particle events. They are plots of the daily solar proton fluences measured by the Interplanetary Monitoring Platform-8 (IMP-8) and Geostationary Operational Environment

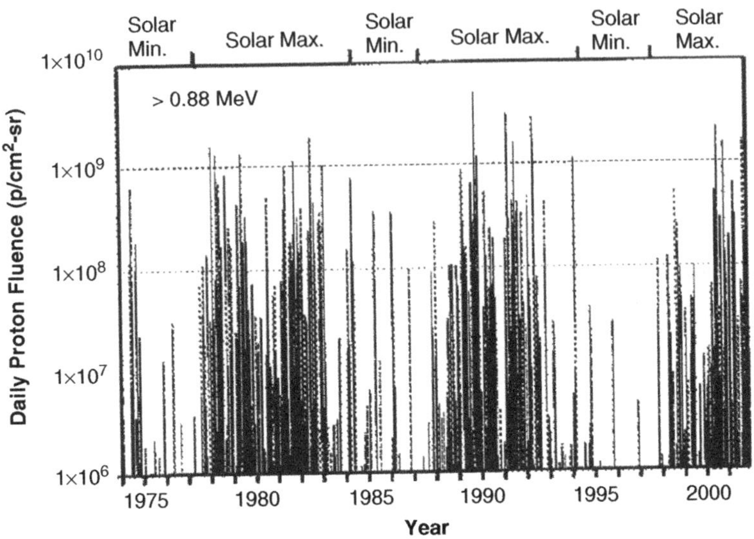

Figure 2-48 Daily fluences of >0.88 MeV protons due to solar particle events between approximately 1974 and 2002 [Bourdarie 2008]. Copyright © 2008 IEEE.

Satellites (GOES) over an approximately 28-year period. Figure 2-48 shows >0.88 MeV fluences while Figure 2-49 shows >92.5 MeV fluences. The solar maximum and solar minimum time periods are shown in the Figures 2-48 and 2-49 to illustrate the dependence on solar cycle.

Figure 2-50 shows the most intense portion of the event in October 1989. Upsets were observed in a number of spacecraft and tracked the proton emission, not the X-ray emission. J. Allen of NOAA determined that problems were observed on several Intelsat, Magellan, DNPS, TDRS-A, TRDS-C, D, GEOS-5, GEOS-6, and Pioneer Venus.

Figure 2-51 shows the heavy ion worst day fluence based on the October 1989 event. Notice that the flare spectrum is soft so that shielding is relatively effective.

2.3.6 Ionization in the Atmosphere

Ionization in the atmosphere was first observed in 1901. Initially it was believed that the ionization was from natural radioactivity of the earth. However, by 1911 it was clear that the intensity of the ionization increased with altitude. By 1912 it was clear that the sources were

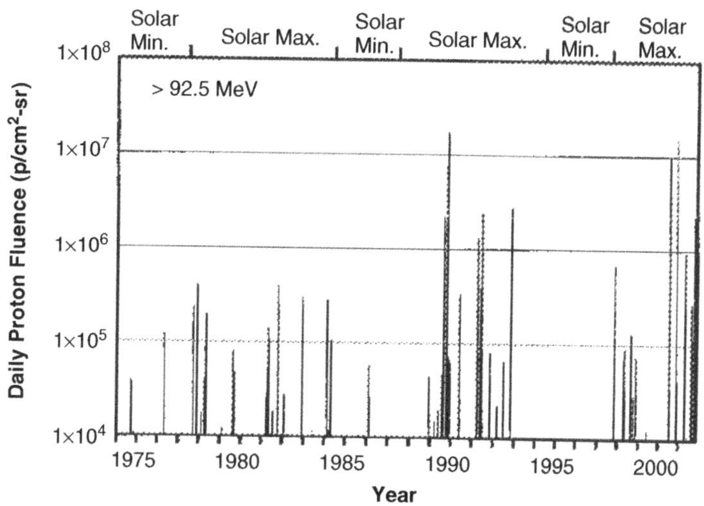

Figure 2-49 Daily fluences of >92.5 MeV protons due to solar particle events between approximately 1974 and 2002 [Bourdarie 2008]. Copyright © 2008 IEEE.

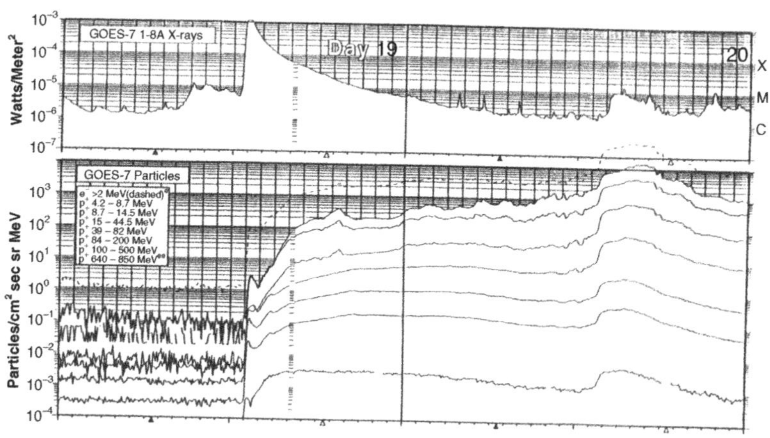

Figure 2-50 The solar terrestrial environment of October 1989 as measured on GEOS-7. Data collected by J. Allen of NOAA and discussed by Ritter [Ritter 1996]. Copyright © 1996 IEEE.

outside the atmosphere. Hundreds of papers explored the phenomena between 1910 and 1940 [Stranathan 1942]. Note that neutrons weren't identified until 1932.

The number of cosmic rays approaching the Earth's surface is initially modified by the Earth's magnetic field. This was discussed

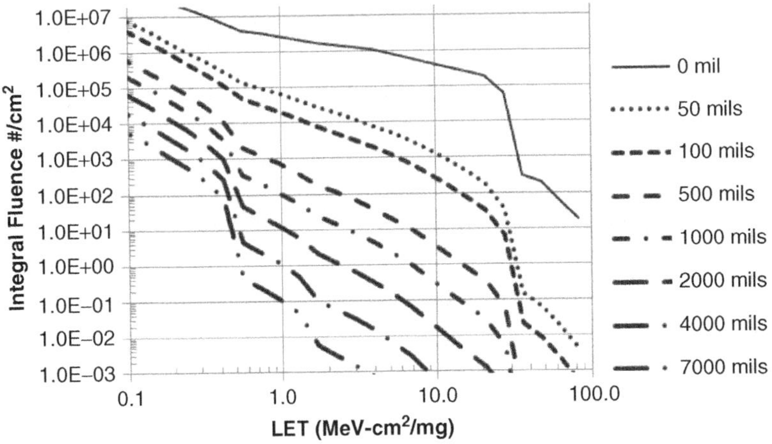

Figure 2-51 Worst day heavy ion fluence based on the October 1989 event.

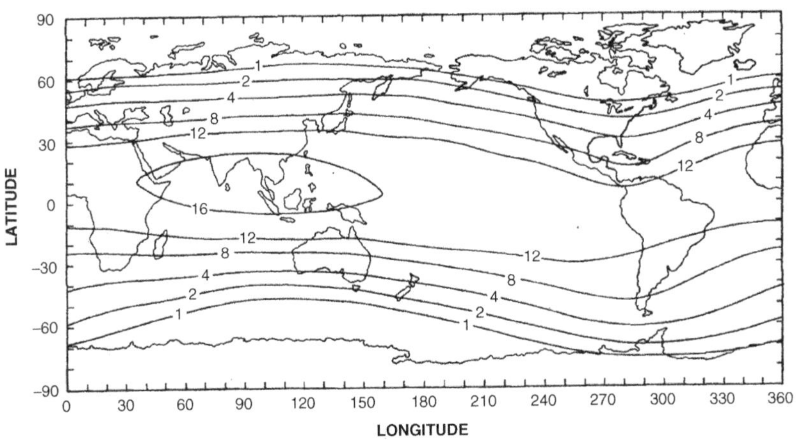

Figure 2-52 Geomagnetic cutoffs at altitudes of 20 km as calculated by Shea and Smart [Tsao 1983]. Copyright © 1983 IEEE.

in Section 2.3.2. Figure 2-52 shows the vertical geomagnetic cutoffs at 20 km [Tsao 1983]. When the particle rigidity is above cutoff, the particle trajectory leaves the Earth without intersecting the Earth's surface. More than half of the surface of the Earth has a cutoff rigidity >4 GeV/c. Hence, most cosmic rays are deflected from the Earth by its magnetic field. Tsao also discusses how the rigidity depends on the angle of arrival, leading to the east–west effect.

As the particles approach Earth, their intensities are modified by nuclear interactions and ionization loss. The heavy nuclei are the dominant source of single event upsets down to roughly 85,000 ft and should be considered down to 60,000 ft [Tsao 1984, Hands 2009]. Figure 2-53 shows the attenuation of the cosmic rays as they enter the atmosphere. The heavy ions with large cross sections are rapidly attenuated. Figure 2-54 shows how the number of heavy ions varies with altitude and cutoff rigidity. The top of the atmosphere is at approximately 150,000 feet.

Figure 2-55 shows the general nature of cosmic ray showers. As the heavy ions are rapidly eliminated, and as there are many more protons than heavy ions, the high energy protons dominate the showers. By the time that the cosmic rays reach the earth there are two components: the relatively low energy components that are of interest to us, up to several hundred MeV, and a small high energy component that can still penetrate several hundred meters of rock.

Although we will discuss the ionization as due to neutrons, Tsao and Silberberg have shown that the relative contributions of primary

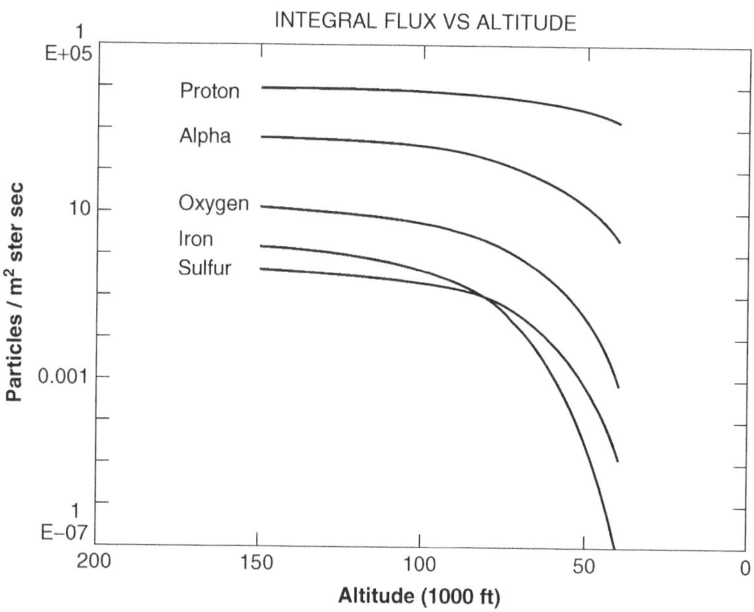

Figure 2-53 Integral flux as a function of altitudes (Rc = 1 GV) for p, a, O, S, and Fe [Tsao 1984]. Copyright © 1984 IEEE.

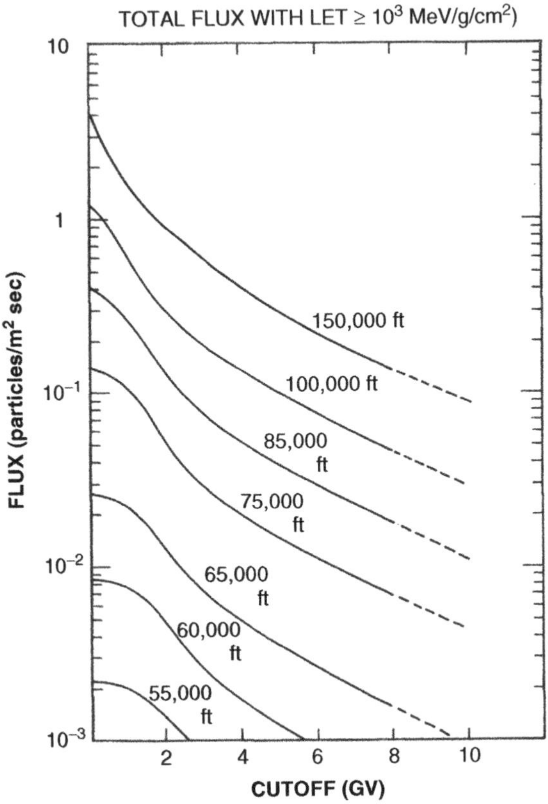

Figure 2-54 The total cosmic ray flux in the atmosphere as a function of altitude and cutoff rigidity [Tsao 1983]. Copyright © 1983 IEEE.

cosmic rays and secondary neutrons and protons vary with altitude [Tsao 1983, Tsao 1984, Silberberg 1984]. These were the first papers from the radiation effects community to point to potential avionics upsets.

The intensity of the cosmic rays can be determined by measuring the resulting ionization as a function of altitude. Figure 2-56 shows the amount of ionization as a function of depth into the atmosphere. The depth in the atmosphere is measured in terms of a corresponding depth of water. A depth of 10 meters of water corresponds to sea level (1038 g/cm^3). These measurements were obtained by Millikan and others in the years 1933 to 1937. These are a sample of a large number of papers that examined the variation with altitude and latitude. Later

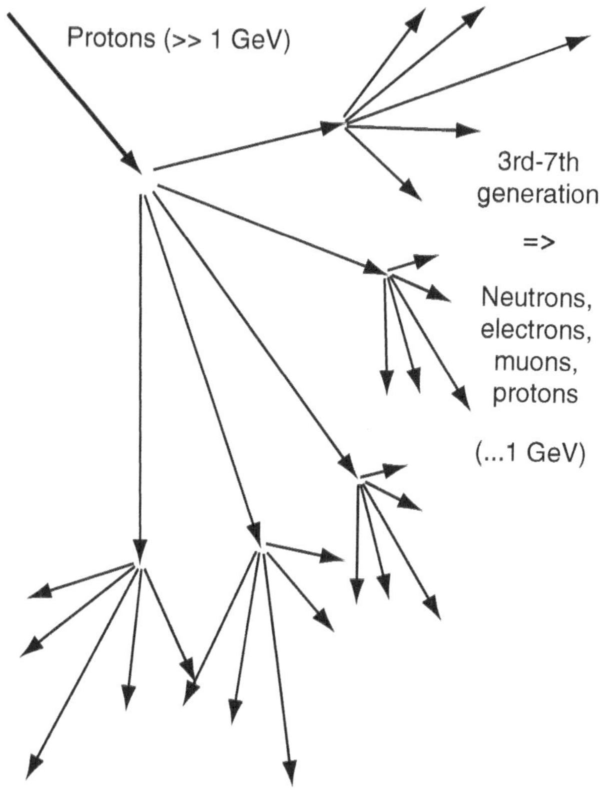

Figure 2-55 Structure of cosmic ray showers [Leray 2007]. Copyright © 2007 J. L. Leray. Used with permission of author.

measurements at latitudes through 60 degrees extended up through 98.8% of the environment continued to show the ionization maxima shown in Figure 2-56. The ordinate in Figure 2-56 represents the number of pairs of ions formed per cubic centimeter per second in an ionization chamber in which the pressure is 1 atmosphere. The relative behavior of ionization with altitude or latitude will be the same whether the ionization is measured in an ionization detector or in a silicon electronic device. The relationship between the mass of the overlaying atmosphere and the latitude are shown in Figure 2-57.

Despite the predictions of Ziegler [Ziegler 1979], little attention was paid to single event effects in the atmosphere until 1993 when Tabor and Normand replaced anecdotal discussions of single event

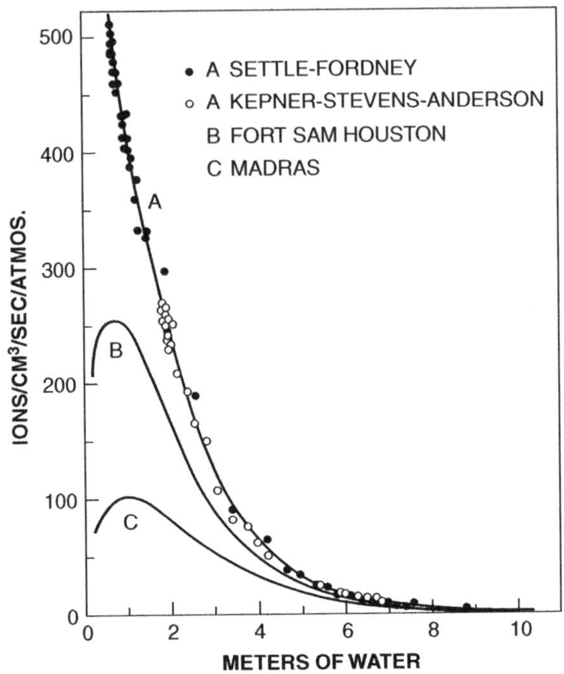

Figure 2-56 A comparison of the variations of cosmic ray intensity at Madras, India (magnetic latitude 3 degrees north), Fort Sam Houston, Texas (magnetic latitude 39 degrees north), and on two stratosphere balloon flights (magnetic latitude 53 degrees north) [Stranathan 1942]. Copyright © 1942 reprint January 1948, The Blakiston Company, Philadelphia, acquired by Doubleday, sold to McGraw Hill, not renewed, so now in public domain.

effects in airplanes with a report on an extensive series of measurements from experimental flights that IBM sponsored in 1988 and 1989 combined with Boeing data on military flights [Tabor 1993]. There were a large number of studies of cosmic rays near Earth during the International Geophysical Years (1958–1967). Ziegler has presented a major review with numerous references for cosmic rays in the atmosphere [Ziegler 1996b].

Normand and Baker presented a detailed discussion of altitude and latitude variations of the 1–10 MeV neutron flux, based on a series of measurements made by geophysicists in the 1960s [Normand 1993, Normand 1996]. They also presented evidence that these results reflected the behavior of the entire neutron spectrum. The results are shown in Figures 2-58 and 2-59. Both papers demonstrated that the

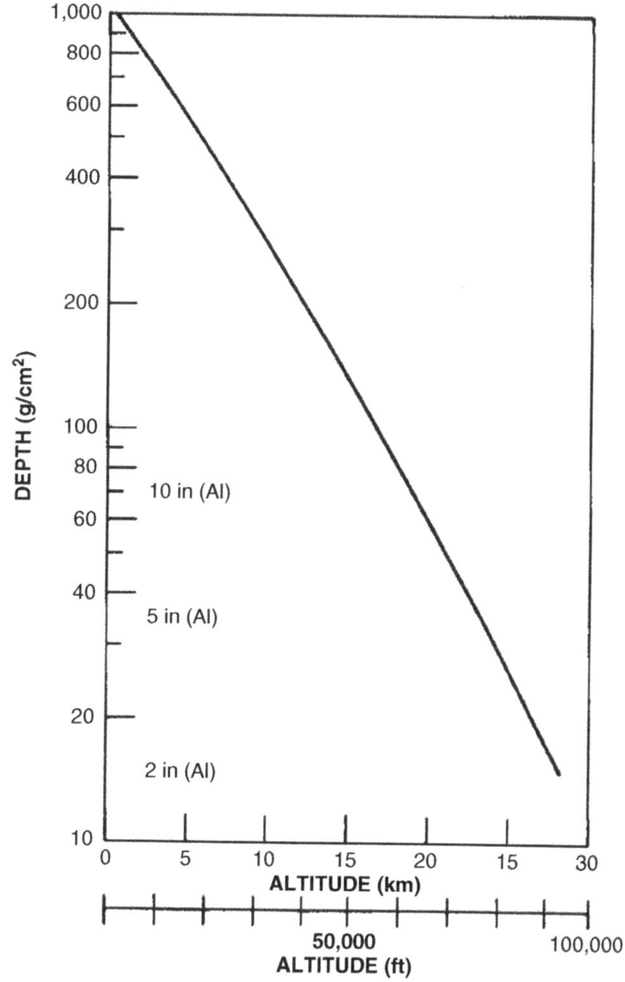

Figure 2-57 The mass of the overlaying atmosphere as a function of altitude.

flight data followed the environmental data. Figure 2-60 shows the energy spectra obtained by the NASA Ames group. The Texas balloon flights shown in Figure 2-56 have essentially been repeated with emphasis on the protons and are shown in Figure 2-61 [Normand 1996]. This high energy portion of the proton spectra is 16% of the neutron flux, so that the total proton contribution is more than 20% of the neutron flux.

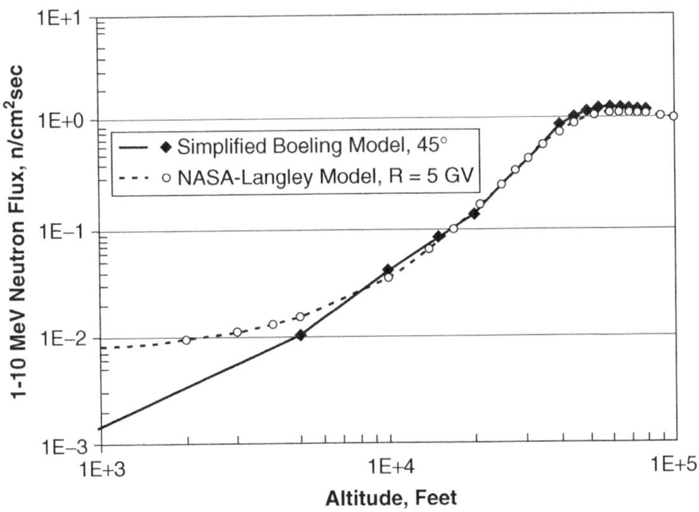

Figure 2-58 The 1–10 MeV atmospheric neutron flux as a function of altitude [Normand 1993, Edwards Edwards 2004]. Copyright © 1993, 2004 IEEE.

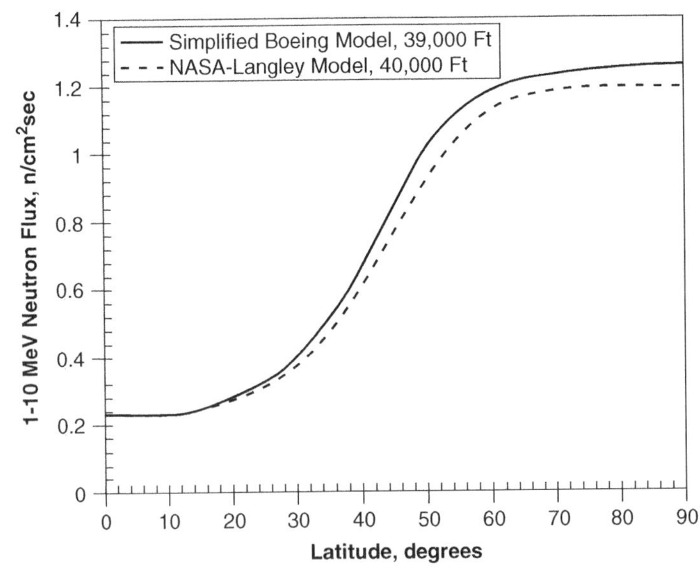

Figure 2-59 The 1–10 MeV atmospheric neutron flux as a function of latitude [Normand 1993, Edwards Edwards 2004]. Copyright © 1993, 2004 IEEE.

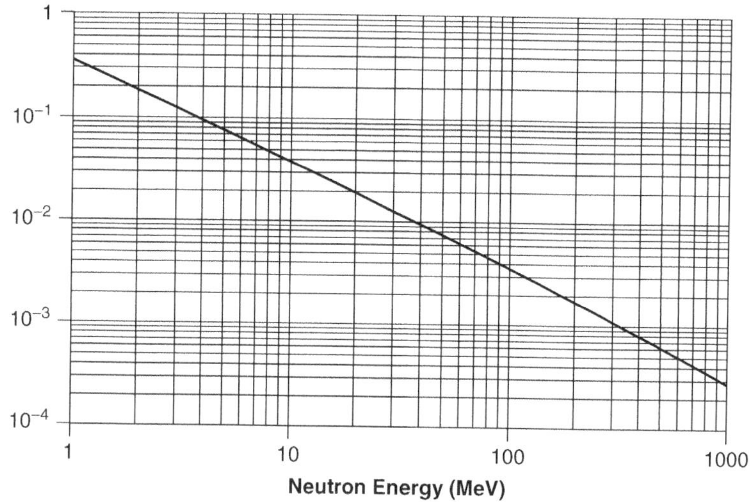

Figure 2-60 Spectrum of the average differential neutron flux in atmosphere at 40,000 feet and 45 degrees latitude, based on NASA Ames data and normalized to a 1–10 MeV neutron flux of 0.85 n/cm^2-s (after Hewitt) [Normand 1996]. Copyright © 1996 IEEE.

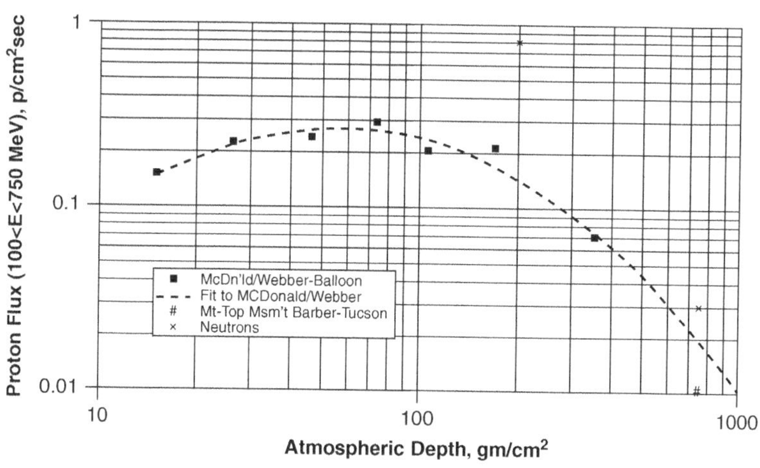

Figure 2-61 Comparison of measured atmospheric proton flux (100 < E < 750 MeV) as a function of atmospheric depth. Note that this portion of the proton flux is approximately 16% of the neutron flux at this location (after Webber) [Normand 1996]. Copyright © 1996 IEEE.

2.4 CHARGE COLLECTION AND UPSET

The ion track leaves a dense plasma of electron–hole pairs along its path. If the electron–hole plasma is generated in a region with an electric field—at a pn junction, for example—electrons and holes are separated and a current spike is observed at the circuit node as free carriers are collected, as shown in Figure 2-62. This current spike has two components: a prompt component that lasts for the order of hundreds of picoseconds after the ion strike, and a delayed component that may last hundreds of nanoseconds. The prompt current is due to charge collection in the original depletion region and the funnel region [Hsieh 1981a, Hsieh 1981b, Hsieh 1983, McLean 1982, Messenger 1982, Gilbert 1985, Murley 1996]. The delayed component of current is due to carrier diffusion up to the depletion region where it is quickly collected by the junction electric field.

We will not have any discussion of the funnel effect. It has been discussed in the articles mentioned above and summarized by Massengill and Oldham [Massengill 1993c, Massengill 1996, Oldham 2003]. It is an important effect in some devices. We will later examine how it is included in upset rate calculations (Section 5.3) and how it is sometimes evident in upset rate measurements (Section 4.1).

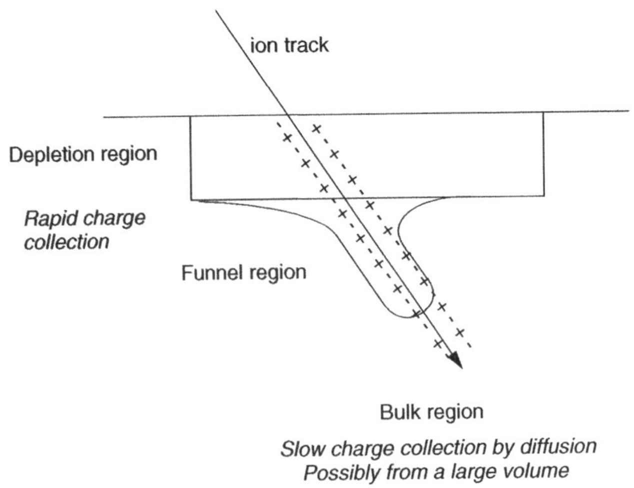

Figure 2-62 SEE charge collection regions in a device.

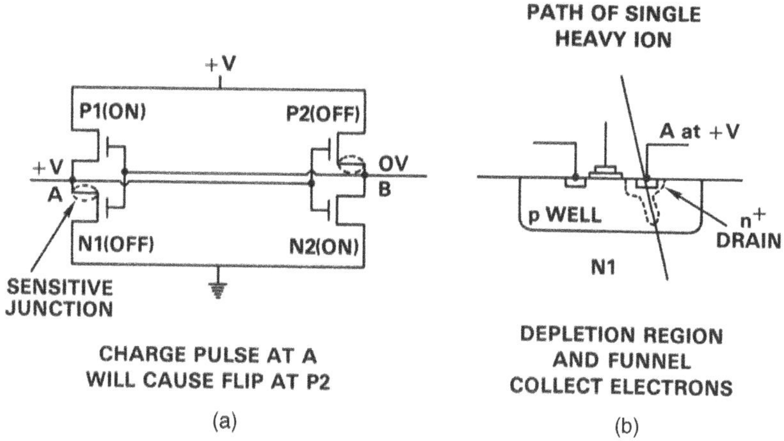

Figure 2-63 Single event upset mechanism in a CMOS static RAM. There are sensitive junctions at A in N1 and B in P2 [Petersen 1983b].

The pulse generated at a junction, as in Figure 2-63, can lead to a change of state in a flip-flop, or propagate down a string of inverters in a logic circuit. In other cases it can lead to latchup or the other single event effects. See the discussions in Sexton, Massengill, and Dodd for a review of the details of these effects [Sexton 1992, Massengill 1993c, Massengill 1996, Dodd 2003].

Figure 2-63 illustrates a SRAM cell held in a valid logic state. Figure 2-63b depicts a single ion strike to the reverse biased drain junction of N1 (node A) within the inverter formed by P1/N1 of the SRAM cell. For the inverter formed by P1/N1, with input node B held low, output node A is pulled to VDD by P1. Similarly, for inverter P2/N2, with input node A forced high, output node B is pulled low by N2 and the SRAM is held in a stable state. If an ion strikes and enough charge is collected at node A to lower its potential below the threshold of inverter P2/N2, the logic state stored in the SRAM may change logic state (flip). The amount of charge necessary to cause a flip is called the critical charge (Q_{crit}). The flip occurs when the input to P2/N2 is driven low enough that P2 turns on and N2 turns off causing the voltage on node B to rise from low to high. If node B crosses the threshold of inverter P1/N2 while node A is still below the threshold of inverter P2/N2, the feedback loop is completed and the logic state stored in the SRAM cell will flip [Hafer 2008].

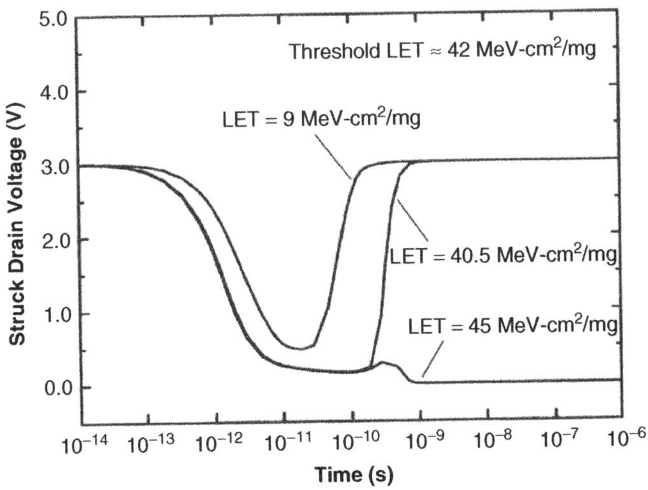

Figure 2-64 SRAM struck drain voltage transients for ion strikes with LET well below, just below, and just above the SEU threshold. Even the ion strike with LET well below the actual SEU threshold is sufficiently ionizing to momentarily flip the struck node voltage [Dodd 2003]. Copyright © 2003 IEEE.

A "transient" logic state change can occur due to an ion strike with the LET below the upset threshold as shown in Figure 2-64 [Dodd 2003]. Below the upset threshold this momentary upset is not sustained. The response time of the opposite inverter is competing with the duration of the drain voltage drop. Above the upset threshold the drain voltage drop is long enough for the opposite SRAM inverter to lock in the new and incorrect logic state. Both the quick charge collection drift process and the slower diffusion process are responsible for the struck drain voltage as a function of time [Hafer 2008].

2.5 EFFECTIVE LET

The fact that the single event effect depends on device thickness, and that the amount of charge deposited depends on the total path length in the device, leads to simplification in device testing. One wants to determine the device sensitivity as a function of the amount of charge deposited. As the charge deposited in a thin lamina depends on the angle of incidence of the particle, it is possible to measure over some range of charge deposition by varying the angle of the beam. However, these results need to be normalized to the effects of the beam at normal incidence. That is, what is the equivalent amount of charge deposition for a beam perpendicular to the face? In early experimental work the idea of linear charge deposition (LCD) was introduced [Koga 1984],

$$\mathrm{LCD} = L_0 \sec\theta$$

where L_0 is the LET of the incident ion. See Figure 2-65

This idea has developed into the concept of effective LET:

$$\mathrm{LET(Eff)} = \mathrm{LET}(0^\circ)\ \sec\theta \tag{2.2}$$

Device measurements are then expressed in terms of the upset cross section as a function of effective LET. The cosine dependence introduced here arises from the charge deposition as a function of angle between parallel surfaces, where the increased path length and charge deposition follow the cosine of the angle. We will see later, in our discussion of geometrical effects in small devices, that the concept of effective LET sometimes is not appropriate or has to be used with care.

2.6 CHARGE COLLECTION VOLUME AND THE RECTANGULAR PARALLELEPIPED (RPP)

The charge collection volume is assumed here to be a plane lamina. This simplifies the discussion of the basic effects. The upset rate calculation depends on the paths available in the actual charge collection region. Ordinarily this region is treated as a rectangular parallelepiped. This has a reasonable relationship with the actual volumes expected. It

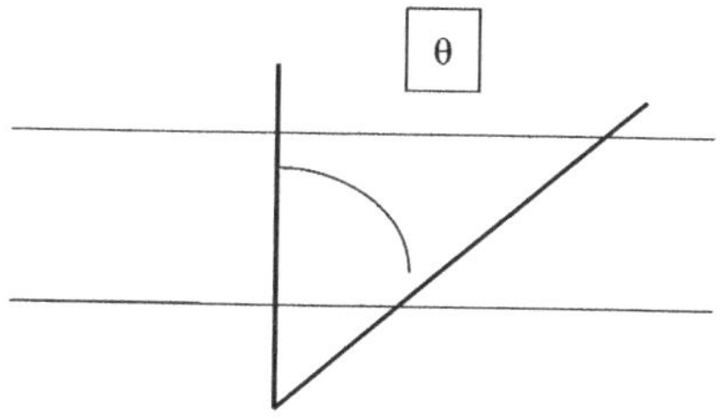

Figure 2-65 Increased path length at an angle leads to effective LET concept.

simplifies the calculations because there are well-known relationships for the distribution of path lengths that omnidirectional particles would encounter in a RPP.

It is possible to do the calculation with other shapes such as a cylinder or a hemisphere and its variations, but these do not seem to have the general application of the RPP. The path length distribution can be expressed in either a differential form or an integral form. As we will see, either form can be used for the upset rate calculations.

2.7 UPSET CROSS SECTION CURVES

We think of the circuit as turning on with the deposit of some unique amount of energy or charge. We also think of the sensitive volume as having a constant depth. Hence there is a unique effective LET at which the circuit will turn on. If we perform an experiment looking for turn-on as a function of LET, we expect that we will see no reaction until we reach a certain critical charge or LET. At this point all of the circuits on the chip should flip, and we will see a step function in the response, or cross section. This expected behavior is shown in Figure 2-66. In practice, we never see this behavior. Instead, we see the type of curve that is shown in Figure 2-67. This curve is replotted in Figure 2-68 so as to see the characteristic "S" shape. These curves show a gradual increase in the area of the device that upsets. The next section will discuss the source of this effect.

2.8 CRITICAL CHARGE

The device characteristic that determines the upset sensitivity of a device is its critical charge. This is the amount of charge that must be collected at the terminal of the device to cause the single event effect. The critical charge depends primarily on the circuit characteristics. The concept of critical charge underlies the entire area of single event upsets. With this basic definition laid down, we still find that the term critical charge has been used in a number of different ways. As we continue on into the subject, and as we read the literature, we need to be aware of the way the term is being used. The following sections will briefly discuss the various usages of the concept of critical charge.

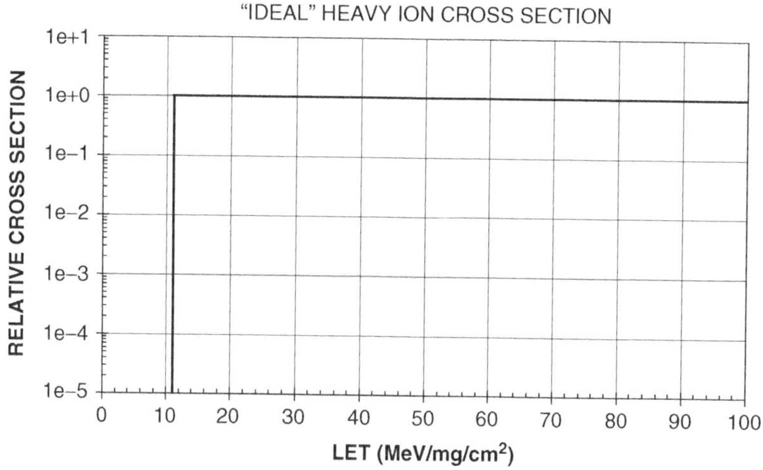

Figure 2-66 The cross-section curve expected if all parts of all cells had the same sensitivity.

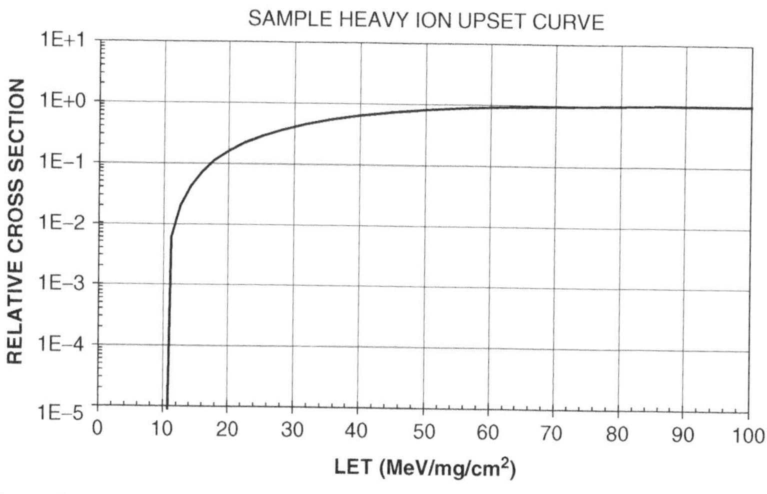

Figure 2-67 Representative heavy ion cross-section curve with cross section expressed as a function of effective LET. This curve has Weibull parameters: $L_0 = 10$ MeV/mg/cm^2, $W = 30$ MeV/mg/cm^2, shape parameter = 1.6.

2.8.1 Critical Charge and LET Threshold

The critical charge is ordinarily assumed to be indicated by the curve of device cross section versus LET. This curve is plotted as an integral curve indicating the relative fraction (cross section) of the total

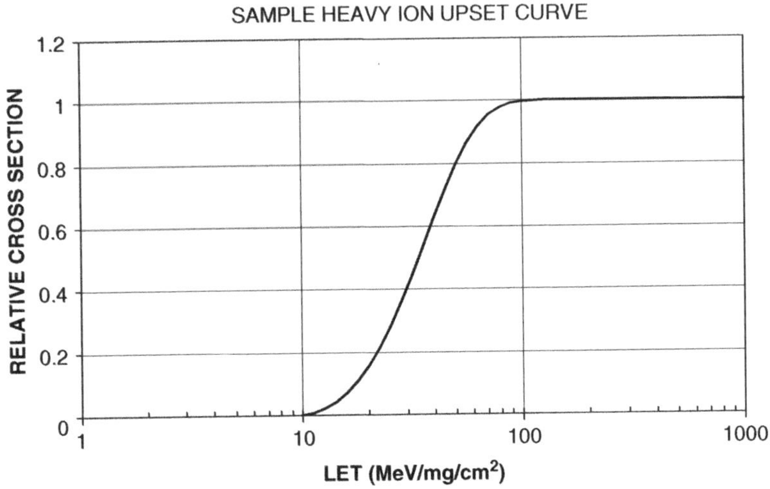

Figure 2-68 The representative cross-section curve of Figure 2-67 plotted so as to show the characteristic shape of the Weibull function. This curve will be used as the basis for a number of examples as we proceed.

device that will upset for any energy deposition equal to or greater than that produced by this effective LET. This was demonstrated in Figures 2-67 and 2-68. The critical charge is determined by multiplying the chosen LET by the depth of the device. The initial workers in the field assumed that the LET threshold was the appropriate LET for calculating the critical charge. This leads to an assumed simple relationship between onset threshold, critical LET, and critical charge. This assumption corresponds to the assumption that that all of the devices on a chip have the same sensitivity, and that the integral curve of cross section should really be a step function. This concept is still often used in the field. The author will attempt to avoid this usage, as he believes that it is wrong. The threshold LET corresponds to the situation of a very small area on a device that first collects enough charge to cause the upset. This is not representative of all parts of the array.

2.8.2 Critical Charge of an Individual Transistor, Two Transistors in a Cell

The early authors were very aware that in some type of circuits, such as CMOS flip-flop memory elements, there are two transistors that are sensitive to upset. These may have different critical charge, different areas, and different depths. The two parts of the circuit have different electrical characteristics. There may be different sensitivities for a flip $0 \rightarrow 1$ versus $1 \rightarrow 0$. These differences need to be considered when measuring upset rates, when performing circuit modeling of the chip, and when making upset rate predictions for the chip. There has been a recent tendency of the community to speak of the critical charge and upset rate of a bit, and to ignore the details of the electronic design of the bit. This is sometimes valid. At other times it can lead to confusion. As we continue, we need to keep the possible importance of this distinction in mind.

2.8.3 Critical Charge from Circuit Modeling Studies

The critical charge of a memory unit depends on the circuit characteristics. It is clearly straightforward to use standard circuit modeling techniques to determine what current pulse injected at a charge collection node will lead to an upset of the memory cell. The charge in this current pulse is then the critical charge. Unfortunately, as one examines this in more detail for a variety of circuits, the situation becomes more opaque. The results depend on the relative time constants of the charge generation and collection, and of the circuit time constants. Circuit modeling has been discussed in detail by Sexton and Massengill and will not be discussed here [Sexton 1992, Massengill 1993c, Black 2006].

2.8.4 Sensitivity Distribution Across the Device

At one time it was believed that the shape of the cross-section curve was primarily determined by a spread of critical charges from cell to cell inside a device [Petersen 1992b]. In this interpretation the cross-section curve reflects the fraction of cells that upset at any given LET. This implied that a point on the low cross-section portion of the curve would correspond to only a few cells upsetting. This was shown not to be true by the experiments of Cutchin and of Buchner [Cutchins 1993, Buchner 1994, Petersen 1993]. Cutchins's experiment showed that a point on the low cross-section portion of the curve corresponded to all of the cells having the same low cross section, while Buchner's experiment showed that all cells had the same critical charge for upset.

Two early papers examined variation across the chip directly. Kohler examined hardened CMOS RAMs [Kohler 1989]. He pointed out that the statistical distribution of hardening resistor values will introduce a spread of critical charges, in addition to the normal width factor of unhardened RAMs. Massengill examined the combined effects of bipolar gain variation in a SOI transistor and parameter variation from cell to cell across the device [Massengill 1993a, Massengill 1993b]. The primary effect is the distributed bipolar gain, which varies with position across a single transistor. In both situations, the better process controls on later chips will reduce the contribution of the variation across the chip [Massengill 1993a, Buchner 1994].

It now appears that there is no support for the concept of critical charge variation from cell to cell across the chip being primarily responsible for the shape of the upset cross-section curves.

2.8.5 Intracell Variation

A great deal of evidence has accumulated in favor of the concept that the variation of cross section with LET commonly observed is due to variations inside the cell. Much of this evidence was obtained while pursuing other questions and was not viewed in this light. We will discuss representative results for a number of technologies [Petersen 1996a, Petersen 1996c].

It appears that all semiconductor technologies involve some combination of geometry, charge collection, and charge amplification such that the device sensitivity varies across an individual transistor. This appears to be the main reason for the variation of the curve of cross section with LET. There appears to be very little contribution from a possible range of sensitivities among the cell population. The spectrum of LET sensitivities is due to intracell variations, not intercell variations.

2.8.6 Summary Discussion of Critical Charge

The charge that should be used in upset rate calculations is that which corresponds to a product of the LET of an incident cosmic ray, and a defined charge collection distance. The cross-section curves indicate upset over a range of LETs. Therefore, upset rate calculations need to take this into account. The general practice is to assume that the device has a single depth and a range of collected charges, and then sum the contributions according to their fractional input into the total upset rate. This is the integral rectangular parallelepiped (IRPP) method.

If we are discussing the critical charge of a typical cell or transistor in the device, we will define critical charge operationally as the value calculated from the LET at which 50% of those transistors lead to upset, combined with the charge collection depth of that transistor. This will correspond to the value obtained from circuit analysis if all of the physical and electronic effects in the device are properly modeled.

2.9 UPSET SENSITIVITY AND FEATURE SIZE

As devices have gotten smaller and faster, they store less charge on circuit nodes. The energy to change the state of a node likewise decreases. This has led to circuit vulnerability to upset from highly ionizing particles increasing as the feature size decreases [Sexton 1992]. This is shown in Figure 2-69, which shows the variation of the SEU threshold for upset as a function of device feature size [Petersen 1982, Petersen 1983b, Petersen 1988]. One of the remarkable things about this figure is the lack of dependence on device technology.

2.10 CROSS-SECTION CONCEPTS

The concept of cross section is basic to our understanding of single event phenomena. It is the most used, and the most misused, concept in the field. It has been developed in analogy to the concept of cross section as used in nuclear physics. The analogy is not complete, so there have been numerous misunderstandings. Let us first review cross section as used in nuclear physics.

2.10.1 Nuclear Physics Cross-Section Concepts

Nuclear physicists are attacking the problem of measuring the probability of something happening when one set of particles is incident on another set of particles. The individual particles are atomic nuclei

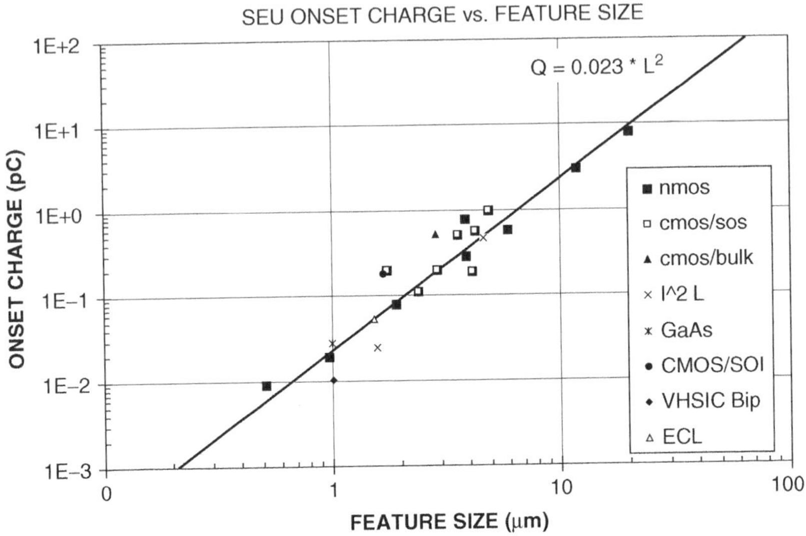

Figure 2-69 The generally observed variation of upset threshold with feature size for memory cells.

and are very very small, on the order of 10^{-26} cm^2. Both the incident particle and the target particle are assumed to have spherical symmetry. The spacing of the target particles is very large compared to the size of the particle. Therefore, the incident particles have a small probability of hitting anything while passing through the assembly of target particles. A common goal is to determine the probability of a given nuclear reaction as a function of the incident energy.

The probability of the target being hit is measured by its apparent cross-sectional area,

$$\sigma = \frac{N}{F} \tag{2.3}$$

where N is the number of times that an event is observed, and F is the incident particle fluence, usually expressed as particles/cm^2. σ is the number of events per unit fluence and has the dimensions of area.

It is not feasible to isolate a single target, so a measurement in the laboratory has to measure the total probability for a large number of target nuclei, and then determine the individual cross section. There are three different experimental setups commonly used. We need to consider these cases separately.

2.10.1.1 Broad Beams

In this situation, shown in Figure 2-70, a small object is obtained consisting of the target nuclei. The object may have any shape. It is placed in a beam of incident projectile nuclei. The cross section is then operationally

$$\sigma = \frac{N}{F \times M} \tag{2.4}$$

The number of target nuclei M is determined from the number of nuclei per unit volume and the total volume of the target. N is now the total number of events observed from the bombardment of the total volume of the target. F is again the total fluence of particles and σ is the measured cross section. The measurement assumes that the cross-sectional area of the beam of incident particles is much larger than the cross section of the volume being irradiated, and that the beam particles are distributed uniformly.

The results do not depend on the shape of the test object, or on its orientation, but only on the element in the sample. The sphere and the two rectangular bars will produce the same cross sections, depending on the number of events observed, the fluence, and the target density and volume, but not on the different shapes or orientations.

2.10.1.2 Uniform Isotropic Exposure

This situation corresponds to an irradiation inside a reactor. There is a uniform field of fast neutrons, going in all directions. A field of

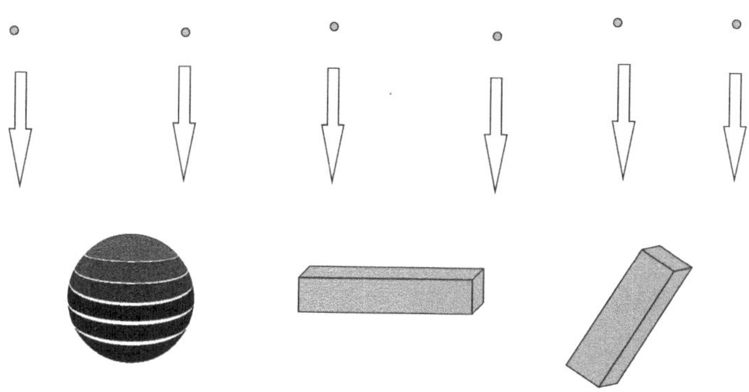

Figure 2-70 Nuclear cross-section measurements in a broad beam.

this type is measured in terms of the number of particles hitting a 1 centimeter circle in the center of a sphere. The units of flux are number/steradian-cm^2-second. If an object is exposed for a time, then the fluence is $4\pi \times$ number/steradian-cm^2. The cross section is calculated exactly as in the case of a broad beam exposure:

$$\sigma = \frac{N}{F \times M} \tag{2.5}$$

In this case, Figure 2-71, as in the broad beam exposure, the cross section does not depend on the shape or orientation of the target. This geometry also corresponds to the geometry in space, where an object is exposed to a field of uniform isotropic cosmic rays.

2.10.1.3 Narrow Beam

In many experimental situations, such as those using a low energy accelerator, it is not possible to use the broad beam, thick target setup described above. In these cases, the beam has a small unknown area and is not uniform, as in Figure 2-72. The target is ordinarily a thin foil whose area is much larger than the area of the beam. It is convenient to express the density of the foil in terms of its areal density, m, that is, the number of target nuclei per cm^2. The total number of nuclei in

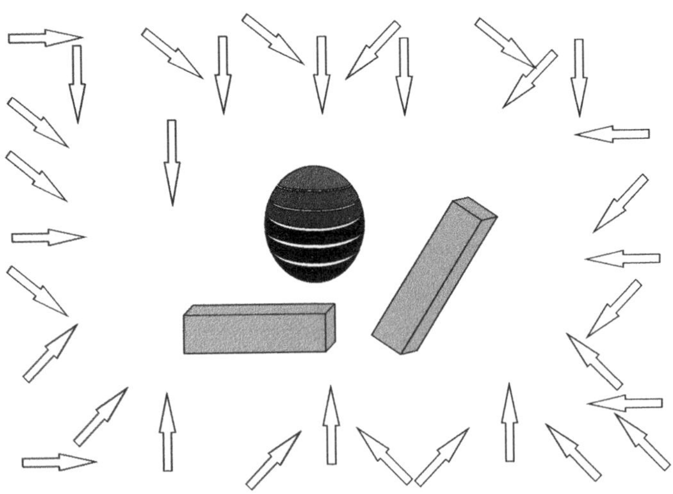

Figure 2-71 Nuclear cross-section measurements with a uniform isotropic exposure.

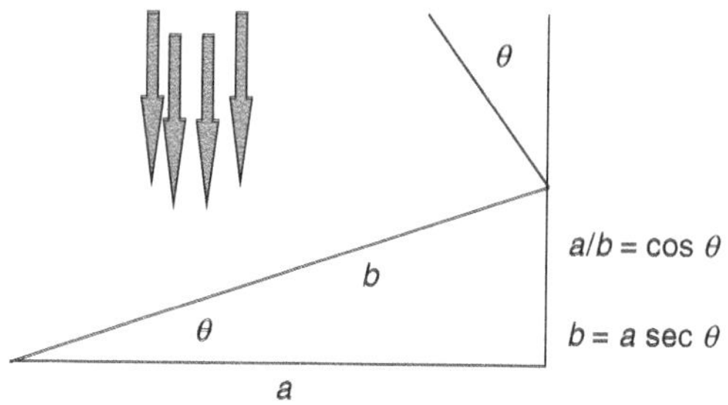

Figure 2-72 Nuclear cross section measurements in a narrow beam.

a unit area of the foil is the areal density times the unit area. The areal density is determined from the product of the volume density and the thickness of the foil. In the narrow beam case we observe the total number of particles passing through the foil, the total fluence f. We can again determine the probability of an event happening, σ, which will have units of cm^2.

$$\sigma = \frac{N}{f \times m} \tag{2.6}$$

It is often necessary to change the angle of the foil relative to the beam during a set of experiments. As the beam has a constant area, the total area of the foil being exposed is changing, and therefore the number of target nuclei exposed is changing. Therefore, the number of observed events depends on the angle of the foil. The angle of the foil is expressed as the angle between the beam direction and a line normal to the foil surface. As the angle increases, the area increases as $1/\cos\theta$, or as secant θ. Therefore, the number of target nuclei is increasing as secant θ. This factor has to be included as one calculates the cross section from the number of events observed:

$$\sigma = \frac{N}{f \times m \times \sec\theta} \tag{2.7}$$

Note that we have identified the increased number of observed events with the increased number of target nuclei exposed to the beam, not with the increased path length of the beam in the foil. For both broad beam and narrow beam experiments, the cross section depends on the ratio of the observed number of events to the number of possible collisions or events. Both types of experiments will yield the same cross sections.

2.10.1.4 Calculation of Expected Number of Events

Now let us consider the calculation of the number of expected events if we know the cross section. The traditional unit of cross section in nuclear physics is the barn, which is 10^{-24} cm^2. More common is the millibarn (mb), which is 10^{-27} cm^2. Consider a nuclear reaction in silicon that has a cross section of 200 mb. First consider the broad beam environment with 10^{10} p/cm^2 on a 10 mg target:

$$\begin{aligned} N &= \sigma \times F \times M \\ &\quad 200 \text{ mb} \times 10^{10} \text{ p/cm}^2 \times 10 \text{ mg} \times 6 \times 10^{23}/28 \text{ g} \\ &\quad \times 10^{-24} \text{ cm}^2/\text{b} \times 10^{-3} \text{ b/mb} \times 10^{-3} \text{ g/mg} \\ &= 4.28 \times 10^5 \text{ reactions} \end{aligned}$$

The $F \times M$ terms could have been expressed in other units for other geometries. The $F \times M$ term in the omnidirectional case would be $4\pi \times 7.96 \times 10^8$ p/ster-cm^2 $\times$ 10 mg. The $F \times M$ term in the thin target case would be 10^{10} p $\times$ 10 mg/cm^2. In all three cases we assume 10^{11} p-mg/cm^2.

2.10.2 Single Event Cross-Section Concepts

The measurements of single event cross sections are conducted using methods parallel to those used in nuclear physics. There are a number of parallel aspects, but the overall concepts and interpretations are different, depending on the single event situation. There are again two different cases: (1) effects due to ionization produced by reaction products after a nuclear reaction in the silicon and (2) effects due to direct ionization by highly ionizing heavy particles. The measurements appear similar, but the interpretations are entirely different.

2.10.2.1 Proton Single Event Cross-Section Measurements

Proton upset and nuclear reaction cross-section measurements essentially correspond. The reason is that the proton upsets depend on nuclear reactions in the device, not on the path length in the device. The experimental geometry corresponds to the broad beam nuclear physics experiment. The beam cross section is much larger than the individual sensitive regions, or even of the devices themselves. The number of upsets depends on the sensitive volume and the device sensitivity. The proton upset measurements usually have the goal of determining the proton cross section as a function of proton energy.

The cross-section equations correspond to the broad beam nuclear physics experiment. The cross section per chip is

$$\sigma = \frac{N}{F} \tag{2.8}$$

The probability of having an upset on the chip is given by σ—which will have units of cm^2. N is the number of upsets observed, and F is the fluence in particles/cm^2. If one wants the upset cross section per bit, then the total cross section is divided by the number of bits M:

$$\sigma = \frac{N}{F \times M} \tag{2.9}$$

In this situation, there is a true correspondence with the broad beam experiments. The upsets depend on nuclear reactions in and near the sensitive volume of bit. Neither the sensitivity nor the volume change as the device is rotated. The situation corresponds to Figure 2-70. There need be no reference to the changing area exposed to the beam, or to the change in path lengths. Neither of these is relevant to the number of upsets observed. The equations just presented apply independent of angle. The cross-section calculation assumes that the beam is uniform over the surface of the target. If you want a higher upset rate and ask the operator for more beam, be careful that he does not get it by focusing the beam.

Normally proton upset experiments are performed only at normal incidence. There should be no angular dependence to first order. However, there are possible second order effects due to details of the nuclear reaction processes [Reed 1994a, Reed 1995].

Note that the proton cross-section measurements are performed at facilities that are normally used for narrow beam nuclear physics experiments. However, the experimental setup is entirely different for upset measurements. The beam is broadened so that it is as uniform as possible. Dosimetry apparatus is set up so that the total fluence on a defined area from the uniform collimated beam is measured. Therefore, the experiment corresponds to the broad beam situation. It is not appropriate to use the target angle cosine factor in calculating the cross section, even though it is necessary when doing narrow beam experiments.

The calculation of proton upset rates in space also corresponds closely to the calculation of nuclear reaction rates in space. This situation corresponds to Figure 2-71. The number of events is calculated from the upset cross section and the total fluence. Actually both are energy dependent, so that the number of events has to be summed over the product of the energy dependent fluence and cross sections. This is discussed in Chapter 8.

2.10.2.2 Heavy Ion Cross-Section Measurements

The typical test situation is to use a beam of heavy ions to illuminate a memory device, with the beam normal to the surface of the device. The device has a number of transistors near its surface; see Figure 2-73a. The ions that pass through the transistors deposit energy and produce a track of ionization. Some of the ionization can be collected by some devices, and the resulting signal can produce an upset in that memory cell. The experimenter is interested in determining the probability of a single event as a function of the energy deposited in the device. If we consider a simplified device as in Figure 2-73b, the device has a

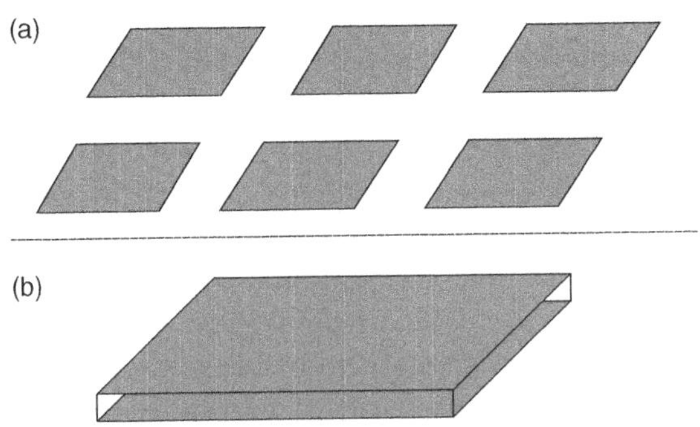

Figure 2-73 (a) A number of sensitive regions in a plane. (b) The assumed geometry for a single sensitive region.

constant thickness, and the energy deposition is

$$E = \frac{dE}{dX} l = \frac{dE}{dX} \frac{h}{\cos\theta} = L_{\text{eff}} h \tag{2.10}$$

Consider an experiment in which we can continually increase the LET of the ions incident normal to the surface. If there was a single well-defined critical energy for upset, and the devices have uniform thickness, than all of the devices will upset at a particular LET, and the measured probability of upset, or upset cross section, will correspond to the total area of the sensitive devices on the chip.

The experiment apparently corresponds to wide beam experiment discussed above.

$$\sigma = \frac{N}{F} \tag{2.11}$$

The probability of having an upset on the chip is given by σ, which will have units of cm^2. N is the number of upsets observed, and F is the fluence in particles/cm^2. If one wants the upset cross section per bit, then the total cross section is divided by the number of bits M:

$$\sigma = \frac{N}{F \times M} \tag{2.12}$$

It is not feasible to continually change the LET at normal incidence. However, it is feasible to rotate the target. This leads to longer path lengths in the target and increased energy deposition. This approach can be described using the concept of effective LET described above.

Now when the experiment is performed, the apparent cross section corresponds to the distance a, not the desired distance b; see Figure 2-74. There may be small end effects, which will be discussed in a later chapter. The cross section per chip or cross section per bit has to be calculated allowing for apparent change in area:

$$\sigma = \frac{N/\cos\theta}{F} \quad \text{or} \quad \sigma = \frac{N/\cos\theta}{F \times M} \tag{2.13}$$

We observe that although the experimental situation corresponds to the broad beam experiment, the cross sections must be calculated

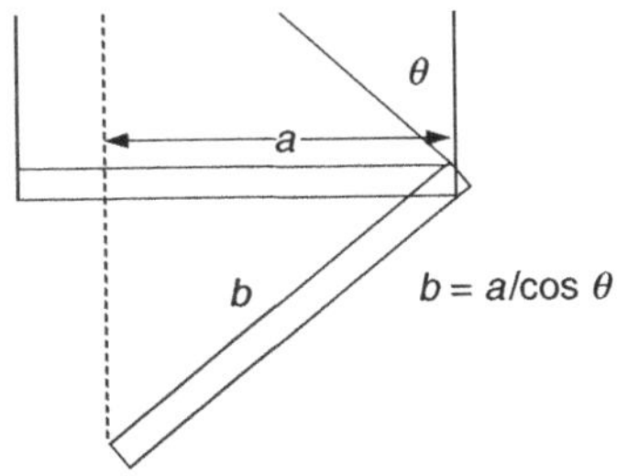

Figure 2-74 The cross-section geometry as the effective LET is varied.

allowing for the dependence of the measurement on the orientation of the target, which was not necessary in broad beam nuclear physics experiments. The results seem to correspond to the narrow beam experiments, but the reason for the cosine θ factor is entirely different. In this case there is very poor correspondence between the nuclear physics and the single event cross-section concepts.

The lack of correspondence between nuclear physics and single event effects is even more pronounced when the measured cross-section results are used to calculate an upset rate in the space environment. We will discuss upset rate calculations in more detail later. We can briefly summarize the situation here. The upset occurs only if a critical energy is deposited in the device. This depends on the ionizing path length in the device. The upset rate is determined by the sum of possible paths that can lead to an adequate energy deposition. Therefore, the calculation uses the combination of the flux in terms of its energy deposition $\Phi(L)$ and the path probability $f(s)$.

$$N(E_c) = A_p \int_{s_{min}}^{s_{max}} \Phi(L_{min}) f(s) ds \tag{2.14}$$

This equation does not correspond to the isotropic nuclear physics case. Although we use the concept of cross section in both the fields of nuclear physics and of single event effects, the correspondence between the two cases is poor and cannot be trusted in detail. There is not a single universal concept of cross section.

CHAPTER 3

Optimizing Heavy Ion Experiments for Analysis

The goal of SEE modeling is to obtain good rate predictions. The success of this effort in turn depends on developing the appropriate high quality data. High quality data is obtained only if the experimenters take careful note of the needs and goals of the program. This in turn depends on carefully thought out test protocols. An early test protocol was ASTM standard F1192 for "Measurement of Single Event Phenomena Induced by Heavy Ion Irradiation of Semiconductor Devices." This was superseded by the EIA/JEDEC Standard 57 "Test Procedures for the Measurement of Single-Event Effects in Semiconductor Devices from Heavy Ion Irradiation." That document lacks guidance on some more recent considerations [Hafer 2008]. This section outlines some of the ideas that need to be considered in designing an experiment.

The goals of the program can be specified in terms of the desired accuracy and the desired precision. If an experiment has small systematic errors, it is said to have high accuracy. If an experiment has small random errors, it is said to have high precision [Beers 1958]. Both of these factors always need to be considered in designing single event experiments.

Single Event Effects in Aerospace, First Edition. By Edward Petersen.
© 2011 the Institute of Electrical and Electronics Engineers, Inc.
Published 2011 by John Wiley & Sons, Inc.

3.1 SAMPLE HEAVY ION DATA

A single event is obtained by measuring the cross section as a function of effective LET. Figures 3-1–3-4 show four views of a set of sample data taken by Sokol [Sokol 1987]. Figure 3-1 shows a log-linear plot and is the type of display that is often reported. This is useful for indicating to the sponsor that you have made SEU measurements. Unfortunately, it hides problem data and can mislead about the saturation cross section, as discussed later. The experimenters should always also plot the data in the other three modes. These plots help you get a much better understanding of the data. The linear cross-section plots in Figures 3-2 and 3-3 give you a much better idea of the saturation cross section and of the spread in the data. Figure 3-4 indicates the threshold region.

3.2 TEST REQUIREMENTS

The upset rate is determined by the folding together of the upset rate as a function of effective LET and the cross section (upset probability) as a function of effective LET using the IRPP approach. This is discussed in much more detail in Section 7.3. We give an example for a

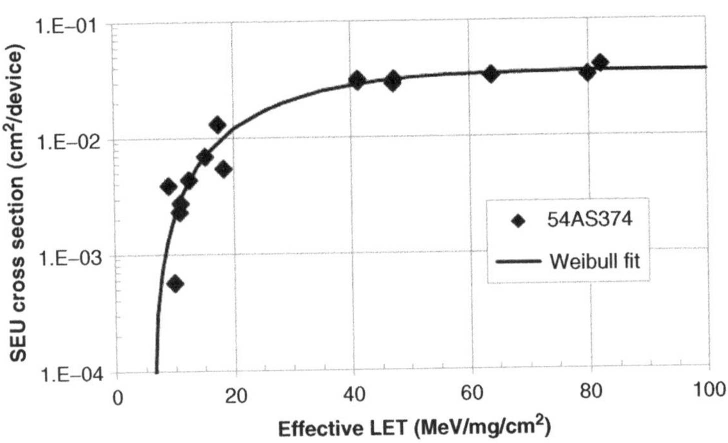

Figure 3-1 Cross section for a octal flip-flop taken from [Sokol 1987]. Copyright © 1987 IEEE. This display is sometimes useful for hiding problem data, although not in this case.

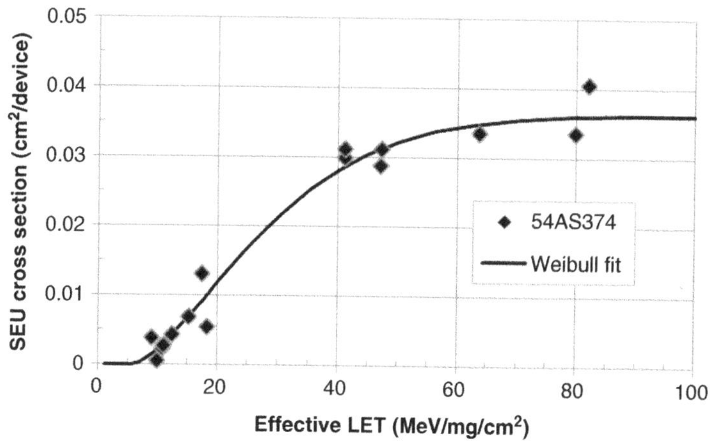

Figure 3-2 The data in Figure 3-1 plotted with a linear cross-section scale. This helps indicate the saturation cross section and the spread in the data.

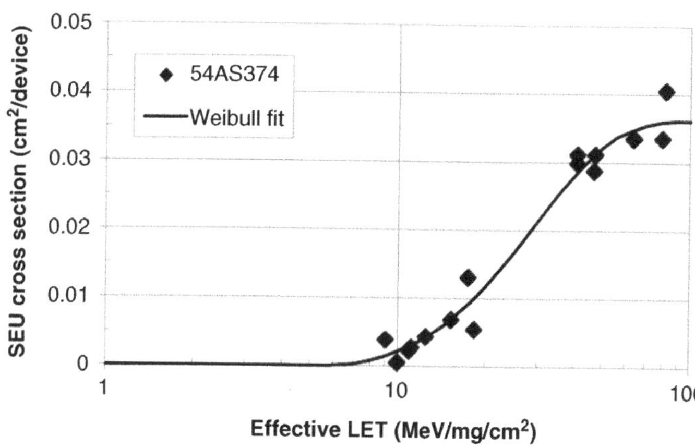

Figure 3-3 The data in Figure 3-2 plotted with a log LET scale. This plot indicates the importance of the midrange data for determining the shape of the cross-section curve.

device with a threshold at 1.0 $MeV/mg/cm^2$. These results are shown in Figure 3-5. The upset rate is determined by the lower-middle of the cross-section curve. The detailed location of the threshold section of the curve has little impact. This is an important point and we will discuss it later with other examples.

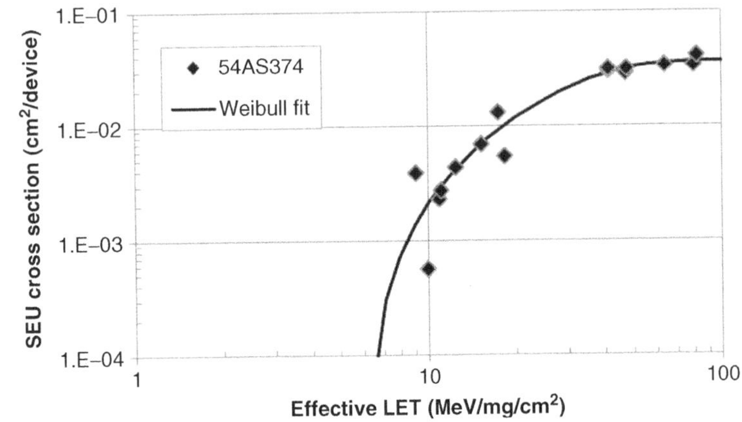

Figure 3-4 The data in Figure 3-3 plotted using the log of the cross section. Each type of plot emphasizes different aspects of the data. Three devices are being tested, leading to a slight spread in the threshold region.

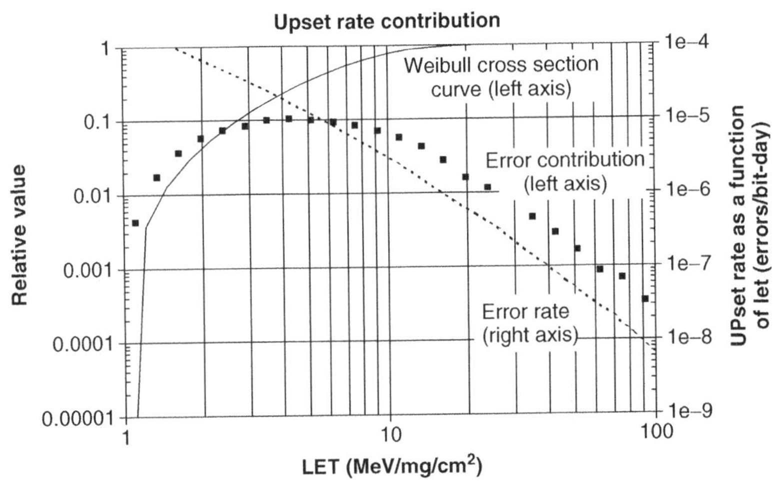

Figure 3-5 Logarithmic view of the combination of cross-section curve and upset rate curve to obtain the total error rate. Note that the major contribution to the upset rate comes in the 5% to 90% regions of the upset cross section.

3.3 CURVE PARAMETERS

The upset rate calculation depends on knowing the LET dependence of the device sensitivity. The LET dependence of the sensitivity seems to follow an integral Weibull or lognormal curve. It is common to use a few basic ion beams, and several angles with each beam to obtain a range of effective LET values. We will see that there are a number of ways in which the data obtained away from normal incidence will not fall on the basic upset rate curve. It is possible and necessary to correct for these effects, but the "purest" and most important points are those obtained using the normal incidence beams. The test facilities normally have a number of standard beams designed to step at approximate steps of a factor of 2 in LET. The basic curves normally have widths (from 0.05 to 0.95) that cover a factor of 8 to 16 in LET. Therefore, it is necessary to use three or four beams to trace out the curve. This will produce three to four points on the rising portion of the curve and give adequate information so that the curve shape is well defined. If the part is well behaved, this is all the data that is required. Even for well-behaved parts, the effective LET should be extended as high as possible using the 60 and 65 degree points with the high LET beam. The other angles in this sequence will give you additional information about the possible angular dependencies. A series of angles, taken with one of the ions whose zero degree LET is in the middle of the curve, will give additional information about the possible angular dependence that does not follow the 1/cos behavior that is expected.

Figure 3-6 indicates use of the normal incidence points to determine the basic curve shape. Two representative cross-section curves are shown together with the normal incidence points that can be obtained to determine the basic curve parameters. If the data points are obtained with very good statistics, they closely define the curve parameters.

The nature of the Weibull and lognormal curves is such that each of the curve parameters is determined by a different portion of the curve. Thus, a point that determines the middle and a point that determines the width basically define the curve. We carry this to an extreme in Figure 3-7. Here we assume that we have only three points and are attempting to fit the data with the four parameter Weibull function. Of course, we obtain a family of curves, not a unique curve. However, we can still use this family of curves in error rate calculations. Figure 3-8 shows the error rates obtained as a function of the threshold. A variation of possible threshold values between 0.5 and 3.0 MeV/mg/cm^2 leads to variation of the upset rate by only a few percent. Furthermore,

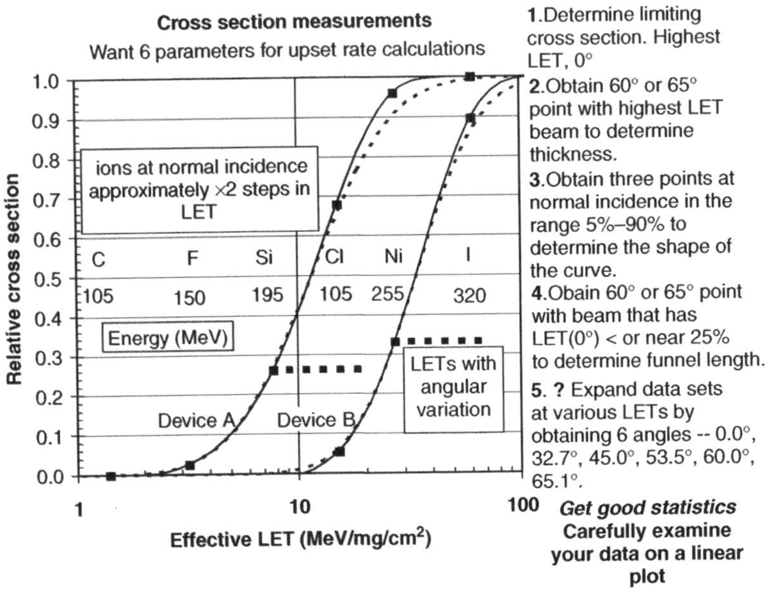

Figure 3-6 Experimental heavy ion cross-section measurements. The representative ions are for the Brookhaven accelerator. The solid and dashed curves are for Weibull and lognormal curves.

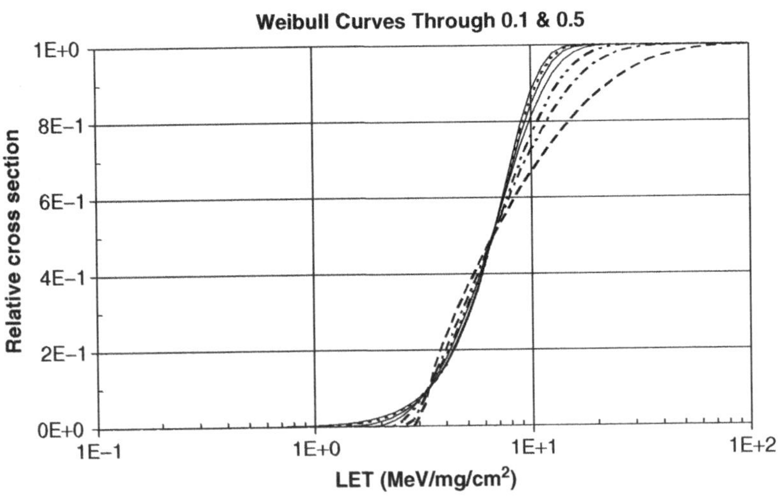

Figure 3-7 The family of Weibull curves that is obtained with knowledge of the limiting cross section and of the 10% and 50% points. Another point at relatively high LET will specify the curve.

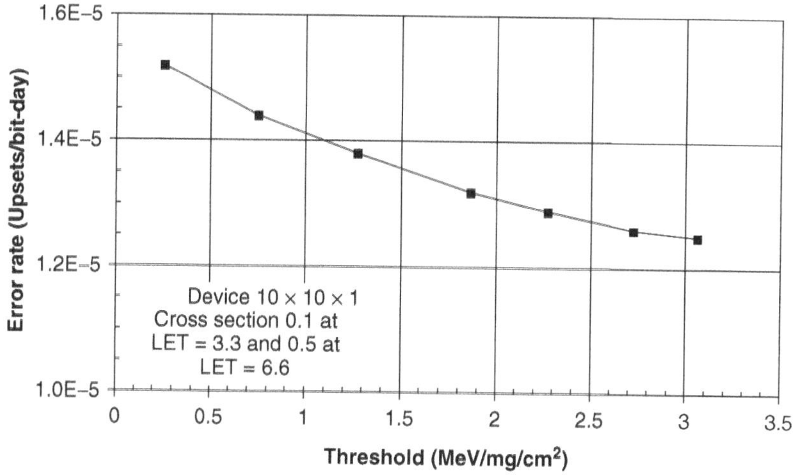

Figure 3-8 Error rate as a function of threshold for the family of Weibull curves shown in Figure 3-7.

the exact shape of the curve can be determined by a cross-section measurement in the LET range above the 50% point. There is no reason to waste time and money by obtaining low cross-section points. If you believe that the curve can be described by the lognormal function, there is little reason to take any more points (assuming that the curve is well behaved with angle). Figure 3-9 shows Weibull and lognormal curves as determined by points at 0.1, 0.5, and 0.9.

Actually the case in Figure 3-9 is an extreme example. This shows a relatively narrow cross-section curve, so that the Weibull and lognormal function differ significantly at low LET. If we calculate the upset rate using the method employed for Figure 3-5, we obtain upset rates of 1.16×10^{-5} using the Weibull fit, and 1.25×10^{-5} using the lognormal function, in conjunction with the $10 \times 10 \times 1$ device used previously. The 8% difference indicates that in some cases we should use a low cutoff with the lognormal function as suggested in [Petersen 1996c]. In this situation a low LET point to determine the cutoff, or to verify that the Weibull function is a significantly better fit, would be valuable.

The basic objective for upset rate predictions is that the parameters describing these curves be obtained as easily and simply as possible. Weibull curves are determined by four parameters: threshold, shape,

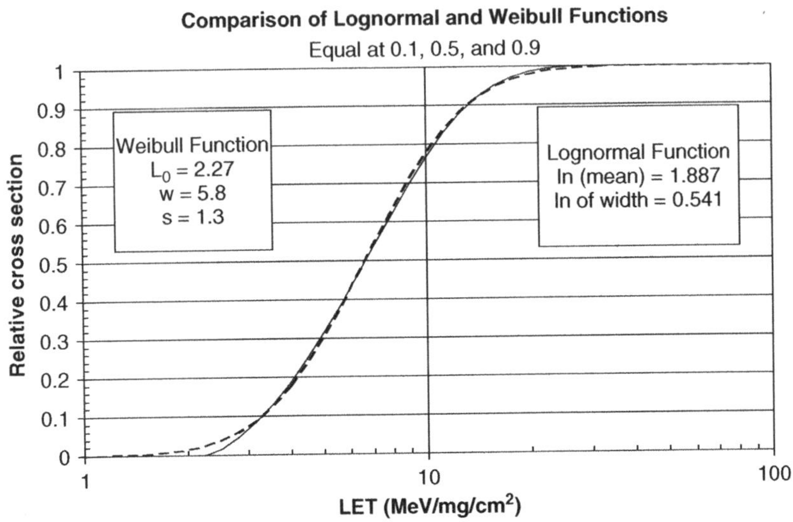

Figure 3-9 The lognormal function that fits the 0.1 and 0.5 points and the Weibull function that matches the lognormal function at 0.9.

width, and limiting cross section. It is in principle possible to determine these from four or five measurements. The limiting cross section is determined by a cross-section measurement at very high LET. The width (and device critical charge) is determined at 0.5 to 0.6 of the limiting cross section. The shape and threshold are determined by the values at 0.1–0.25 and 0.8 of the limiting cross section. You can obtain data at 10^{-3} to 10^{-4} of the limiting cross section to tie down the threshold, but is it not really necessary. Figure 3-6 shows how the appropriate data points might appear on a representative Weibull curve. The upset rate is dominated by the cross-section values in the range of 0.01 to 0.8 relative cross section. The shape should be well specified in this region. Figure 3-5 showed an example of the relative contributions to the upset rate.

In practice, it is often necessary and desirable to take more data. Initially the shape of the curve is not known, so the available points are determined by the available ions. The curves shown have full widths of 0.5 to 2 orders of magnitude. An individual ion will cover a factor of 2 in LET with an angular variation of 0–60 degrees. (If we are lucky, the device mount will allow angles larger than 60 degrees.) It is then necessary to use three to five ions to trace the entire curve.

If all the data falls on a smooth curve, the depth of the device is not significant and the charge collection processes are most likely simple. In this case the standard upset rate calculation should be adequate. A careful examination of the minimum discontinuity that could be observed in the experiment can lead to an estimate of the maximum possible device depth. If the data has discontinuities, then the depth of the device is important. The charge collection processes may also be important in this case. Careful angular data is necessary so the data can be used to improve the upset rate predictions

3.4 ANGULAR STEPS

We have mentioned the importance of observing the angular dependence of the cross section. The measurements at a series of angles enable you to determine the thickness of the device and the length of the funnel. If there is little observable variation from the smooth Weibull curve, that gives you information. More often there is a variation evident in careful measurements. Figure 3-10 indicates the two different effects that can be observed. In one case the data at large angles is above data at the same effective LET obtained with a heavier

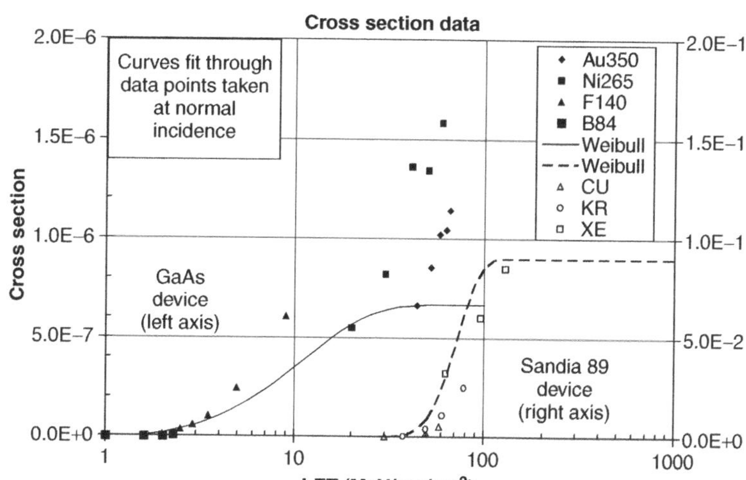

Figure 3-10 Data showing possible effects observed when data is taken at several angles for different ions.

ion. In the other case, the data at large angles is below data at the same effective LET obtained with the heavier ion. We will examine possible causes for these effects.

It is easy and necessary to vary the relative angle of the device and the beam in order to change the effective LET. It is common to step the angle in even angle increments: 0, 15, 30, 45, 60, 65, and 70 degrees. However, the data is of interest as a function of effective LET, not angle. Another choice is in steps linear with effective LET: 0, 37, 48, 55, 60, 63.6, 66.4, 68.7, 70.5 degrees.

The upset data is best represented when plotted versus *log LET* as in Figure 3-6. In this case the data should be taken in *even steps of log LET: 0.0, 32.7, 45.0, 53.5, 60.0, 65.13, and 69.3 degrees*. Therefore, we recommend that this last series be used. If the data indicates that smaller intervals should be used, consider the angles from the series: 0, 23.5, 32.7, 39.5, 45.0, 49.6, 53.5, 57.0, 60.0, 62.7, 65.1, 67.3, and 69.3. This is likely to occur for deep devices such as those using deep trench isolation (DTI). In this case, the medium and large angles do not contribute at low LETs.

The caveat to these sets of angles is that is some situations the ion energy loss is important so that ions at large angles loose appreciable energy and the LET is not the value expected from simple geometrical arguments. The experimenter needs to consider this case in planning the experiment.

3.5 STOPPING DATA ACCUMULATION WHEN YOU REACH THE SATURATION CROSS SECTION

It is sometimes the policy to acquire data until the saturation cross section is reached. This can cause problems, because the often used plot of logarithmic cross section versus linear LET gives a misleading visual saturation. The normal process of fitting with a Weibull distribution does not cure the problem. We give an example in Figure 3-11. This shows data for a part, where the highest data point was for a LET of 18.3, corresponding to a 195 MeV Si beam at 65 degrees. It appears that the cross-section curve, on the normal log-linear scale, is reaching saturation. If we perform a least squares fit using the Weibull function, we obtain a saturation cross section of $4.8\,\mu m^2$, consistent with the visual fit.

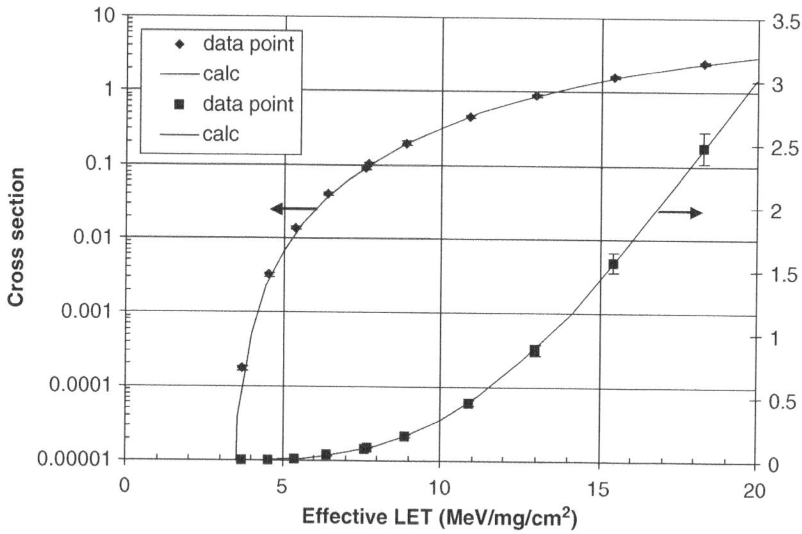

Figure 3-11 Logarithmic and linear fits to a low LET data set. The log-linear plot appears to be reaching saturation. A least squares Weibull fit confirms a saturation cross section of 4.8. However, the linear plot does not seem to be at saturation and a least squares lognormal fit gives a saturation cross section of 22.

However, there is some question about this fit, as the linear-linear plot does not seem to approach saturation and a lognormal fit leads to a saturation cross section of 22 (although that is not well defined, depending on the initial value). Either of these results should make us question our initial conclusion that the data has reached saturation.

Now let us consider another set of data on the same part. In this case we spend less time obtaining data in the threshold region, and take more runs, although not necessarily more time, in the saturation region. The data is shown in Figure 3-12. In this case there is a clear saturation in both the logarithmic and linear plots.

There are several effects that influence the appearance of the cross-section curve in the saturation region. There are geometrical effects that can lead to saturating cross section that do not appear to do so. These are laboratory artifacts and are discussed in later sections. They arise from the fact that the device has a significant depth, so that the cross section apparently increases as it is exposed at increasing angles.

The device may have multiple upsets per event. The *event* cross section may not be increasing, while the upset cross section is

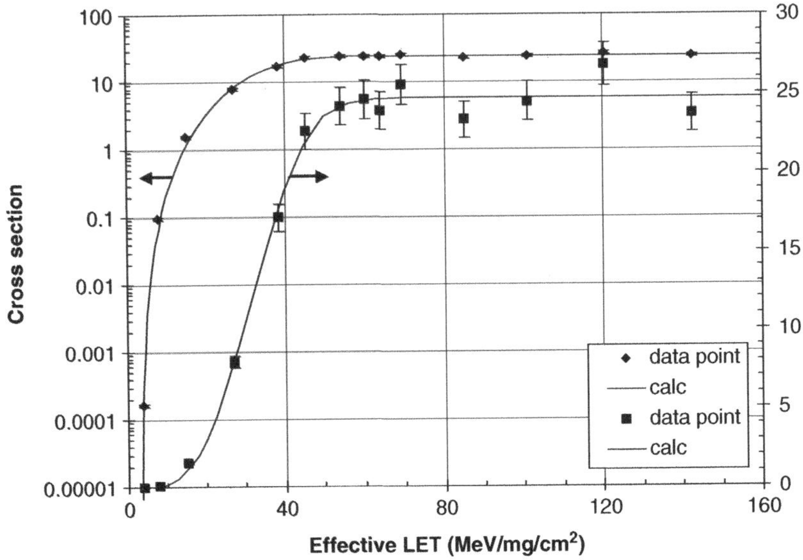

Figure 3-12 A different set of data for the same part used in Figure 3-11. In this case the experimental data has been extended so that saturation is shown clearly on both the logarithmic and linear plots. Fits with the Weibull function and the lognormal function will produce essentially the same saturation cross section.

increasing. One has to carefully examine the data so see if multiple upsets are properly accounted for.

There appear to be cases for single event latchup where the cross section does not saturate. This may be due to diffusion type effects, or perhaps the measurements were not extended to high enough LETs.

3.6 DEVICE SHADOWING EFFECTS

If possible, the angles should extend past 60 degrees. This is an easy way to extend the effective LET range. The experimenter must be careful that the device mount allows the larger angles. If there are mounting limitations, these can cause a decrease of measured cross sections at large angles that are not characteristic of the device itself. Figure 3-13 shows what can happen in an extreme case. The points show the situation when several series of runs are taken, each starting with a different LET beam at normal incidence. That beam is then used at a series of angles so as to obtain data at larger effective LETs.

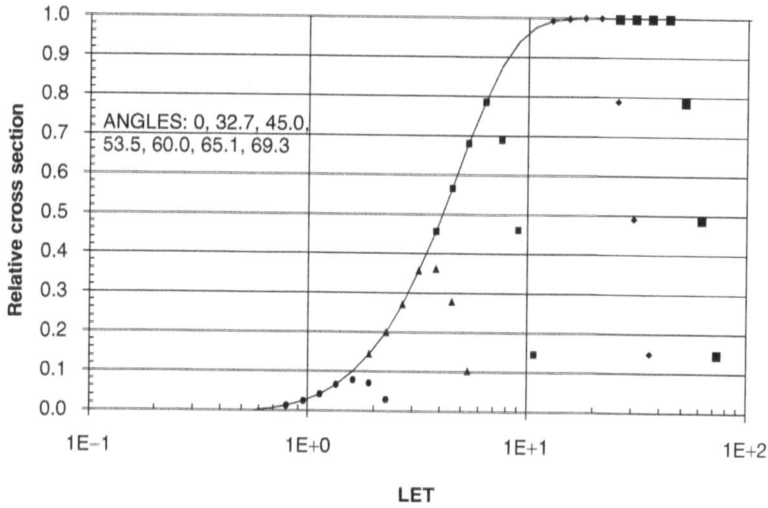

Figure 3-13 The effect of physical shielding on the angular variation of the data. This plot assumes that the shadowing starts at 55 degrees and is complete by 70 degrees.

The figure shows the results if shadowing starts at 55 degrees, with the device being completely shadowed at 70 degrees. The measured cross sections at large angle show a sharp decrease. The angular dependence in this case is different from that in the cases where the deviations from the Weibull curve are caused by the device geometry.

3.7 CHOICE OF IONS

Various ions are available at different facilities. We will discuss use of the ions available at the Brookhaven single event upset test facility (SEUTF). The general considerations for choice of LET apply to any facility. The main problem with the beams at Brookhaven is that they lose appreciable energy passing into the sensitive region, so that the LET has to be corrected for the actual energy at the sensitive region. As devices become more complex, with more layers above the sensitive volume, it may be necessary to use other accelerators. This is discussed by Hafer [Hafer 2008]. If you use other accelerators, you must be aware there are different spreads in the statistical errors of the measured cross sections: they have different standard deviations (see Section 5.2.1.4).

The initial measurements should be made with an ion of high LET. We suggest iodine with energy 320 MeV. This will produce a LET range of 60–120 over 0 to 60 degrees. The first reason for starting at high LET is the very practical one of checking the experimental equipment and verifying that it actually detects upsets. If devices upset, they ordinarily do so at a LET of 120. The second reason is that this range should determine the limiting cross section for most devices. Take measurements at 0, 45, and 60 degrees. If the values are consistent, then you have determined the limiting cross section, and some of the possible geometrical effects are eliminated as confusing factors. If the values are not consistent, then data should be taken at the additional angles of 32.7, 53.5, 65.1, and 69.3 degrees.

It is important not to use gold ions at Brookhaven. At first glance they cover an interesting range of LETs, but they have so much energy absorption that it is difficult to determine the actual LET in the sensitive region. Data at several angles, taken with expectations of different LETs, will fall in a small range of LET after the proper corrections are made. Even in this case there may be questions if the proper corrections were used due to uncertainties in the thickness of the absorbing layers on the device.

After obtaining data at the maximum LET range with iodine, it is desirable to step down in LET steps to thoroughly measure the device characteristics. The second ion suggested is 255 MeV Ni with a LET range of 27–54. This could be skipped if the device is expected to be very soft and if the cross-section results with iodine were well behaved in their angular dependence.

The total set of ions that we suggest is the following:

I	$E = 320$	LET 60–120
Ni	$E = 255$	LET 27–54
Cl	$E = 105$	LET 15–30
Si	$E = 195$	LET 7.7–15
F	$E = 150$	LET 3.2–6.4
C	$E = 105$	LET 1.4–2.8
B	$E = 90$	LET 1.0–2 or
Li	$E = 60$	LET 0.36–0.72 as necessary to determine threshold

Note that there will generally be three normal incidence points in the regions that do a good job of determining the Weibull parameters. If the angles can be extended beyond 60 degrees, the overlaps will be improved. Figure 3-6 shows how the normal incidence points might appear on a representative cross-section curve. The normal incidence points should be obtained with care and good statistics.

3.8 DETERMINING THE LET IN THE DEVICE

Often heavy ions produced at accelerators are at relatively low energies relative to cosmic ray heavy ions. Whereas the cosmic ray ions have very little energy loss as they pass through the device, and therefore maintain the same LET, the accelerator heavy ions can have appreciable energy loss and corresponding changes in LET. Figure 3-14 shows the energy as a function of depth in silicon for 330 MeV iodine ions.

Figure 3-15 shows the corresponding change in the ion LET. These changes are significant as all devices have several layers above the sensitive region. These can be plastic pacification layers to protect the surface, metal layers, or silicon dioxide layers. The depths may range from 2 to 5 microns of effective silicon depth. In addition, the sensitive layer may be one to several microns deep, so the change of LET across the layer may be significant.

The cross-section measurements in the laboratory are obtained as a function of effective LET by measuring at a series of angles. This means that the effective LET of the ion must be calculated allowing for

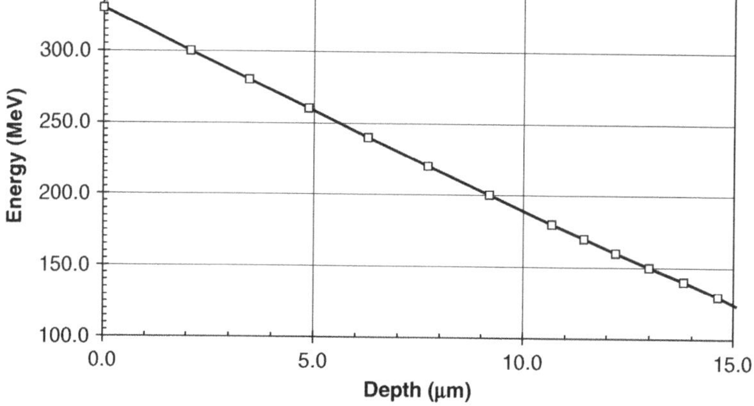

Figure 3-14 The change of energy with depth of 330 MeV iodine incident on silicon.

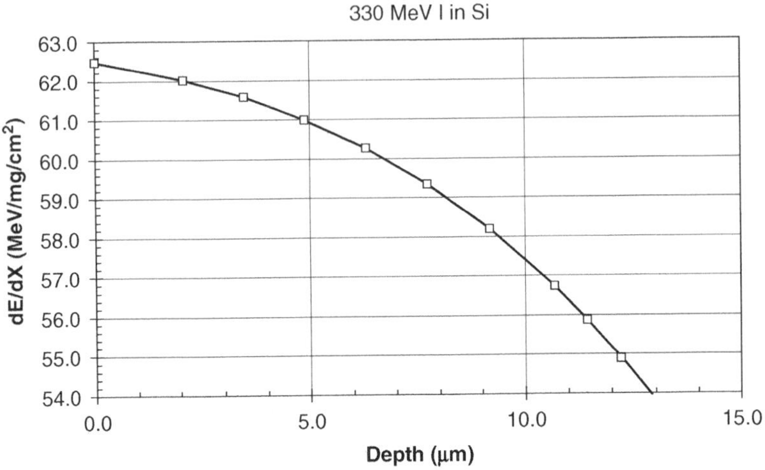

Figure 3-15 The energy loss of incident 330 MeV iodine in silicon as a function of depth.

a changing dead layer above the sensitive region. The effective LET should reflect the actual energy deposition of the ion in the sensitive region. Therefore, it needs to be corrected for the effect of the dead layer. Figure 3-16 shows the effective LET that would be expected from the LET of the ion at the surface of the device, and the actual

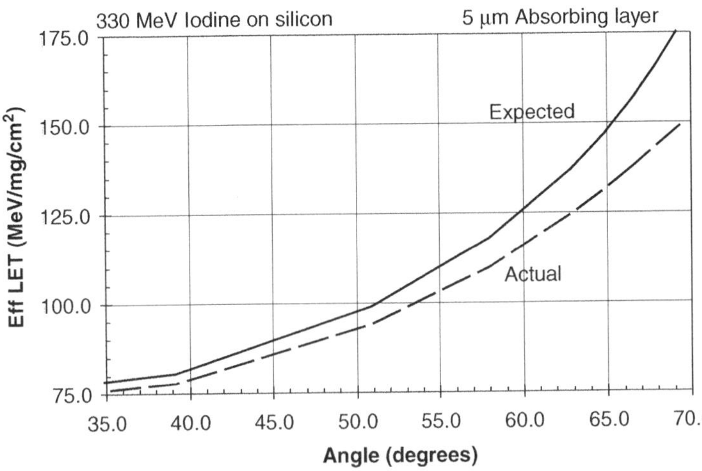

Figure 3-16 Effective LET correction that must be made for 330 MeV iodine in silicon as a function of angle at a depth of 5 microns.

effective LET if the center of the sensitive volume is 5 microns below the surface.

The previous example is for a heavy ion. There is the opposite effect for some of the lighter ions that are used in SEE measurements. These ions have an increase of LET as they lose energy. Figure 3-17 shows the changes of LET with depth for six different ions that might be used in a series of measurements.

Figure 3-17 supports the recommendation that gold not be used. If there is uncertainty about the depth of the dead layer, the gold ions at 60 degrees may be passing through 10 or 15 microns. The effective LET may then be somewhere in a range from 100 to 130 MeV/mg/cm^2. The uncertainty will be even worse at larger angles. There is less uncertainty with an iodine beam, and the apparent high effective LET that one might expect from the surface LET of gold is not actually obtained. Figure 3-18 shows that one also has to consider LET variation with depth at other facilities [Hafer 2008]. The TAMU facility clearly has an advantage with beam penetration. However, there is a disadvantage in that it has a very large dosimetry systematic error (see Section 5.2.1.4). You need to examine the factors important in your experiment.

If the device has heavy overlayers, the effects are even more important. See Figure 3-19. The equivalent thickness is defined as the Si thickness with the equivalent ion stopping power of the overlayer of interest [Hafer 2008].

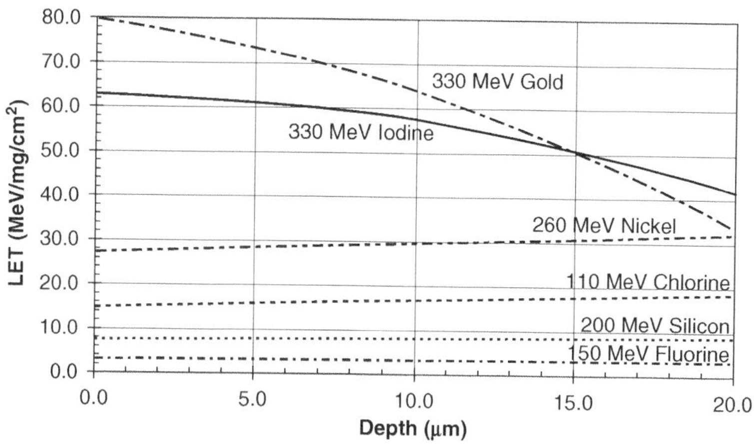

Figure 3-17 LET of various ions as a function of penetration depth in silicon.

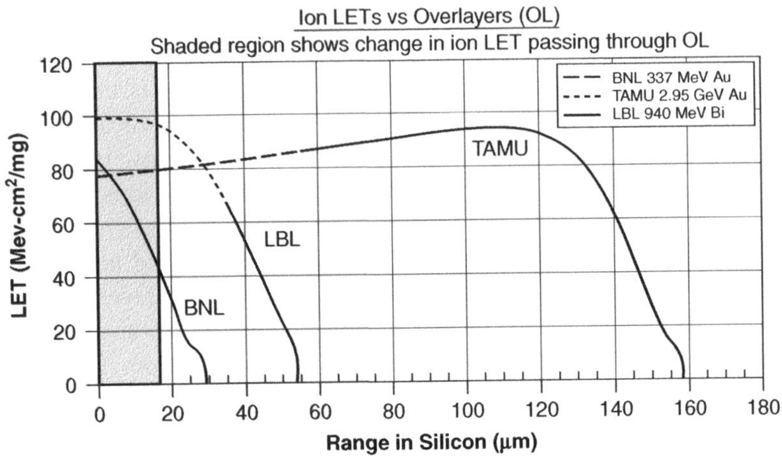

Figure 3-18 LET versus ion range in silicon for the heaviest ion at three test facilities—Brookhaven National Laboratory, Texas A&M University, and Lawrence Berkeley Laboratory [Wert 2005, Hafer 2008]. Copyright © 2008 IEEE.

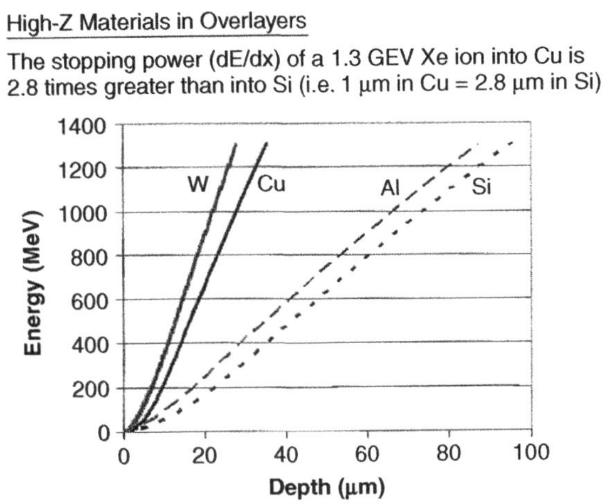

Figure 3-19 The ion stopping power (dE/dX) of Cu and W is 2.8 times the stopping power of Si for a 1.3 GeV xenon ion [Wert 2005, Hafer 2008]. Copyright © 2008 IEEE.

3.9 ENERGY LOSS SPREAD

One obvious contribution to the shape of the cross-section curve is that of the statistical distribution of energy loss by high energy particles in thin layers. This has been examined in detail by Xapsos and shown

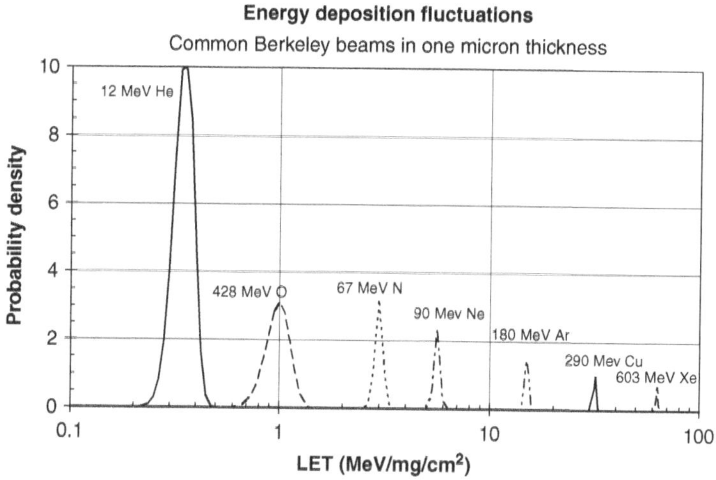

Figure 3-20 Calculated LET spread for monoenergetic beams in a 1 micron distance for Berkeley beams.

to play a very small role [Xapsos 1992, Xapsos 1993]. Figures 3-20 and 3-21 show the results of calculations using the Xapsos approach for the commonly used heavy ion beams at Brookhaven and Berkeley. There are spreads in the energy loss, but they are small compared to the widths ordinarily observed in single event cross-section curves.

3.10 DATA REQUIREMENTS

The experimental team should specify, before the experiment, the desired accuracy and precision of the experiment. The precision of the experiment is determined by the spread of values due to random errors. If the team acquires large numbers of errors, they have good statistics and good precision. The accuracy of the experiment is determined by systematic or constant errors. In this case, individual values are in error by the same amount for a test with one ion and energy. The most likely source of systematic errors in single event experiments is dosimetry.

3.10.1 Desired Precision

Historically, SEE data has emphasized the low LET points to determine the threshold. This wastes much of the time used in obtaining the

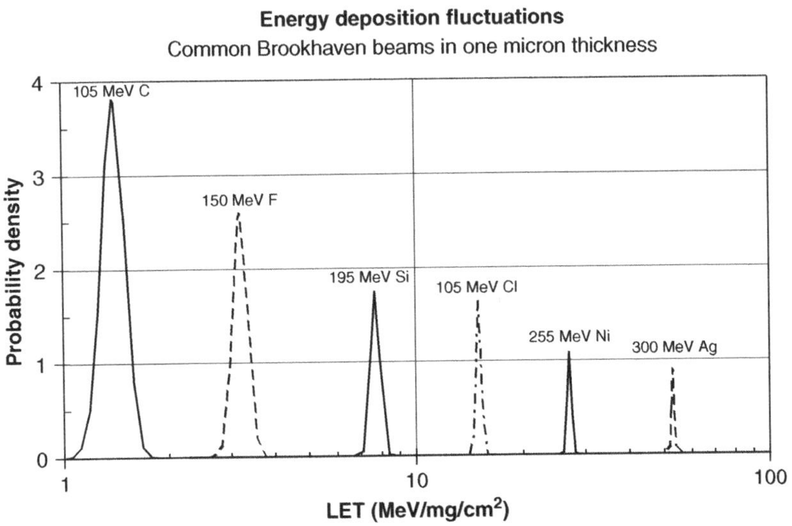

Figure 3-21 Calculated LET spread of monoenergetic beams in a 1 micron thickness for Brookhaven beams.

data. The previous sections have pointed out that the significant data is primarily in the regions of LET that have high cross sections (>5% of limiting cross section). The previous sections have also indicated how significant information is available in the data at high angles. In order to use this information, the data has to be obtained with good statistics. However, as the data is in the high cross-section regime, there is not a significant time penalty incurred. The statistical uncertainties of the points can be assigned in terms of the number of upsets observed. If N upsets are observed, the uncertainty is $\sqrt{N}/N$. A thousand upsets will correspond to 3.16% statistics. We recommend the following goals:

3% statistics for cross sections >5% of limit

5% statistics for cross sections 1% to 5% of the limit

25% statistics for cross sections <1% of limit

If the data is not smooth with changes of angle and ion, it might be desirable to obtain 1 or 2 points in the region where the cross section is <1% of the limit. One of these should be in the region 10^{-4} to 10^{-3} of the limiting cross section. A complete angular distribution is desirable at low LET if there is an interesting deviation from the expected Weibull curve. This can be checked by using several angles. If the data rises and then falls again, a complete angular distribution should be obtained. Of course, near the threshold, there may not be any upsets at small angles, so the experimenter has to take her data with care.

A more sophisticated approach to determine the desirable statistics is given by Ladbury [Ladbury 2007b]. This approach is useful for situations in which it is difficult to obtain the statistics recommended above.

3.10.2 Desired Accuracy

The desired accuracy is determined by the goal of the experiment. If you are interested primarily in the order of magnitude of the event rate in space, you may not need high accuracy. The event rate may vary over twelve or more orders of magnitude depending on the environment and the part. You need to know that your error handling techniques are adequate and that the probability of catastrophic error is tolerable. If you do not need high accuracy, you may decide to live with 30% systematic error and 10% random error, for a total 32% uncertainty.

If the goal of the flight is to determine the accuracy of rate prediction techniques, then you may desire an accuracy of 5% to 10%. You also may desire high accuracy if the event rates depend strongly on the geometry of the device.

3.11 EXPERIMENTAL STATISTICS AND UNCERTAINTIES

The previous section recommended that the experimenter obtain 3% statistics (>1000 upsets) for data points in the range from the limiting cross section down to 5% of the limiting cross section. This is a more stringent requirement than that often used, with many groups stopping when they obtain 10% statistics. We need to examine the feasibility of our requirement. ASTM document F1192 states that nominal beam intensities range from 10^2 to 10^5 ions/cm^2/s. We have previously seen that the limiting cross sections are normally in the range of 10^{-7}

to 10^{-5} cm^2/bit. At the time of these standards, devices often had 0.256×10^6 bits.

If we assume a beam intensity of 10^4 ions/cm^2/s, a cross section of 10^{-7} cm^2/bit, and 0.256×10^6 bits, the observed rate will be

$$\text{Rate} = 10^4 \times 10^{-7} \times 0.256 \times 10^6$$
$$= 256 \text{ upsets/second}$$

Three percent statistics require 10^3 upsets. We see that it is very feasible to reach this goal. Furthermore, as most experimental setups allow very quick changes of angle, it is also very feasible to obtain the five to seven angles that were suggested above.

The problem is different in modern devices that may have 10^9 bits. In these cases, a total exposure of 10^7 ions/cm^2 will expose only a small fraction of the total number of cells. This does not satisfy a conservative approach to radiation assurance [LaBel 2008]. We discuss systematic errors in Section 5.2 and point out that they are rarely as good as 5%. The experimenter has to decide what his goals are before deciding the desired balance of systematic and random errors and the amount of data required.

3.12 EFFECT OF DUAL THRESHOLDS

A CMOS flip-flop memory cell has two transistors that can contribute to upset in each memory state. Thus, there are two critical charges, two depths, and two cross-sectional areas. This has been discussed by various authors from the very earliest [Pickel 1980, Pickel 1982]. There have been few if any cases in which the cross-section data has clearly indicated the two contributions. This is partially because of the normal emphasis on the low cross-section data, not the high cross-section data that might show this effect. However, it is probable that in many cases the data would not clearly show two contributions. Figure 3-22 examines how two separated Gaussians can appear as a single Weibull. In one case we combine two separated Weibull function with a Gaussian shape factor ($s = 4$). The combination is nearly impossible to distinguish from a single Weibull with a shape factor of 3. It is only when the two Gaussians are widely separated that the combination is clearly not a single Weibull function.

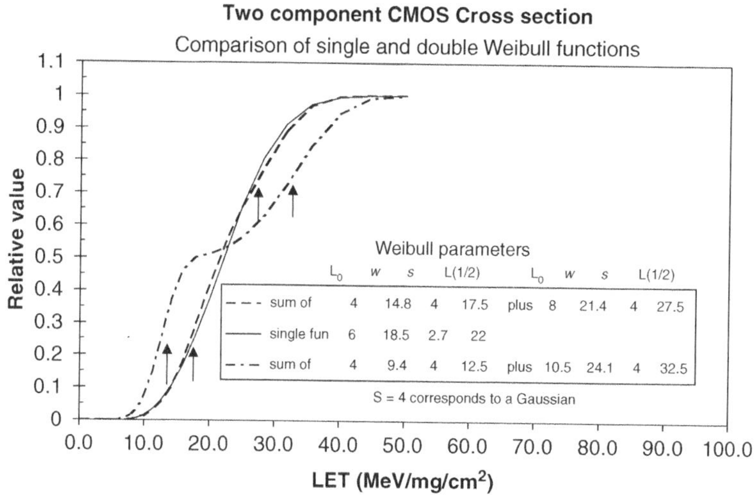

Figure 3-22 The single Weibull curve and two possible combinations of double Weibull curves. Two Gaussian Weibull functions ($s = 4$) can combine to resemble a single Weibull with $s = 3$. If the Gaussians are well separated, this does not occur. The arrows indicate the centers of the various distributions.

Most of the cells that are susceptible to upset have two sensitive locations at two different transistors with different characteristics. Therefore, most of the early calculations carefully calculated for the two different characteristics and summed the two upset rates. This seems to be the right thing to do. However, in practice, although the two areas can be distinguished in experiments carefully designed to see them [Zoutendyk 1984], the cross-section curves have never been observed to have two different plateaus.

3.13 FITTING CROSS-SECTION DATA

Modern experimentalists commonly use very complicated codes for analyzing their data. These are not always necessary. The fitting of heavy ion experimental data to the functions described here can be done with ordinary spreadsheets. The basic functions are available as standard functions, at least in Microsoft Excel.

To use the Weibull distributions $= \text{IF}(\text{L} - \text{L}_0 > 0,$

$\text{Weibull}(\text{L} - \text{L}_0,\text{s},\text{w},\text{true}),0)$

To use the lognormal distribution = lognormdist(L,C,W)

To use the gamma distribution = gammadist(x,alpha,1,true)

You can then fit the data to either these functions, or these functions combined with the corrections described in the RPP geometry sections. You set up the calculation of the sum of the squares of the differences between the data and the calculated values. If you have the errors on the points (and you should have) you can use a chi square fitting procedure. The Excel solver can then be used to find the curve and correction parameters that lead to the best (least squares or chi square) fit.

The fitting routine cannot be used blindly. The data often extends over several orders of magnitude. Therefore, a simple fit using the linear differences between the observed and calculated values may not fit the threshold region, where the differences are always small. Thus, it is always useful to also make a fit using the differences of the logarithms of observed and calculated values. This will improve the fit in the threshold region, but not necessarily give a good fit in the saturation region. You can now repeat the linear fit, forcing the threshold to the value found with the logarithmic fit.

The fits should always be examined visually using the four different displays described in Figures 3-1–3-4. You want to have good fits at the threshold, at the middle of the curve, and at the limiting cross section. The fitting values should also be examined to make sure they are physically reasonable. The threshold should not be negative. In fact, it should be slightly above zero if you believe that the part does not upset for stopping protons. Often, where there is inadequate high LET data, the fitting routine will assume an unphysical wide distribution, with a corresponding unphysical large limiting cross section. You should always estimate the limiting cross section from your knowledge of the device layout before you start the fitting process. It is sometimes necessary to place limits on one or more of the fitting parameters. It sometimes takes five to ten iterations to obtain a good fit to the data.

The results of the fit will be used to calculate the event rate in space. Therefore, you should be sure that you are fitting the observed event rate, which may be different from the total upset rate due to multiple upsets of one kind or another.

Most modern small scale devices have depths that are significant relative to the areal dimensions so that the geometrical corrections discussed in Section 5.3 should be used in the fitting process. The Spaceradiation code will also do this, in conjunction with the RPP geometry model. The fitted curve parameters and device depth can then be used to calculate upset rates

3.14 OTHER SOURCES OF ERROR AND UNCERTAINTIES

There are a number of other possible sources of errors and uncertainties. These have been well presented by Koga in a review article and by Sexton in a NSREC short course [Koga 1996, Sexton 1992]. See also the material on rate prediction in Chapter 16. Articles by Hafer, LaBel, and Ladbury discuss some of the problems introduced with modern small scale technology [Hafer 2008, LaBel 2008, Ladbury 2009].

Some of the possible complications are:

Multiple upsets

Role of charge sharing in multinode effects

Beam uniformity

Part to part variation

Nominally same lot

Different lots, different times, different experimenters

Sample size consideration

Variation with time due to total dose effects

Testing algorithms

Static versus dynamic testing

Pattern sensitivity

Misalignment during rotation

Complicated overlayers

Approaches to die access

Ion penetration and range issues in power and packaged components

Complications at high speeds

Device complexity

Varieties of packaging

Large variety of operating modes

Cross-contamination of various error modes

Error latency

Repeatability

The report on the experiment should address each of these possible problems. This book does not specifically discuss hardness assurance issues. That is, do the parts that you are flying have the same single event characteristics as the parts that you tested? However, we can see from this list of potential problems that hardness assurance is not trivial.

CHAPTER 4

Optimizing Proton Testing

4.1 MONITORING THE BEAM INTENSITY AND UNIFORMITY

Many aspects of single event testing with protons are similar to single event testing with heavy ions [Koga 1996]. There are two important differences. The primary one is the problem of monitoring the beam intensity. The beam intensity necessary in order to make measurements in a reasonable time is in an awkward regime. It is higher than that used for heavy ion testing, so that single particle monitoring cannot be used. It is lower than that traditionally used for nuclear physics experiments at the same accelerators, so direct beam current measurements cannot be used. A common technique uses a secondary emission monitor [Murray 1989, Foster 1996]. This consists of a foil structure placed in the beam in front of the target in which electrons knocked out of a foil lead to a measurable current in pickup foils mounted after the front foil. Sometimes several of these structures are cascaded. Another technique is the transmission ion chamber. This consists of a chamber with a high voltage centered foil between two thin signal foils [McMahon 2008].

The second problem involves obtaining and monitoring the beam uniformity across the single event target. Normal nuclear physics

Single Event Effects in Aerospace, First Edition. By Edward Petersen.
© 2011 the Institute of Electrical and Electronics Engineers, Inc.
Published 2011 by John Wiley & Sons, Inc.

beams are well focused on the target. The SEU beams want to be broad and uniform. This is ordinarily obtained by defocusing the beam or by using a thick beam scatterer. In either case, the beam uniformity should be carefully observed and factored into the results if necessary. The experimenters should be aware of any problems of beam scattering or beam divergence or convergence.

We have previously mentioned that it is necessary to know the beam energy spectrum. If the beam is attained through the use of energy absorbers, there may be a great deal of energy straggling so that there is a wide energy peak.

The beam intensities can lead to problems with total dose damage as discussed in Section 4.2. A related problem is that of activation of the target and surrounding materials. These can become radioactive, so that care in approaching and handling everything in the area is necessary. Since the half-lives involved are normally short, the problems are not serious.

An ASTM standard is being developed for proton testing [McMahon 2008].

4.2 TOTAL DOSE LIMITATIONS ON TESTING

Figure 4-2 shows typical cross sections in the region of 10^{-14} cm^2/bit. In the past, due to historical accident, any cross section less than 10^{-14} cm^2/bit was assumed to be zero. This point of view was adopted because initial proton upset measurements were made on 1K parts with a hardness of 10^4 rads (Si). As approximately 10^7 protons/cm^2 deliver a rad, the parts die from total dose at the same flux that would produce one upset. Measurements by Pickel and co-workers on CMOS/SOS parts that were assumed to be immune to proton induced upsets revealed upsets with cross sections of approximately 10^{-15} cm^2 [Pickel 1989]. It is now clear, as we move to high bit-count parts or closer to the heart of the proton belts, that we will encounter parts for which low cross sections are important.

Total ionizing dose effects become an important limitation in testing for soft parts. Figure 4-1 shows dose delivered as a function of proton energy. Consider a maximum allowable upset rate of less than 10^{-6} upsets per bit-day in a 64K part in the heart of the proton belts. If the part has a hardness of 10 krads, a single part will often die from

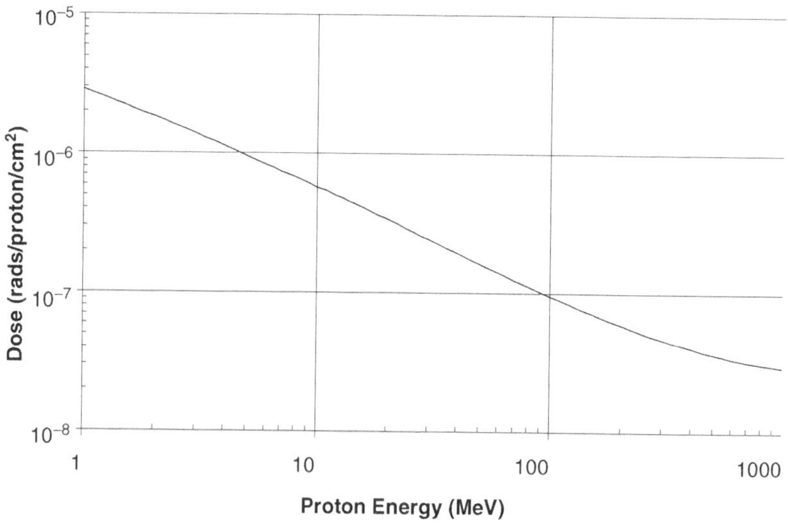

Figure 4-1 Proton dose deposition in silicon.

total dose before it is possible to expose it to enough beam to verify the SEU hardness [Petersen 1988].

4.3 SHAPE OF THE CROSS-SECTION CURVE

Initially the Bendel–Petersen two parameter semiempirical equation was used to describe the cross-section curve. This is described in Appendix B at the end of the book. This curve was based on data that had very few low energy points. However, good low energy data [Duzellier 1997, Harboe-Sorensen 1998, Poivey 1998] show that the threshold region is better described by the integral lognormal distribution [Petersen 1998]. If adequate data is available, the Weibull distribution can also be used.

Figure 4-2 shows some of the data plotted on a linear-log plot that shows the details of the threshold region for several devices. The form of the plot in Figure 4-2 is best for showing the characteristics form of the integral lognormal and Weibull distributions. The solid lines are the lognormal fits. The curve parameters are the mean, the logarithmic standard deviation, and the limiting cross section. The Weibull fits, indicated by only the symbols, lay on top of the lognormal fit. In

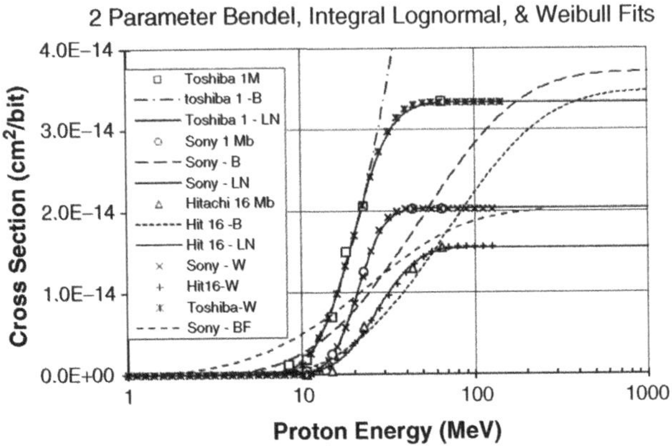

Figure 4-2 Proton upset cross-section curves and fits with the Bendel, integral lognormal, and Weibull functions. The data is from references [Duzellier 1997, Harboe-Sorensen 1998, and Poivey 1998].

general, we say that the Weibull curve should fit any data that the lognormal curve fits. However, this may not be true for proton data on sensitive devices; when there is not adequate low energy data to determine the three low energy Weibull parameters. In this case you need to use the lognormal curve and then find a matching Weibull curve.

Figure 4-2 also shows the results of attempting to fit this data with the two parameter Bendel curve. It cannot simultaneously fit the threshold region and the limiting cross section. The three curves labeled B in the insert represent an overall fit to the data and fit the threshold, but have the wrong limiting cross section. If we force the Bendel curve to fit the limiting cross section, shown for the Sony data and labeled BF, then the threshold region is approximated. The early version of the CREME code only had the Bendel fits for calculating proton upset rates, so this type of fit has been widely used. As the high energy protons dominate the rate, this approach works very well, especially for devices with low thresholds. The CREME96 and CREME-MC codes allow input using the Weibull parameters, not the lognormal parameters, so that Weibull fits should be used for proton rate calculations.

When we want to show a family of curves we resort to the log-log plot. Figure 4-3 shows a number of other devices from these studies. We see that a variety of devices are fit using the lognormal distribution.

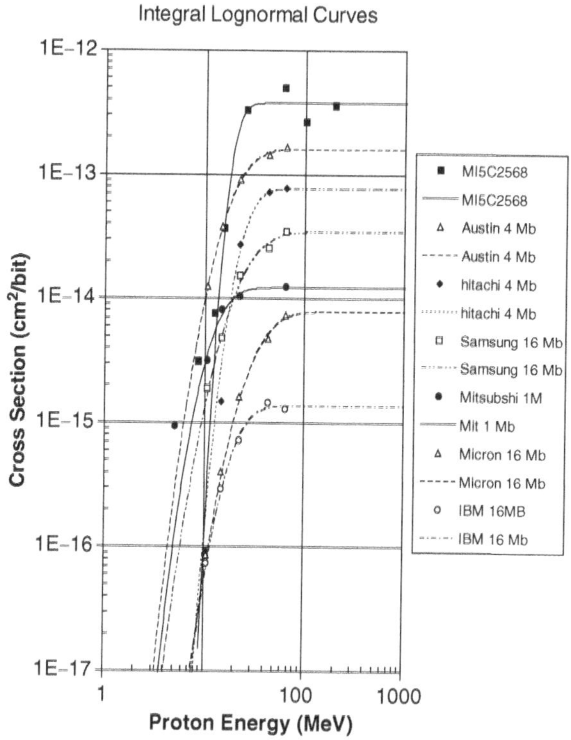

Figure 4-3 Proton upset threshold cross sections for several devices from references [Duzellier 1997, Harboe-Sorensen 1998, Poivey 1998]. The data have been fit with the integral lognormal function.

Table 4-1 lists the lognormal and Weibull parameters for these devices. We observe that the limiting cross sections do not agree for two of the devices. In these cases, the maximum measured energy is 64 MeV, not high enough to define the limiting cross section. Later we discuss using measurements in the range of 100–200 MeV to characterize device limiting cross sections. For the rare devices with very high thresholds, the upper part of this range should be used.

Data at higher energies assembled by Harboe-Sorensen [Harboe-Sorensen 1997] supports the use of the lognormal function. Examining parts in [Harboe-Sorensen 1997] for which a number of cross-section points are listed, we obtain a ratio of cross sections $200\ \text{MeV}/100\ \text{MeV} = 1.03 \pm 0.18$ (58 parts) and $300\ \text{MeV}/200\ \text{MeV} = 1.17 \pm 0.07$ (29 parts). Above 300 MeV there

Table 4-1
Integral Lognormal and Weibull Device Parameters

Device	Mean (MeV)	Standard Deviation (log MeV)	σ_{PL} (cm^2/bit)	L_0 (MeV)	w (MeV)	s	C_s (cm^2/bit)
Toshiba 1M	19.7	0.41	3.3×10^{-14}	8.00	14.68	1.67	3.34×10^{-14}
Sony 1M	21	0.29	2.0×10^{-14}	10.16	12.91	2.09	2.03×10^{-14}
Hitachi 16M	27.1	0.42	1.55×10^{-14}	10.50	21.76	1.53	1.56×10^{-14}
MI5C2568	20.3	0.2	3.7×10^{-13}	5.62	16.97	4.28	3.73×10^{-13}
Austin 4M	21.1	0.49	1.62×10^{-13}	8.98	18.36	1.09	1.70×10^{-13}
Hitachi 4M	26.1	0.33	7.73×10^{-14}	10.40	19.25	1.93	7.72×10^{-14}
Samsung 16M	26.3	0.55	3.47×10^{-14}	9.51	40.29	0.94	4.65×10^{-14}
Mitsubshi 1M	13.3	0.41	1.28×10^{-14}	5.00	10.31	1.62	1.22×10^{-14}
Micron 16M	35.2	0.51	7.92×10^{-15}	10.19	62.90	1.27	1.32×10^{-14}
IBM 16M	21.1	0.42	1.38×10^{-15}	0.00	24.53	3.07	1.37×10^{-15}

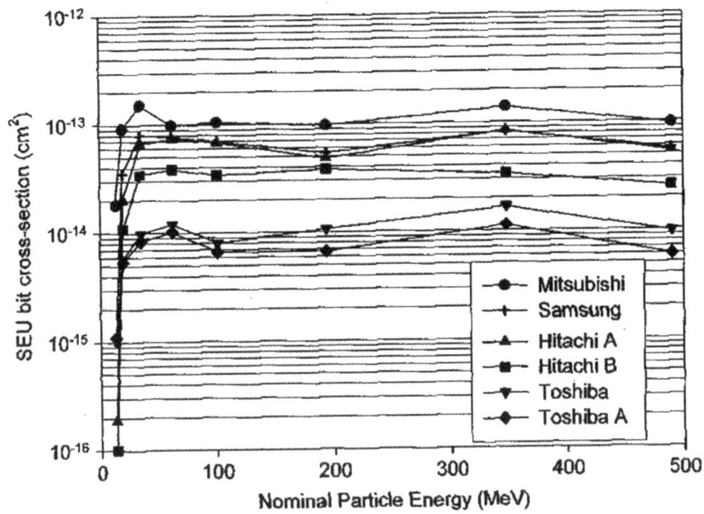

Figure 4-4 Proton induced SEU cross sections for 4 megabit SRAMs [Dyer 2004]. Copyright © 2004 IEEE.

may an increase in cross section due to meson production, but that is not important for our purposes. Figure 4-4 shows data obtained at TRIUMPH indicating essentially constant high energy proton cross sections [Dyer 2004].

We will observe that heavy ion upset rates can be determined from the figure of merit and that the figure of merit can be obtained from the limiting proton cross section. We now observe that the proton cross section is usually constant from 200 MeV to 100 MeV (a distance of 85 mm in Al). This means that we can test a part inside an electronics box in a 200 MeV proton beam and use the resulting proton upset cross sections to determine the part's heavy ion upset rate in any orbit. If we want the proton or heavy ion rate *in this box*, we need to consider the effects of shielding. See Section 11.6.

CHAPTER 5

Data Qualification and Interpretation

5.1 DATA CHARACTERISTICS

Before we can analyze SEE data, we have to understand the basic nature of the data. That means that we need to recognize both the normal and abnormal errors that may show up as the data is obtained. SEE measurements are a fertile ground for all types of error as we will show with several examples [Petersen 2008a].

5.1.1 Illegitimate, Systematic, and Random Errors

Illegitimate errors are likely to occur even in the best of experiments and should be discussed in the written report. The most likely type of avoidable error in SEE experiments is the *blund*er. This is an outright mistake in reading instruments, adjusting the conditions of the experiment, or performing calculations. One example of a blunder is forgetting to reset an upset counter after a run. Another example is to have one program to record errors and run information and a separate program to log the data, with the logging program set for a large number of runs. If the number is set for 100 during the equipment testing procedures, and then that number is not reset before making 140 data runs, data for forty runs may be lost. That is a blunder! These may be

Single Event Effects in Aerospace, First Edition. By Edward Petersen.
© 2011 the Institute of Electrical and Electronics Engineers, Inc.
Published 2011 by John Wiley & Sons, Inc.

largely eliminated by care and by repetitions of the experiments and calculations. If the effects of the disturbances become unreasonably large—that is, large compared with the natural random errors—they are called *chaotic errors*. In such situations the experiment should be discontinued until the source of the disturbance is removed [Beers 1958, p. 5].

Systematic errors normally arise due to errors in the calibration of instruments. The most likely source in single event experiments is beam dosimetry. Nuclear physics experiments ordinarily involve only one set of beam conditions, so there is a single systematic dosimetry error in the results. When the same beam is used repeatedly over many months for ordinary nuclear physics experiments, it is routine to maintain 3% absolute errors [Petersen 1967]. However, SEE experiments usually involve a series of different accelerator beams (different ions or energies), each of which requires good calibration of the instruments used to measure the beam intensity. Therefore, a systematic error for one beam shows up as an inconsistency either from one beam to another or for repeated measurements with the same nominal beam when measurements with other beams are interspersed. In this case, it is very difficult to obtain accelerator beam dosimetry to better than 5% for a set of repeated runs. However, one should expect the beam dosimetry to be in the range of 5% to 8%. Observations of this type should be made during the experiment and the written report should discuss the systematic errors.

Another type of systematic error that may occur in SEE experiments is a mistake in device readout. You may be reading only 32K of a 64K device. Systematic errors can also arise if an instrument is used under constant experimental conditions different from those for which it was calibrated, and if no correction is made, a systematic error results.

Random errors arise from several sources. A common source is fluctuating conditions such as temperature or voltage. It may be that your particular SEE is very sensitive to one of the voltages applied to the device. Do you know that? How well regulated are your power supplies? Another source is small disturbances. An accelerator vault usually is very noisy electrically. Are you adequately shielded? Are there any ground loops?

In SEE experiments there is one source of random error that is inherent in the measurement. The SEEs occur at completely random intervals of time, whether in space or in accelerator beams. Therefore, repeated measurements will fail to agree. The number of counts observed in a given time obeys the Poisson distribution, and, as the number of counts becomes large, it approaches a Gaussian distribution. We need to understand these distributions if we are to understand our results.

5.1.2 Inherent Random Errors

Single event data normally has a Gaussian (normal) distribution about a best fit line through the data. The standard deviation of N events is $N^{1/2}$. The fractional standard deviation is $N^{1/2}/N = 1/N^{1/2}$. Table 5-1 shows the fractional standard deviation for various numbers of events.

Figure 5-1 shows a plot of the normal distribution and of its cumulative value. For this distribution, 1 in 3 points is outside one standard deviation. Points more than three standard deviations from the average are very unlikely. Data points are normally plotted with a bar or flag indicating the one standard deviation value. This is very useful for quickly judging the data. If all of the data points are within one standard deviation of the average, someone is not calculating the errors properly. If the spread in the data is consistently more than one standard deviation, the errors may be governed by systematic errors, not the random counting errors. If there are one or more points more than three standard deviations from the average, there is a major problem with the experiment. Data is sometimes plotted with a two standard

Table 5-1
Fractional Standard Deviations Corresponding to the Number of Observed Upsets

N	% Standard Deviation
30	18.26
100	10.00
300	5.77
1000	3.16
3000	1.83
10000	1.00

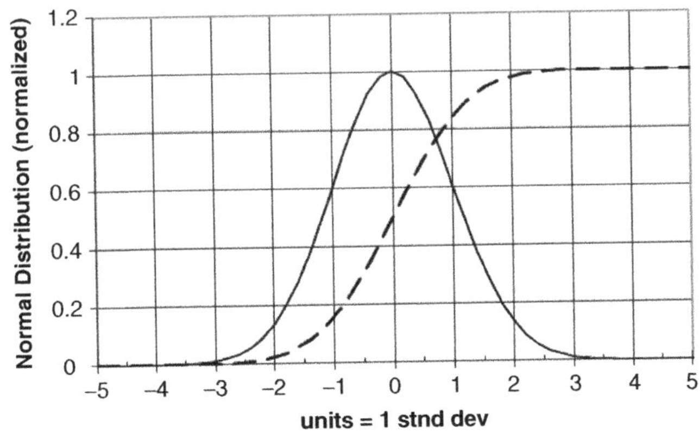

Figure 5-1 The normal distribution.

deviation bar, but that does not appear to be as informative as the one standard deviation bar.

When we are observing only a few counts, the probability distribution is Poisson rather than Gaussian. Consider an experiment in which you observe ground level cosmic rays with a Geiger counter on a desktop. A single detector will also detect products of the decay of natural radioisotopes in concrete. It is necessary to use a counter telescope to detect only cosmic rays. The average rate is four counts per minute and you make 100 one minute runs. The results may be as shown in Figure 5-2. In this case we only observed a total of

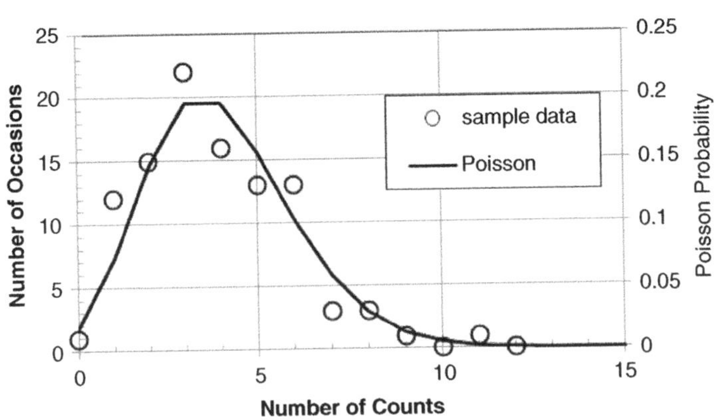

Figure 5-2 Sample data reflecting a Poisson distribution when observing 100 one minute runs expecting an average of four counts per run.

380 counts. Note the expected asymmetric distribution and that we may observe zero or eleven counts. Eighteen cases were two or less; twenty-one cases were six or more. There is a significant probability that a single run can be very misleading.

When the result of the measurement of a physical quantity is shown as $x \pm \sigma$ without further explanation, it is implied that the probability distribution of the measured quantity is a Gaussian (normal) function. Thus, the probability that the true value of the measured quantity is within the confidence interval $(x - \sigma,\ x + \sigma)$ is 68.3%. When N is small, the distribution is Poisson rather than Gaussian and is no longer symmetrical. We would still like to quote the upper and lower values of the confidence limits for 68.3% probability corresponding to the standard deviation and 95% corresponding to two standard deviations. Table 5-2 shows the corresponding confidence levels for small numbers of events described by the Poisson distribution [Helene 1984]. Note the error ranges that we would quote for the extreme cases in Figure 5-2. A single measurement of a small number should always be treated with some mistrust.

The cross-section curve is best fit using a weighted least squares approach. For least squares fitting you calculate and minimize the quantity S:

$$S = \Sigma \Delta y^2 \tag{5.1}$$

For linear least squares fitting Δy is the difference between the measured and calculated cross section while for logarithmic least squares fitting it is the differences of the logarithms of measured and calculated cross sections. The Excel Solver routine is useful for calculating the minimum by varying the parameters in the fitting equation.

For weighted least squares fitting, you calculate and minimize the quantity S_W [Beers 1958, p. 37]. This places more weight on the points with small statistical errors and less weight on the points with large statistical errors.

$$S_W = \sum \left(\frac{\Delta y}{s_y^2} \right)^2 \tag{5.2}$$

Δy is defined above. For SEU analysis, the weights s_y are the fractional standard deviations of the data points. Table 5-1 showed the fractional standard deviations, or weights, for large numbers of upsets. The last column of Table 5-2 shows the corresponding weights for small numbers.

Of particular interest is the situation with zero observed upsets. The data is normally plotted showing the one standard deviation limits, or for small values, the limits corresponding to the values in Table 5-2. Therefore, points with zero counts should show a zero cross section (or very small value on a log scale) with the error bar extending to the cross section corresponding to 1.15 counts. The least squares weight is 0.87. When fitting the data, only the zero point adjacent to LET points with observed upsets should be used. Any zero points at lower LETs do not belong to this distribution and should not be included in the

Table 5-2
Limits of Confidence Intervals of Poisson Distribution for Confidence levels of 0.683 and 0.95[a]

N	68.3%	± Range	95%	± Range	% Standard Deviation
0	0.000	1.15	0.000	3.00	89.9
1	0.268	2.50	0.042	4.77	66.7
2	0.864	3.86	0.304	6.40	53.7
3	1.55	5.15	0.713	7.95	46.5
4	2.29	6.40	1.21	9.43	41.7
5	3.06	7.63	1.76	10.9	38.0
6	3.85	8.84	2.35	12.3	35.2
7	4.65	10.0	2.97	13.6	33.3
8	5.47	11.2	3.62	15.0	31.25
9	6.30	12.4	4.29	16.3	29.4
10	7.14	13.5	4.98	17.6	28.6
12	8.84	15.8	6.40	20.2	26.3
14	10.6	18.1	7.86	22.7	24.4
16	12.3	20.4	9.36	25.2	22.7
18	14.1	22.6	10.9	27.7	21.7
20	15.8	24.8	12.4	30.1	20.8

[a] This corresponds to one and two standard deviations. The last column shows the weights to be used in weighted least squares fitting (corresponding to the fractional standard deviations).
SOURCE: [Helene 1984]. Copyright © 1984 Elsevier Science Publishers. Reproduced with permission from Elsevier.

fit. The plot of zero cross section with its standard deviation differs from the common practice of plotting zero counts as if one count had been observed, and allows a better understanding of the data.

Some members of the single event community have advocated plotting data using 2σ error bars. We believe that the 1σ error bars are more informative.

1. This is the standard approach in the experimental statistics community.
2. If fewer than one in three points is outside 1σ, there is a problem with the assignment of errors.
3. If more than one in three points is outside 1σ statistical error, there are problems with the data. In SEE testing these are probably due to systematic (dosimetry) errors.
4. Showing 1σ error bars makes it very easy to recognize the points that are 3σ or more away from a smooth curve through the data. These are candidates for rejection.
5. The 1σ error bars give an indication of the relative weight of the points when fitting the data.
6. The fractional standard deviation of the cross section is determined from the combination of the statistical fractional standard deviation and the fractional standard deviation of the dosimetry. Note that the rejection criteria use the total standard deviation, not just the statistical error.

There are always issues about obtaining adequate statistics for devices with few bits, small cross sections, and possible radiation damage. These are discussed elsewhere in the book.

5.1.3 Fractional Standard Deviation of Your Data

You should always characterize your data by determining its statistical characteristics. In particular, you should determine the fractional standard deviation. You fit the data so that you obtain expected values (w) along the curve. You then determine the difference δy between the data point (y) and the expected value. The fractional difference

Dy is then

$$Dy = \frac{\delta y}{w} = \frac{y - w}{w} \tag{5.3}$$

From this you obtain the average absolute fractional deviation, using N as the number of terms.

$$a = \frac{\Sigma|Dy|}{N} \tag{5.4}$$

The fractional standard deviation is 1.25 times the fractional average deviation:

$$S = 1.25 \times a \tag{5.5}$$

Consider the data shown earlier in Figure 3.1. Several devices with slightly different thresholds were tested; so to avoid uncertainty near threshold, we will use the points above an LET of 10 MeV/mg/cm^2. They have an average deviation of 6% so the standard deviation is 7.5%.

We have Equation (2.3),

$$\sigma = \frac{N}{F}$$

where N is the number of upsets and F is the fluence. The fractional standard deviations are related by the expression [Beers 1958]

$$S_\sigma^2 = S_N^2 + S_F^2 \tag{5.6}$$

Therefore, with the total fractional standard deviation 7.5%, if the counting statistics are 3% to 4%, the error in the fluence measurements is 7% to 6%. Single event experiments will always have two error terms.

There are normally several data sets. Each device may be tested at two voltages and with two test patterns. Hopefully there is also a variety of angles used for at least one data set. It is not necessary to have a variety of angles for examination of systematic errors, as seven normal incidence points are adequate, but the more data the better. (Remember the Connell corrections if you are fitting angular data. See Section 5.3.3.) Each set of data should be fit using the least squares approach to determine a best fit line through the data. Points on this line describe the expected data. As we want to determine the dosimetry (systematic) errors for an entire set of runs, we can combine fractional standard deviations for all runs.

Although we have been discussing heavy ion experiments, the same discussion also applies to proton experiments. You fit the energy data with a lognormal or Weibull curve and determine the standard deviation about the curve.

5.1.4 Rejection of Data

This topic rarely arises with well-behaved experiments or in books discussing approaches to data analysis. We will follow the discussion of Beers [Beers 1958, p. 23; Petersen 2008a]. In our SEE experiments, quite frequently most of the resulting values are found to lie close together while one or two others differ very much from this group (outliers). The question immediately arises as to whether these apparently anomalous measurements should be rejected. If we know that there were obvious problems at the time the measurements were made, obviously they should be rejected. In fact, even if the data obtained during a disturbance appear to agree with the others, they should be rejected.

In normal experimental situations where no disturbances were known to be present, it is highly controversial whether the data should be rejected. Some observers contend that all data should be retained. Ideally, so many additional measurements should be made that retaining or rejecting the dubious measurement will have little effect on the average value. Unfortunately, such a procedure is often impractical in SEE experiments.

Another approach to outliers, when one is using regression analysis to fit the data, is to assign outliers lower weights [Hamilton 1992, Chapter 6; Mosteller 1977, Chapter 14]. This uses a weighted least squares fit with outlying cases getting down-weighted gradually: the farther out a case lies, the lower its weight. This appears appropriate when a few cases widen the distribution, but not with the extreme cases often observed in SEE experiments.

We are inclined to reject the questionable measurements on the basis that the probability of obtaining such large deviations is so negligible that they must have resulted from some abnormal cause. This approach does present two difficulties. In the first place, there is no clear-cut basis for deciding just how small the probability must be to be considered negligible. In the second place, if some definite criterion is adopted, and then applied without discretion, it may lead to an unreasonable situation. In applying it to the original set of data, the observer rejects some of the measurements. He/she may then suppose that some of the remaining discrepancies are also due to abnormal causes, and apply the criterion to the remaining data, and reject other measurements, and so on. Obviously repetitive application of a criterion is ridiculous. Such a criterion should only be applied to the original data.

The Gauss error law provides a rational basis for setting up a criterion. If the probability of obtaining deviation greater in magnitude than some limit $|z|$ is negligible compared to $1/k$, where k is the number of measurements, then all measurements with deviation greater in magnitude than $|z|$ are to be rejected. Table 5-3 lists the probability of obtaining a deviation greater than some preassigned magnitude. We still need to define "negligible." If we have ten measurements, and set $|z|$ less than 1/30 of $1/k$ (1/10), the probability is less than 0.003, corresponding to three standard deviations. Note that the probability falls off very rapidly at larger standard deviations. A fraction of 1/1600 would correspond to four standard deviations. This would suggest that a fuzzy criterion in the vicinity of three standard deviations would be appropriate.

There are more elegant procedures that can be applied to various size samples (even very small ones!) that enables you to choose the probability or risk you are willing to take in rejecting an observation that really belongs in the group [Natrella 1963, Chapter 17]. These approaches do not appear necessary in analyzing SEE data.

A common and usable approximation for the criteria for the maximum acceptable spread in the data is to use the value at three standard deviations. In the normal distribution this should occur for only 3 events in 1000. We often have only 5 to 25 data point, so this event is very unlikely. Any data points more than three standard deviations from the average should be considered for rejection (not automatically rejected). And note that the outliers in SEE experiments are often well

Table 5-3
Probability of Obtaining a Deviation from the Average Greater than a Preassigned Value

Values Relative to Standard Deviation	Probability
0.6745	0.5
0.707	0.48
0.8	0.43
1	0.32
1.41	0.16
1.5	0.13
2	0.05
2.5	0.012
3	0.003
3.5	4.6E−04
4	6.3E−05
4.5	6.8E−06
5	5.7E−07

outside three standard deviations. We do not want to automatically reject outliers, as they often give us very important information about our experiment [Draper 1981, Chapter 4]. In all such cases we would like to identify the problem (opportunity?) before continuing.

5.2 APPROACHES TO PROBLEM DATA

There are a number of types of problem data. We will give samples of a few of these and discuss how we examine these.

5.2.1 Examination of Systematic Errors

In most experimental situations the random statistical errors dominate. This is not true in typical SEU accelerator experiments. A group will arrive at the accelerator planning to test three devices, each with seven ions. That means that they plan to do 21 different experiments in 16 hours, each with its own dosimetry. There is normally an abrupt time cutoff, so the group feels time pressure. This is an ideal situation to generate blunders and systematic errors.

There appear to be several types of common problems in SEE experiments. It is normal to have situations in which the dosimetry is controlled, but has a statistical spread significantly greater than the counting statistics. You should immediately determine the standard deviation due to the dosimetry and determine if it is consistent with that necessary for the observation of the effects of interest. There are outliers associated with single runs where the other runs with the same conditions are well behaved so that the outlier is not obviously identified with problems with dosimetry. There are cases in which the dosimetry with a single beam is wrong, but the other beams are well behaved. There are cases of large variations in the dosimetry in which the dosimetry does not appear to be controlled.

We now want to examine approaches to observing the systematic errors and give several examples of data with various problems. The problems may be indicated by lack of internal consistency or by one or more points that are more than three standard deviations from a smooth curve through the data points. Most of these examples use constructed data that reflects the problem. They are constructed for illustrative purposes and are loosely patterned after unpublished data from reputable SEE groups, some of which were taken during a period in which the dosimetry was not under control. The groups later worked with the facility to remove the major problems. We will show the examples using plots of the data. During the data run, the numerical results should be roughly predicted and then quickly examined for reasonableness. They should then be plotted for comparison with the other data.

We are making the point that the problems should be recognized during the runs, and are suggesting methods for doing so. The constructed data does not reflect later facility capabilities or any actual part response. There are several complementary approaches for observing these problems [Petersen 2008a]. The experimental team should use as many as possible.

5.2.1.1 Golden Chip Approach

This approach uses a standard chip that has a repeatable upset cross section. The device is inserted into the beam for the initial measurements with a new heavy ion beam or with new proton energy [Harboe-Sorensen 2008]. It is possible to quickly observe if the response is not the expected response. The chip is mounted on its own test board so that it can be introduced into and taken out of the beam without interfering with the beam or the test object. It is then possible to cancel an exposure to a very expensive test object if there are questions about the dosimetry.

5.2.1.2 Examination of Internal Consistency of Data

The most common type of problem arises due to dosimetry problems at the accelerator. It is difficult to do dosimetry on any single beam. SEE experiments normally involve four to eight different beams. It is therefore important that the lead experimenter carefully monitor the beam dosimetry. One of her tools is the examination of the internal consistency of her data.

Figure 5-3 shows an example with test data for a small unhardened memory cell. The Weibull logarithmic fit wants the threshold to be below zero. We therefore use a threshold of 0.001. Even this is questionable, as any threshold below 0.04 indicates that the device may be sensitive to directly ionizing protons. More data is needed at low LETs. The final fit is done linearly, using this threshold. There is more spread in the data than would be expected from the small error bars. This figure used the traditional log-linear plot, which hides the error spread. Figure 5-4 shows the same data on linear-log scales. Here it is very clear that the scatter in the data is much larger than the

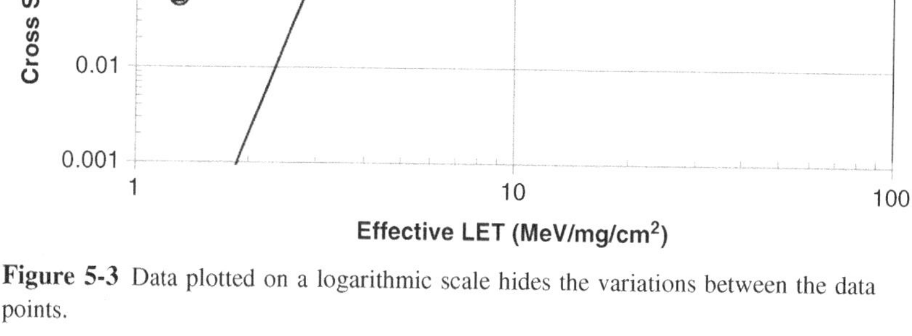

Figure 5-3 Data plotted on a logarithmic scale hides the variations between the data points.

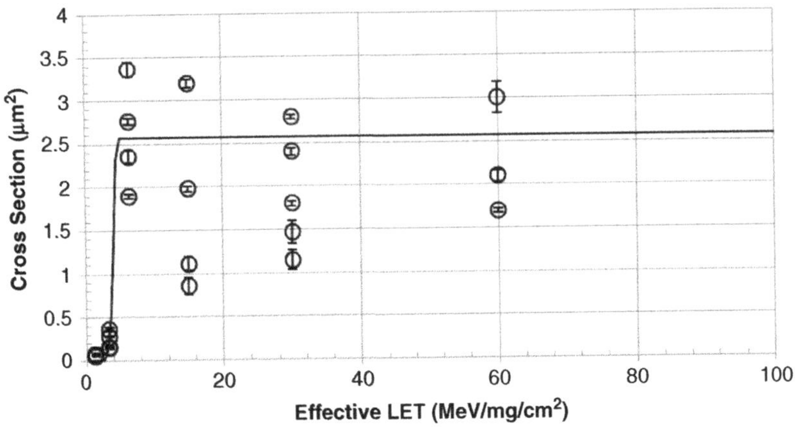

Figure 5-4 The spread in the data of Figure 5-3 is obvious on a linear scale.

error bars. The solid curve is the fit shown in Figure 5-3. It appears that the dosimetry is not under control.

Most of the data is more than three standard deviations from the fitted curve. If we arbitrarily removed points, we could obtain a curve through the lower points, with more points near the curve. However, there is no justification for this. This data is not good enough to use the three standard deviation rejection rule. The only thing that we can do with this data is to recognize that the errors are systematic errors, not the 1% to 2% statistical errors. The weighted Weibull fit gives a limiting cross section of 2.57. The unweighted average of the points above a LET of five is 2.12 ± 0.77. This indicates a 36% systematic error. Therefore, the data is usable, if we are satisfied with 36% errors in the results. But why take time to obtain 1000 to 16000 errors if you don't need the accuracy?

The experimental team leader should be plotting and studying the data as it is obtained. Large systematic errors would immediately be recognized and attempts could be made to remove them. In this experiment the leader is wasting much more time taking bad data, than would be used in tracing and fixing the problems. The first level data analyst should have recognized the problems with the data. The project leader should question the data and never pass it on to the sponsor.

The experimental team should continually check the data for internal consistency. We give a second example in Figure 5-5 based on

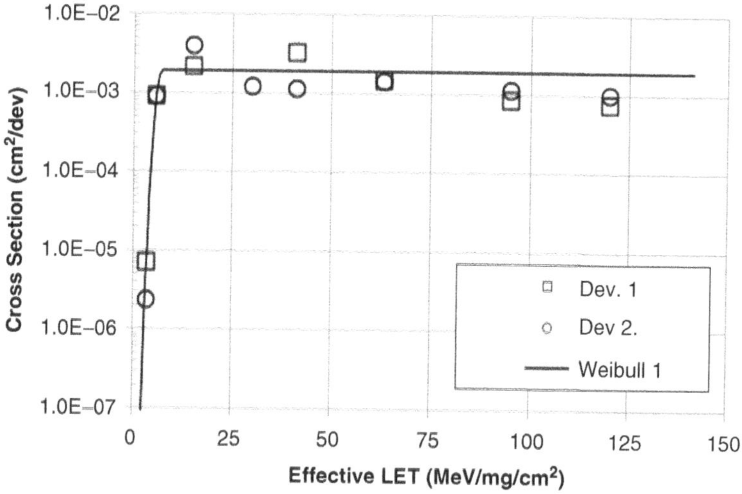

Figure 5-5 Sample SET cross-section measurements with wide data spread.

reported data for single event transients in an op-amp. The two samples appear to be consistent, so that is not a problem and the results can be combined. The plot has included the statistical one standard deviation error bars, but they are basically equivalent to or smaller than the size of the data points. This situation leads to the common practice of not including error bars, and reporting that the error bars are smaller than the data points. That would seem to indicate very small errors, but actually, as in this case, the standard deviations can average 10%, which are not small errors.

We should be bothered about two things in this data plot. First, the data is plotted on log-linear scales, which is very good at concealing data problems. Second, the vertical spread in the data does not seem consistent with the reported errors.

The situation is made clearer in Figure 5-6. Here the data are replotted on linear-log scales. We now observe that there are three points that are much more than three standard deviations from the other points. There appears to be a major problem with the dosimetry for the beam with a LET of 15 MeV/mg/cm^2. One of the points at a LET of 41 is inconsistent with the other data. This type of problem is rare, but it happens. Normally the experimenters would observe this discrepancy and redo their data. It is not clear, after the fact, what happened. It may well be that they only plotted the data on the

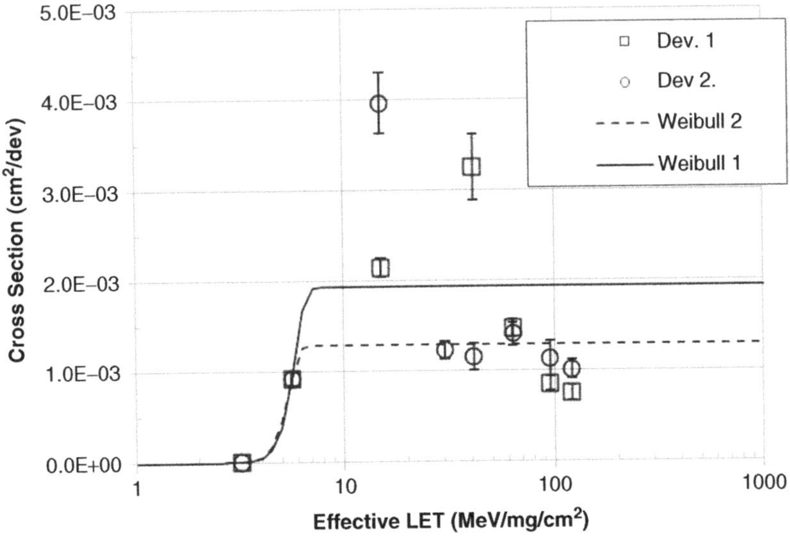

Figure 5-6 The data from Figure 5-5 replotted to emphasize the problems with the data.

log-linear scales and hid the problem from themselves, or worse, that they were not examining the data as they obtained it. The dosimetry appears to be under control at about the 10% level for the remainder of the points.

In most experimental work, it is very unusual to have outliers. The experimenter considers his options thoroughly before discarding data. After all, in a well-designed experiment, only one population is being sampled and the outliers should follow the Gaussian probabilities. However, in this example, there are eight different experiments, each with its own dosimetry. Even though a cocktail beam speeds up the experiments, it does not remove dosimetry questions. Furthermore, the dosimetry is set up by technicians who are not members of the experimental team. In this situation, there is more room for Murphy to operate, and, if experiments are not carefully monitored as they are performed, there are more occasions in which data must be discarded. The author would recommend trying to avoid this situation.

5.2.1.3 Repeated Runs

A primary method of testing for systematic errors is to repeat the testing with the same device and test conditions at various times during

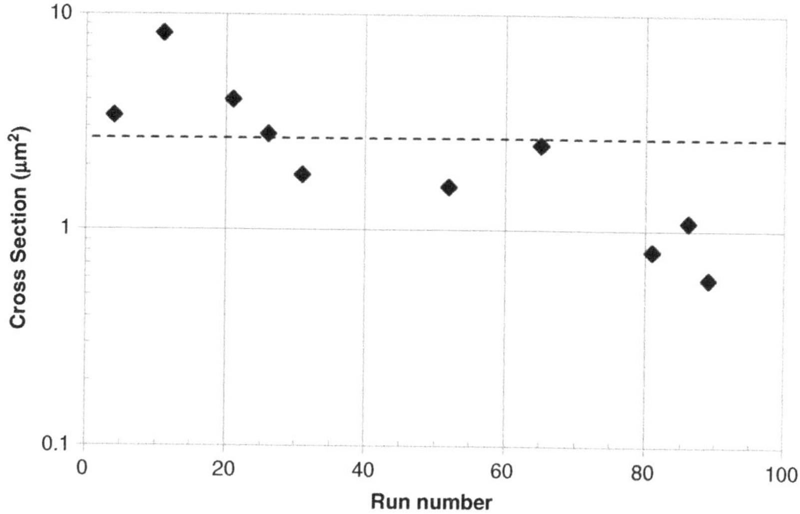

Figure 5-7 Data repeated in occasional runs in a large series of runs. Unlike Figure 5-8, the experimenters did not notice that their data were inconsistent.

the data run. Figure 5-7 shows a sample of repeated data that is part of a large group of experimental runs. The tests are all nominally under the same conditions with very good statistics on each individual run. The systematic error in this case appears to be 83%. The experiments are continually changing ions, so it is very possible that this reflects dosimetry errors. With the ratio of the maximum measurement to the minimum measurement of 13.7, it appears that the dosimetry is not under control. It does not appear that the experimental group was paying any attention to what it was doing.

The data in Figure 5-8 is a small sample from a sequence of 89 runs, in which ions, temperatures, and voltages were repeatedly changed. There were 18 changes of ion during the total run. The authors were conducting 19 different experiments in a short time frame! This plot represents runs taken at various times during the experiment using the same ion, angles, temperatures, and voltages. Therefore, they are measuring the same cross section. The data is plotted using a logarithmic cross-section plot, common with SEE data. The error bars are smaller than the symbols. We observe one point well outside three standard deviations. It is almost exactly a factor of 2 more than the other points. This suggests that someone forget to hit a reset button. Clearly the experimental team recognized a problem, as

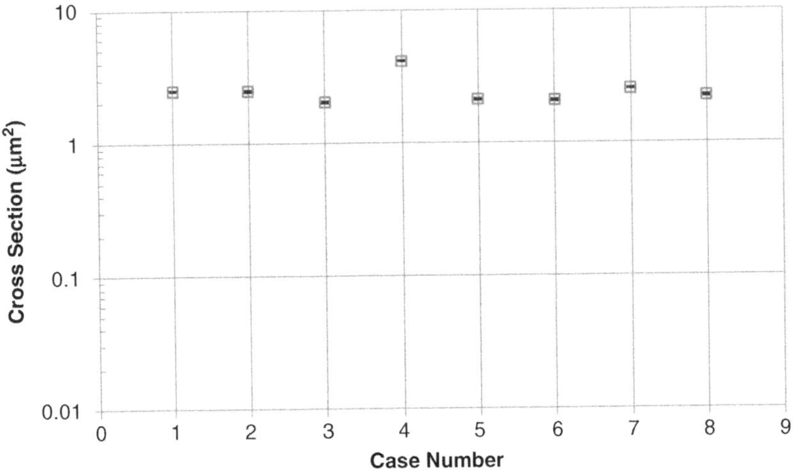

Figure 5-8 Sample data with outlier. The outlier is a factor of 2 higher than the other points. This is clearly a blunder and the point should be rejected.

they then repeated the point. In this case, it is clear that it is legitimate to discard the problem data. Appropriate notes should be added to the logbook and the experimental write-up.

One has to look very carefully at the plot to observe that there may be more scatter in the measured points than expected. We need to replot the data, without the outlier, on a linear scale to observe its detailed behavior. This is shown in Figure 5-9. We expect that two-thirds of the data points should have error bars that overlap the average value. None of the error bars overlap the average value! Clearly there are systematic errors much larger than the statistical random errors. One clue is given if we plot the occurrences of one or more beam changes. The data jumps each time. The standard deviation of the data set is 10%, not 1.5%. The experimental team needs to decide before the experiment what quality of experiment they are performing, and then carefully observe that quality. In this case they could get by with 150 errors per data point rather than the 4000 errors they took the trouble to obtain.

5.2.1.4 Use of Fractional Standard Deviation

We have treated systematic errors as if they were abnormal. Actually, as mentioned above, there are always systematic errors in dosimetry.

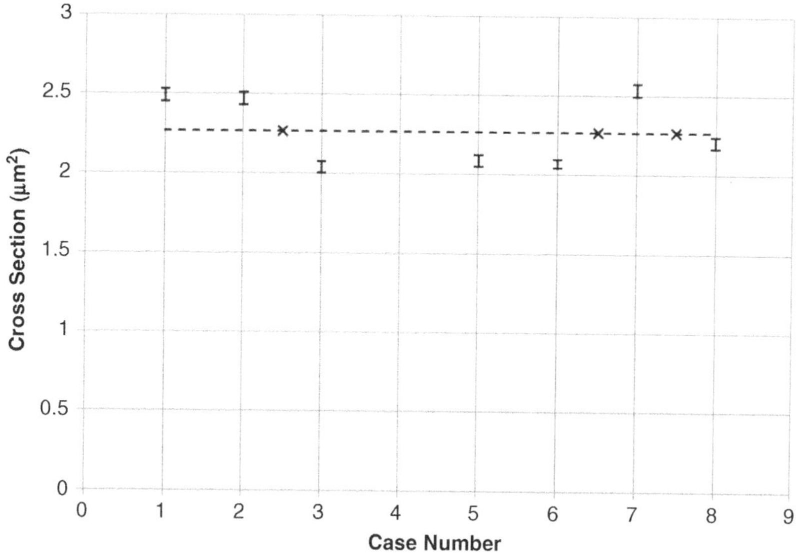

Figure 5-9 This data shows that there are systematic errors much larger than the statistical errors. They appear to occur at beam changes (×), which indicate dosimeter problems.

Even if there are no abnormal points such as those discussed above, the systematic error shows up as deviations from the best curve fit to the data. A powerful approach to systematic errors is to estimate the fractional standard deviation. Figure 5-10 shows an example of data with significant systematic error [Hansen 2007]. This data is also challenging to fit. The low LET points (nonzero) are evidently not in the same population as the remainder of the points. The author does not discuss them. They may be due to either low LET beam impurities or heavy ion reactions discussed elsewhere in this book. The author assumes that he has reached the saturation cross section and averages the high LET cross sections for a value of 6.3 MeV/mg/cm^2. Our plot does not indicate saturation. The R^2 for the authors' fit parameters is 0.68. A least squares fit suggests a cross section of 329 μm^2 with $R^2 = 0.73$. This is unphysical. We use a cross section of 8 μm^2, which gives an R^2 of 0.72. The authors do not discuss the expected cross section and did not extend the cross-section measurements to high LET to be closer to saturation.

The average deviation for the points with LETs greater than 3 MeV/mg/cm^2 is 26%. Therefore, the fractional standard deviation is

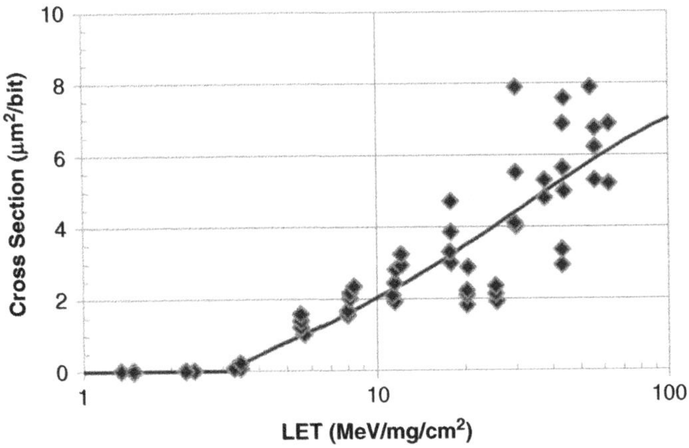

Figure 5-10 Cross section of a 0.25 µm SRAM [Hansen 2007]. This data shows a fractional standard deviation of 32% and systematic (dosimetry) error of 30%. The counting statistics were 10%.

32%! The authors stopped after approximately 100 upsets in each run. In this case the Poisson distribution will be evident in the run time, not the number of counts. A 32% total fractional standard deviation and 10% counting deviation lead to the dosimetry (systematic) error of 30.4%. Are you comfortable working with a 32% fractional standard deviation?

The data can also be examined visually plotting the ratio of the observed data to the expected data, as determined from the fit, at each LET. The ratio of these results can be averaged for the various data sets for a given ion and part (a single experiment). With good experimental statistics, the ratio should always be very close to one. You may not want to include the lowest LET cases, as that ratio is very sensitive to how the threshold is fit. If the ratio is not near one for only a single data set out of several sets in an experiment, a blunder is indicated, not a systematic error.

We have observed errors of 7% and 11% obtained at different times at Berkeley. The Hansen value of 32% was obtained at TAMU. This is consistent with the value calculated for the measurements of Swift, also taken at TAMU [Swift 2008]. Figure 5-11 shows data taken at Brookhaven by Hartwell [Hartwell 2004]. This has a standard deviation of 13%. This data is difficult to fit as the data does not have a clearly defined limiting cross section. The authors assigned a limiting

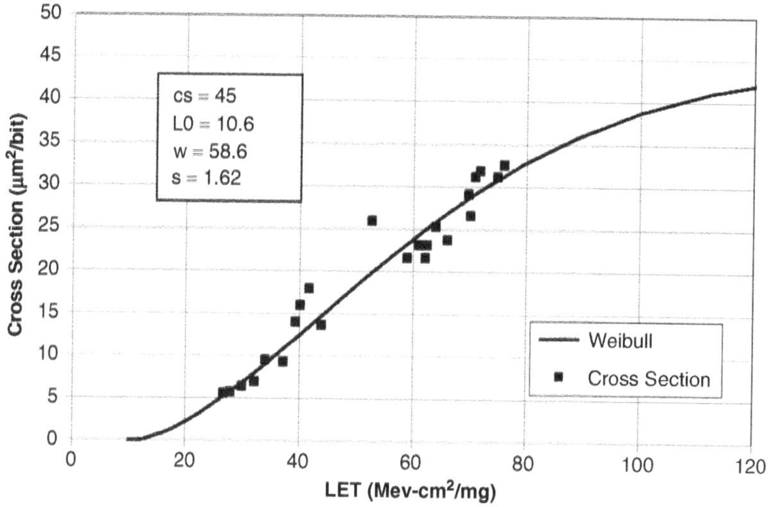

Figure 5-11 Upset cross sections for 0.25 μm ASICs taken at Brookhaven [Hartwell 2004].

cross section of 28 μm^2. Our fit was obtained by minimizing the standard deviation.

We also observe that our examples with 10% to 11% systematic errors still have one to three points that are well removed from the average. In one case, it appeared to be the fault of the experimental team. In the other cases, it is not clear whether these are the responsibility of the experimental team or of the dosimetry team or are an unexpected single event effect. It is clear that they are to be expected and that the cases should be carefully examined as they are observed. It is also clear that you should not attempt to draw any conclusions on the basis of a single point that is significantly removed from the other data.

Table 5-4 shows the variety of deviations that we have observed in a random sample of representative experiments [Petersen 2010]. We observe a wide range of dosimetry errors (average 35%). The errors depend on the operators, the dosimetry system and experts, and the direction of the experimental team leader. Undoubtedly, better and worse examples could be found for all of these facilities. The point is that you cannot assume that there is little systematic error in the dosimetry. Someone on the experimental team needs to be closely involved with the dosimetry. This is a list of examples, and the columns have no specific meaning.

Table 5-4
SEU Data Standard Deviations (%) Observed at Various Heavy Ion Accelerators

Brookhaven National Laboratory (BNL)	7.3	13	24	48
Lawrence Berkeley National Laboratory (LBNL)	7.5	11	67	
Texas A&M University (TAMU)	21	33		
University of Jyvaskyla, Finland (JYFL)	25			
Takasaki, Japan (TIARA)	12	93		
Michigan State University (MSU)	17	24	86	
Universite Catholique de Louvain, Belgium (UCL)	40			
Grand Accelerateur National d'Ions Lourds, France (GANIL)	78			

The standard criterion for considering that data points be rejected is if the points are more than three standard deviations from the mean. This normally refers to the standard deviation of the statistical errors. We see that for SEU testing of this nature, where the errors are dominated by the systematic errors, it should be three standard deviations of the combined statistical and systematic errors.

A systematic error may be acceptable in some experiments and be grossly too much for other experiments. You need to determine acceptable values before performing the experiment. If you want the order of magnitude of the rates in space, you may be satisfied with large systematic errors. If you want to determine the depth of the device, as described in Section 5.3.2, or perform experiments on the quality of space SEE rate prediction, as described in Chapter 16, you probably want the systematic errors to be less than 10%. If you need small systematic errors, your dosimetry expert should consult with the operations staff before you start your run.

5.2.1.5 Study of Systematic Dosimetry Errors

Heavy ion upset experiments are performed in wide beam geometry. The beam is spread so that there is nearly uniform intensity across an area greater than the size of the device being tested. The measurements of beam flux are ordinarily obtained using four detectors continually measuring the beam intensity at the sides of the beam, and a central detector intermittently measuring the intensity at the center of the beam. In principle, this is an accurate system, and there should be no problems. However, as we have seen above, problems often occur.

If you are concerned about systematic errors, the first step is to carefully examine your data, using the techniques above, and calculate and document the systematic errors, based on measured cross sections. This should be part of the experiment plan and be started as early as possible during the experiment [Petersen 2010].

We pointed out that the dosimetry problems were often associated with changes. That is, we make a cross-section measurement, change beams, and then come back to the initial beam and observe a different cross section. A problem could also occur if the device is rotated, and the axis of rotation and the center of the test device are not coincident. We also need to study the ever present systematic error in experiments with well controlled dosimetry.

The beam measurements determine the beam flux, the beam uniformity, the axial gain, and the central shift. These are calculated from the detector counting rates. You also ordinarily have access to the average count rates in the four outside detectors, which can give the direction of the central shifts. The facility may or may not make direct measurements of the initial beam distribution (possibly density in a x-y detector or photographic film). The beam can ordinarily be considered to have horizontal and vertical Gaussian distributions that are truncated at the sides.

Some of the systematic errors studied above seemed to have a random component. The dosimetry depends on detector counting rates, so will inherently have a random component. One should examine the total number of counts in order to see if they are contributing to the systematic errors.

The most common beam change is for the test team to request a change in beam intensity. This is easily done by changing the beam optics so that the Gaussian distribution is slightly sharper. The beam often seems to have a mind of its own, so that it twists and shifts, and will also move the center of the distribution, as well as change the beam uniformity. The operators may believe that the test team expects and plans to live with these changes, while the test team expects no significant change.

If the experimental team considers the accuracy of the dosimetry to be important for their experiment, they should consider recording the beam characteristics mentioned above, the measured standard deviations from the best fit to the data, and the comparison of measurements made with different beam tunes as discussed above. It may be valuable to make a continued study of the correlation between changing beam characteristics and the systematic errors.

Modern technology also introduces another factor. Most early measurements were made on memory chips. These are uniform, so that slight changes in the distribution of the beam across the chip has little effect on the measurements. Many modern measurements are made on some variety of system on a chip. The memory or register portions of the chip may be located on the side or in a corner of the chip. Therefore, two measurements using the same average beam flux, but very different beam distributions, can lead to very different upset cross sections. The modern experimenter needs to be aware of the changing shape of the beam distribution relative to her chip layout.

5.2.2 An Example of Voltage Variation

Now let us consider a pair of runs in which one of two voltages on the device is changed, and we observe a large change in the cross section, a factor of 4.5. Data is shown in Table 5-5. This may be telling us something important about the device or about the experiment.

Devices with feature sizes less than 200 nm can have extremely complicated electric fields. It is very possible for various bipolar gain effects to connect various internal regions of the device, depending on the internal voltages. Therefore, these results could be indicative of differing internal upset mechanisms and possibly indicating a hardening approach. The experimental team should follow up with more measurements and try to get a better understanding of their device.

These results could also indicate the cause of systematic errors. A 7% change in I/O voltage changes the cross section by a factor of 4.6. This implies that a 0.85% change in I/O voltage would lead to a 10% change in the cross section, which was the observed systematic error in this set of measurements. How well regulated is that voltage?

Table 5-5
Experimental Voltage Variation

LET	Angle	Effective LET	Core V	I/O V	Cross Section
60	57	110	1.8	3.5	2.4
60	60	120	1.8	3.75	0.53

Another possible explanation of this type of data is the shadowing effect at large angles. An example of this effect is shown in Section 3.6. The effect here is much more striking than the effect in that example, so more data is required to differentiate the voltage and angle effects.

5.2.3 Data Inconsistent with LET

We have discussed cases of outliers and systematic errors where data are inconsistent between runs or along the cross-section curve. There are other types of inconsistencies. Figure 5-12 shows such data. There is one group of measurements with one ion (Mn) that seems completely inconsistent with the data obtained with the four other ions. There is either a new SEE effect, or there was a blunder. The most probable blunder is that the operations crew misidentified the ion to the experimental team. It appears that this was really a Kr beam.

5.2.4 Beam Contamination

There can also be similar problems if the beam is not thoroughly understood. Criswell used the BEVELAC high energy beams to study several devices [Criswell 1984, Criswell 1987]. They used two ion beams with beam degradation to vary the beam LET. The very sensitive 93L422 did not show any difference between the two

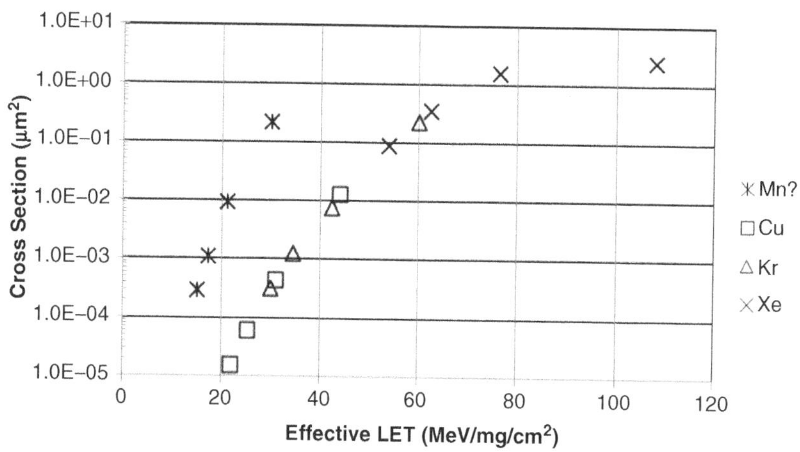

Figure 5-12 Data that appears inconsistent with LET.

ion beams, whereas the higher threshold 27LS00 appeared to show different LET thresholds for the two ion beams. There appeared to be "ion species dependence." This is shown in Figure 5-13. However, later, Koga and co-workers carefully examined the LET distributions of the BEVELAC beams and found that it was probable to have a large amount of high or low LET contamination of the beam [Koga 1990]. The high LET tail on the neon beam could lead to upsets when the nominal beam LET was below the threshold to produce upsets. High energy ion beams may have either high LET contaminant in the primary beam, or very broad LET peaks in degraded beams. One cannot always assume monoenergetic, single species beams. Figure 5-14 shows one of the BEVELAC beams. A degraded neon beam had a similar, but broader, distribution. If you used this beam to test a device with a threshold of 2.5, you would observe upsets at an apparent LET of 1.4. The BEVELAC was closed in 1993.

Problems like this are more common with high energy accelerators. Cyclotron generated beams tend to be nearly monoenergetic and relatively pure, provided that (1) the cyclotron has been properly tuned, (2) the source is carefully maintained, and (3) the beam delivery system is properly maintained [Koga 1990]. The higher the energy, the more likely the possibility of impurities in the beam. The

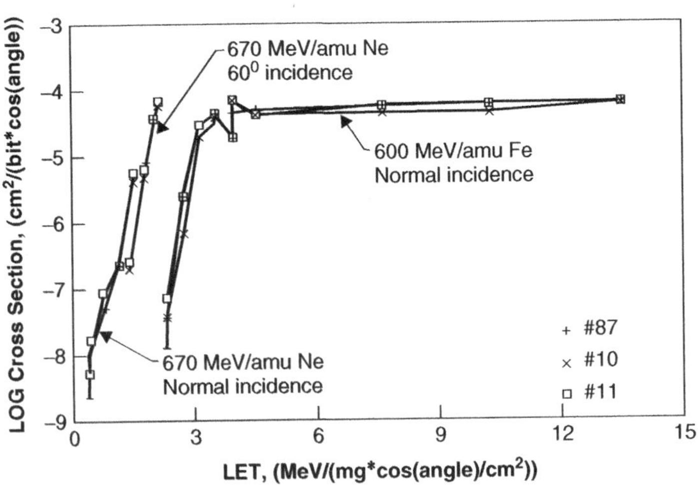

Figure 5-13 Apparent evidence of species dependence in BEVELAC beam [Criswell 1987].

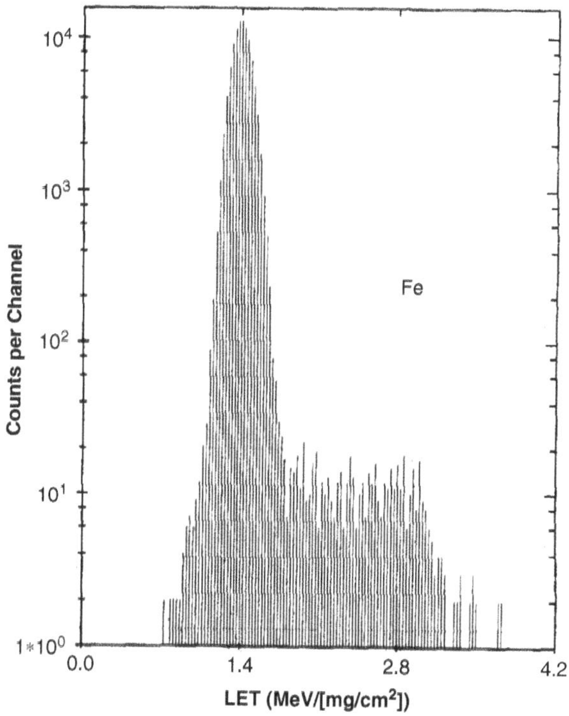

Figure 5-14 LET distribution of primary BEVELAC Fe ion beam [Koga 1990].

impurities at the BEVELAC lead to very noticeable contributions to the upset rate.

Several recent experiments have observed very low numbers of upsets below the normal upset curve. These have very low cross sections and it is not known if the purity of the accelerator beams was monitored. Consider the data in Figure 5-15. The experimenters should determine that the combined Kr, Xe, and Bi contents in the low LET beams are less than 0.005%. That is 1 in 20,000 incident particles.

The beam purity should be examined each time there is a new low LET beam. There can be large differences between the accelerator beam tuned by a senior operator at 8:30 a.m. Monday morning and the beam tuned by a junior operator at 4 a.m. Sunday morning. (Been there, done that.)

In studies of this type, it is important for authors to cross-check their dosimetry and beam purity. As the work of Koga demonstrated,

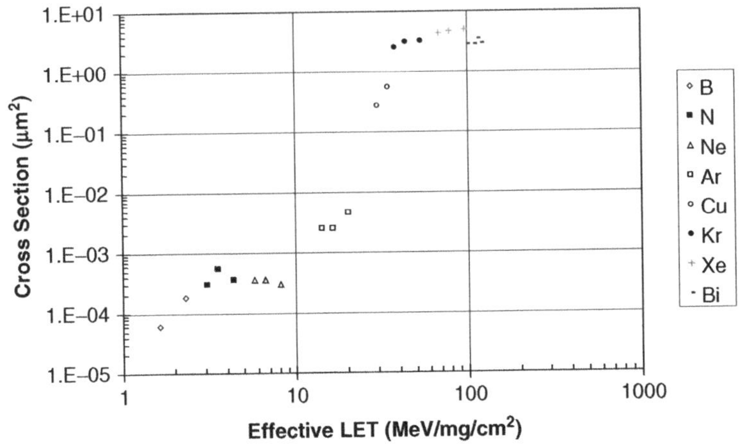

Figure 5-15 A representative set of data showing a low LET shoulder. Is this a SEU effect or a beam effect? In this case we believe that the data represents a legitimate SEU effect [Warren 2005], but all experimenters should be aware of the possibilities of beam contamination when doing this type of experiment. Is the beam pure to 1 part in 20,000?

it is very easy for high energy accelerators to have impure beams and for the electrical engineers measuring upsets to be unaware of the problem. We have also pointed out a number of examples of poor dosimetry leading to systematic errors in one or more beams, and one example of probable beam misidentification. These issues have to be addressed and discussed.

5.2.5 No Event Observed

One repeated problem occurs when no single event effects are observed. As that is the desired result, it is often accepted without adequate checking of the equipment. There are some hardened memories, latches, and other devices that do not upset except at exceptionally high effective LETs, if at all. Even if you are testing one of these, you should check that the equipment can detect upsets. Very often new test equipment is rushed into use without adequate testing. One needs to be very sure that the equipment will report effects if they occur. It is too easy to have a misconnected wire or a switch thrown the wrong way. The group should always have some method of inserting errors or a test device that they know produces errors. This type of problem is very common and has probably been experienced by all major test groups. The test group should always have PLAN B (and PLAN C) available in case their prime test setup does not work.

The experimental groups should always include all of the team members necessary for understanding everything they are doing. One group was monitoring the beam intensity themselves. After a number of successful runs, they decided that they did not need the team member who designed the beam dosimetry. They were using photomultiplier tubes that operate at negative high voltage. Accidentally the voltage was switched positive. When the accelerator crew turned on the beam, the experimental team could not detect it: an unsuccessful (and unpaid for) run with many harsh words and many lasting hard feelings. This nearly led to institutional lawyers.

5.2.6 Sloppy or Wrong Fits to the Data

Sometimes both the experimenters and the analyst give poor fits to data. Figure 5-16 shows two "fits" to a set of data. The solid curve is the Weibull fit using the geometrical correction. This is the best Weibull fit; however, it has a negative threshold and is probably not physical. Furthermore, if the device is sensitive near zero LET, it should show upset to direct proton ionization, and this is not observed.

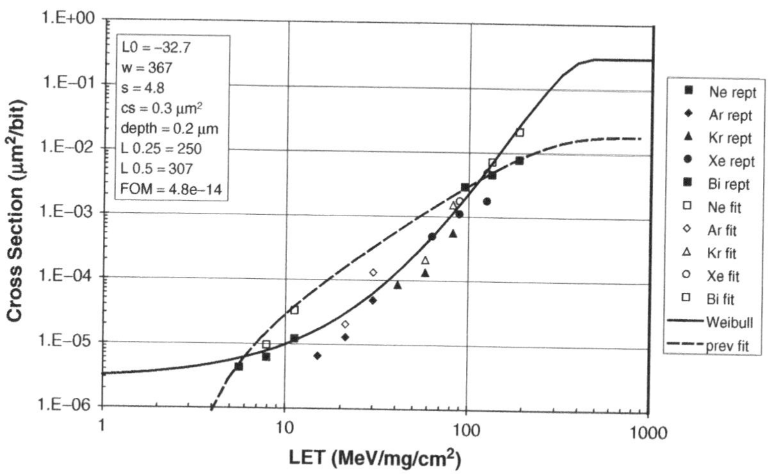

Figure 5-16 Upset data for a hardened 4 Meg SRAM. The solid line is a Weibull fit with a negative threshold. The dashed line is the original fit by the authors [Liu 2004].

The dashed line is the fit by the original authors, and is clearly wrong, although it does not look too bad on a log-linear plot! It is so easy to make least squares fits that one should never be satisfied with an eyeball fit! The fit by the authors indicates a saturation cross section that is much smaller than any physical dimension of the device. This should wave a red flag. Neither of the fits meets the criterion that the fit should be physically reasonable. We will not calculate the upset rate for either of these curves.

Figure 5-17 shows the data fitted with a Weibull function that is forced to be nonnegative and with a lognormal function. The two curves differ more than normal at the peak because of the lack of data in that region. We have previously mentioned that the major contribution to the upset rate occurs in the region of 0.1 to 0.9 of the cross-section curve. That is not true here, as that entire region is above the iron peak of the space cosmic ray spectrum. L0.25 (the LET that corresponds to 1/4 of the maximum cross section) is in the range of 400 to 500.

The upset rate for the Weibull curve is 1.6e−12 upsets/bit-day while the upset rate for the lognormal curve is 4.9e−13 upsets/bit-day—a factor of 3. In this case, the low LET data does have an effect on the upset rate, although they still have the same order of

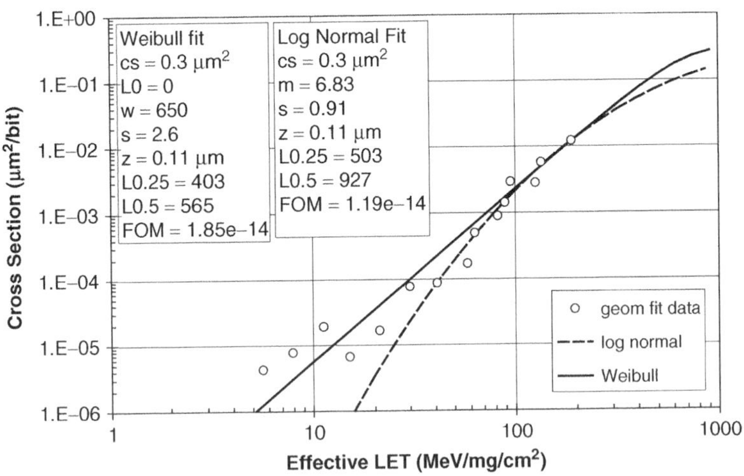

Figure 5-17 The same data fitted with a Weibull curve that is forced to have a zero or positive threshold, and with a lognormal curve. The data points are the points after the geometrical correction.

magnitude. This is an extraordinarily hard device so this situation will be very rare.

The upset analysis here treats the device as if the hardening was done with a simple increase of critical charge. The actual device used a more sophisticated method, and the upset rates need to be calculated using a double hit approach [Pickel 1984, Liu 2004].

5.2.7 Experiment Monitoring and Planning

Data always needs to be carefully examined before one makes any conclusions about the SEE behavior of the parts being tested. We have given a few examples of possible problems. Clearly each experiment is capable of generating new and exciting problems. The data has to be closely observed as it is taken, in real time. Any group that attempts to only record data, leaving examination of the data for later, is asking for trouble. There is some question about the experimental competence of such groups. In particular, each group should plan their experiment with careful consideration of the systematic and statistical errors that they will accept. They should have a feeling before each run about the expected results for that run. We have seen that it is possible to have consistently very small statistical errors, but to ignore the much larger systematic errors. The conclusions that one reaches about SEE behavior are no better than the quality of the data that was used.

We have observed that it is very common to have systematic error in the range of 10% to 40%. In these cases there is no need to have statistical errors smaller than 3% to 5% (400–1000 counts). Special efforts will be required if you need small systematic errors so that you can study the depth and funnel effect in the device.

We have also observed that it is not uncommon to have one data point that is inconsistent with all of the other data points. The authors should not base any conclusions on this point. If it implies anything significant, the authors should perform a thorough study with a number of data points.

In many cases, the conclusions of a single event paper depend on the validity of the dosimetry. In these cases the person performing the dosimetry should be a participating member of the experimental team. He/she should continually be engaged in monitoring the experiment, should be a co-author of the paper, and should contribute a detailed discussion of the dosimetry.

5.3 INTERPRETATION OF HEAVY ION EXPERIMENTS

The device geometry plays a very important role in the calculation of device upset rates. The device geometry is also important in the interpretation of the experimental data. The experimental data for a large flat device will have a smooth cross-section curve that follows the integral Weibull distribution. However, as the depth becomes more important, the experimental data that is obtained as a function of angle will show differences between large angle data with one beam and low angle data with another beam at the same nominal effective LET. Previous experimental data has shown both positive and negative variations. These can be explained by various possible geometrical and charge collection effects.

Note that the geometrical effects are studied in order to learn more about the device structure. We will incidentally learn more about the cross-section curve. The two aspects of the device geometry that are important are the depth of the device and the amount of augmented charge collection (funnel effect) that occurs. In some cases, the upset rate depends greatly on the depth of the device.

We will now discuss several of the effects that can lead to discontinuities in the data when the same effective LET is obtained using two ions at two different angles. These effects can be used to determine the depth of the device, and therefore improve our upset calculations [Petersen 2008b].

5.3.1 Modification of Effective LET by the Funnel

It is possible for the funnel to cause the breakdown of the effective LET concept. Figure 5-18a shows a simple structure with the path lengths corresponding to two angles. The idea of effective LET was introduced to compare the amount of charge produced (corresponding to the energy deposited) by the two tracks. As path x2 (at 60 degrees) has twice the length of x1, for the same LET, there will be twice the charge generated. It is then convenient to use effective LET to represent the charge deposition. Figure 5-18b shows the same volume with an additional length "F" added to the track corresponding to a funnel length. Here we are assuming that the funnel length is independent of angle. The use of effective LET assumes that, at this angle, the

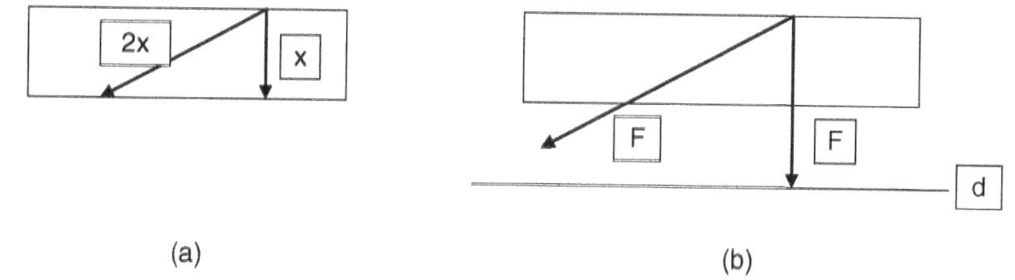

Figure 5-18 (a) No funnel and (b) funnel.

funnel is twice as long as at normal incidence. As that is not true, the effective LET calculation is wrong, and the data needs to be corrected.

Figure 5-18a shows the different charge deposition lengths for beams at an angle, which leads to the effective LET concept. If the funnel is of constant length, as in Figure 5-18b, the effective LET approximation breaks down. There may be a barrier at depth d.

Following Golke, we introduce a corrected effective LET such that the charge deposition is still calculated using the available depth at normal incidence [Golke 1993]. The LET of the ion is normalized for the actual path length relative to the path length at normal incidence.

$$LET_{\mathrm{eff,c}} = LET_{\mathrm{ion}} \times \left[\frac{\frac{x}{\cos\theta} + \min\left(F, \frac{d}{\cos\theta}\right)}{(x + \min(F, d))} \right] \tag{5.7}$$

The energy deposition that is calculated to obtain the charge deposited is

$$E = LET_{\mathrm{eff,c}} \times (x + \min(F, d)) \tag{5.8}$$

It is presumed that you have some information about x, and perhaps about $x + d$. The values of F and d can then be adjusted so that the discontinuities in the data are removed and a single smooth cross-section curve is obtained. Golke interpreted his version of Equation (5.7) so that the funnel has to intercept structure d before a corrected LET could be uniquely determined. We believe that these concepts apply in the more general case. Note that in this interpretation it is possible to calculate the funnel length from an assumed x (if $d > F$).

The data points in Figure 5-19 show the type of data that would be obtained in this situation. The points show the way the data would be quoted assuming the standard effective LET definition. The data

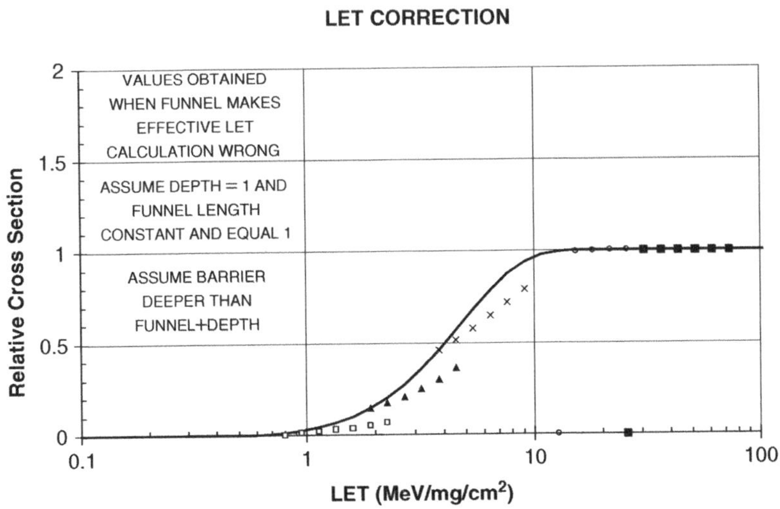

Figure 5-19 Cross-section measurements obtained when the presence of the funnel makes effective LET calculation invalid.

point should be converted using Equation (5.7) so that they move to the left to be on the Weibull curve.

5.3.2 Effects of True RPP Shape

If the charge collection volume has the shape of a true rectangular parallelepiped with appreciable depth, the cross section will have large angle distortions. This was originally pointed out in 1983 [Petersen 1983a]. The distortion will change depending on whether the ions have a very high LET and only need to pass through a corner of the volume to cause upset [Sexton 1989], or if they have a low LET, needing to pass through all of the volume to cause upset [Petersen 1992b]. Figure 5-20 shows the two possible geometries.

Ordinarily we assume that the target is a thin lamina. In that case there is not a geometrical correction except for the change of projected area described earlier in our discussion of cross-section concepts. Let us consider the basic cross section σ_0, the reported cross section is σ_R, and σ_m is the measured cross section:

$$\sigma = \frac{N}{F \times M}$$

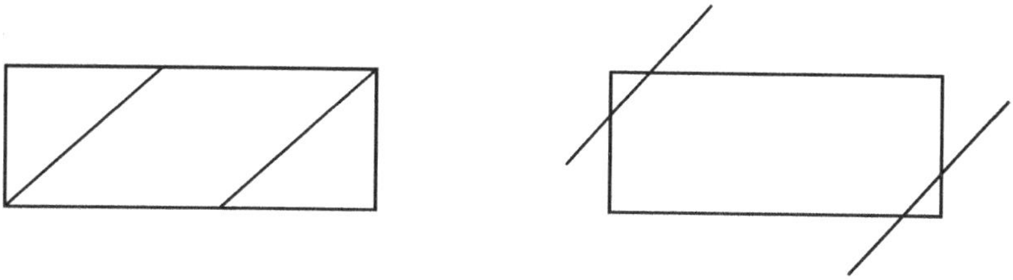

Figure 5-20 Change of apparent cross section in a RPP at large angle for low LET ions and for high LET ions.

If we are given the reported cross section that was obtained using this relationship, we can easily determine the measured cross section.

Now let us consider the situation when geometrical corrections need to be made. The left part of Figure 5-20 corresponds to a situation at very low LET where the ion needs the full path length in order to deposit energy for upset. In this case the standard cosine correction is not appropriate. Instead

$$\sigma_0 = \frac{\sigma_\mathrm{m}}{\cos\theta - (z/x)\sin\theta} \tag{5.9}$$

The situation at the right portion of Figure 5-20 corresponds to very high LETs for which the merest touch of ionization at a corner can cause upset. In this case the geometrical correction is

$$\sigma_0 = \frac{\sigma_\mathrm{m}}{\cos\theta + (z/x)\sin\theta} \tag{5.10}$$

These corrections only apply in extreme situations. For a true RPP geometry there should be a correction $B(L)$ that smoothly bridges these two terms as a function of LET [Petersen 1993]:

$$\sigma_0 = \frac{\sigma_\mathrm{m}}{\cos\theta + (z/x)B(L)\sin\theta} \tag{5.11}$$

Petersen presented an approximate form for the function $B(L)$ [Petersen 1993]. Connell presented a better approximation, based on the Weibull function, in two papers [Connell 1995a, Connell 1995b]. (Beware of misprints in Equations 42 and C1 of [Connell 1995a].)

Connell considers the variation being in energy E, rather than LET L. Consider the subscript (eff) as meaning that the variable is scaled to normal incidence:

$$E = L_{\text{eff}}(h + f)$$
$$F(E) = \sigma_{\text{eff}}/\sigma_{\text{sat}}$$
$$E_0 = L_0(h + f)$$

The Weibull function is

$$F(x) = 1 - \exp(-x^b) \qquad x = (E - E_0)/W$$

The SEU parameters are

$$\begin{aligned} A &= \text{area} \\ h &= \text{depth} \\ f &= \text{funnel length} \\ E_0 &= \text{threshold} \\ b &= \text{shape parameter} \\ W &= \text{width parameter} \end{aligned}$$

Connell also considers the aspect ratio l/w for the area and the azimuth angle φ. We will consider the aspect ratio to be one ($l = w = \sqrt{A}$), and ignore the φ dependence. He demonstrates that the rate is not very sensitive to this variable. He introduces as an integrand variable

$$L'(t) = \frac{t - S_0(L)}{W/L}$$
$$t > S_0$$
$$S_0(L) = \frac{E_0}{L} - f \tag{5.12}$$

Then the bridging function is

$$B(L) = 1 - \frac{2\cos\theta}{h}\bar{S}_c(\theta, L) \tag{5.13}$$

$$\bar{S}_c = S_0(L)\left[1 - \frac{F(L'(0))}{F(L'(S_{mx}))}\right] + \frac{W}{L}\left[\frac{\gamma[\alpha_1, L'(S_{mx})^b] - \gamma[\alpha_1, L'(0)^b]}{F(L'(S_{mx}))}\right] \quad (5.14)$$

Here $\gamma(\alpha, x)$ is the cumulative gamma function,

$$\alpha_1 = 1 + 1/b$$

S_{mx} is the local maximum chord length. θ_c is the critical tilt angle for which the maximum chord pierces sides of the RPP rather than the top or bottom. In Section 5.4.5 we will examine cases in which this is important.

$$\theta_c = a\tan\left(\frac{w}{h}\right)$$

$$\begin{aligned} S_{mx} &= \frac{h}{\cos\theta}, \quad 0 \le \theta \le \theta_c \\ &= \frac{w}{\sin\theta}, \quad \theta_c < \theta \le \pi/2 \end{aligned}$$

We assume that there are two basic scaling relationships: the first is in the effective LET as given in Equation (2.2). The second is in the cross section as given in Equation (5.11). The bridge function necessary in Equation (5.11) is that given in Equation (5.13). If the cross section shows discontinuities, these two scaling relationships can be used to fit the data to a smooth curve, and incidentally obtain the depth and funnel length for the device.

Figure 5-21 shows two examples of the bridging function. We see that it provides a smooth transition from -1 at low LET to $+1$ at high LET. Figure 5-22 shows possible data for the same device. It assumes that beams with LETs = 10, 20, 40, and 80 MeV/mg/cm^2 are used at five angles between 0 degrees and 60 degrees. The hollow circles indicate the data as it would ordinarily be reported. The solid circles indicate the corrected data. The arrows indicate the two corrections that need to be made in the 60 degree data. The horizontal arrows indicate the corrections in effective LET and the vertical arrows represent the correction in the effective cross section. Figure 5-23 presents the data

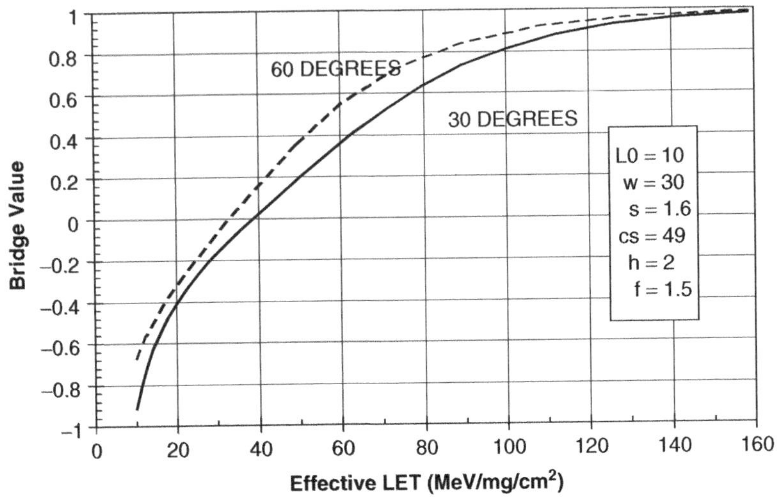

Figure 5-21 The bridging function that allows for the cross-section correction for RPP geometry where the depth of the device is important.

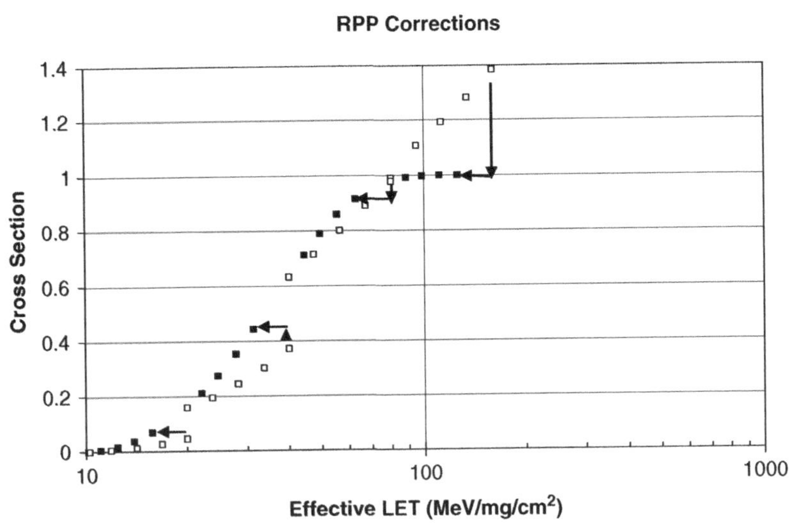

Figure 5-22 Angular data that would be obtained if the volume were true RPP with some charge collected from the funnel. The arrows indicate the two corrections that would have to be made in 60 degree data.

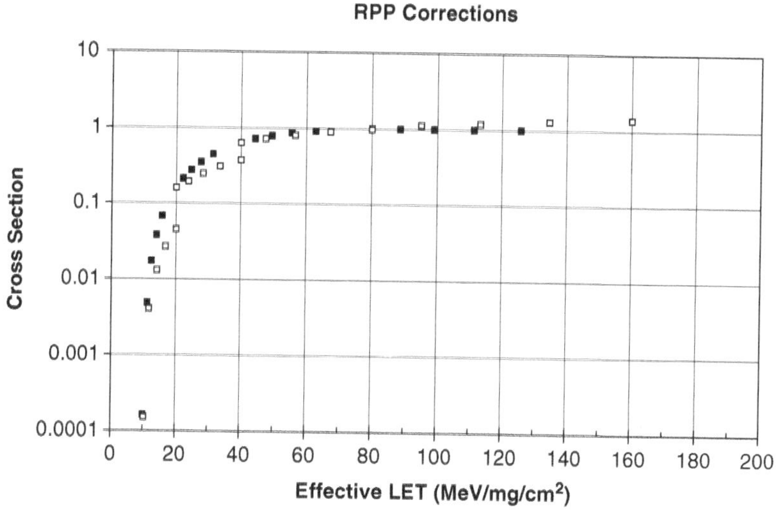

Figure 5-23 The hollow points show data as ordinarily reported and the solid points show the data after being corrected for device depth and funnel effect. The plot corresponds to the plot of Figure 5-22.

as it would ordinarily be presented on a log plot. This plot tends to hide the effects of the depth.

Note a way in which the data may confuse you. The {LET = 40, 60°} data point falls at the same value as the {LET = 80, 0°} data point. Naively, this is where you expect it if the effective LET concept holds. You need information with other ions, and other angles, in order to realize that the data is not falling where you expect it when you change ions.

For very low LET ions, there is the possibility that the measured upsets will increase and then decrease and disappear with increasing angle. This has been observed in some experiments.

5.3.3 Fitting Data to Determine Depth and Funnel Length

The procedure for fitting data with geometrical corrections is straightforward. One assumes that the data is basically determined by a smooth Weibull function. One therefore obtains initial values using a simple fit. You then assume that the data is distorted in LET by a funnel, if you have reason to believe that that is so. Therefore, the LET values for the experimental points are really described by Equation (5.7). For many parts this is not necessary.

If you have geometrical effects, and you usually do, the measured cross-section points are shifted with angle as described by Equation (5.11). These shifts are shown in Figures 5-22 and 5-23. You then calculate the expected locations of the observed data by modifying the "pure" Weibull points by the funnel and the depth contributions. You start with assumed Weibull parameters and assumed funnel length and depth. You then vary these values in a least squares fit comparing the calculated values with the initial data that includes these effects. The least square fit to the data provides the "true" Weibull function, the funnel, and the depth of the device. These parameters in turn allow you to do an accurate calculation of the device upset rates.

The RPP approach has often been criticized due to the often uncertain depth of the device. We have mentioned previously how a conservative approach to avoiding this is to assume that the depth is 1 micron. We now see how the depth can be obtained from an examination of the data. If the RPP device has an appreciable depth or funnel, there will be discontinuities when one compares equivalent LETs obtained at 0 degree and 60 degrees or 65 degrees. The amount of the discontinuity depends on the device geometry. The thorough way to approach the problem is to use a least square or chi square fit of the data and the full set of SEU parameters that includes the Weibull parameters in conjunction with the depth and funnel length.

Many modern devices have little funnel, and so we are primarily interested in determining the depth. The effects of the depth show up primarily at high LET. Figures 5-24 and Figure 5-25 show two examples of the depth effect. Figure 5-24 shows the effect for a $10 \times 10 \times 1\ \mu m^3$ device. Although this is a shallow device, the effect is still noticeable. There will be similar appearance for any device with a depth 1/10 of the surface dimensions. The 60 degree points are more than 10% above the 0 degree points. If the depth was 1/20 of the surface dimensions, the effects would be approximately 5%. Exactly the same effect will be evident if the device is $2 \times 2 \times 0.2\ \mu m^3$.

Figure 5-25 shows the depth effect for a $10 \times 10 \times 2\ \mu m^3$ device. In this case there is a 20% to 30% variation between 0 degrees and 60 degrees. From these examples we can conclude that the depth of a majority of devices can be estimated from the behavior of high LET angular data. There may be no need to do a complete Connell fit to

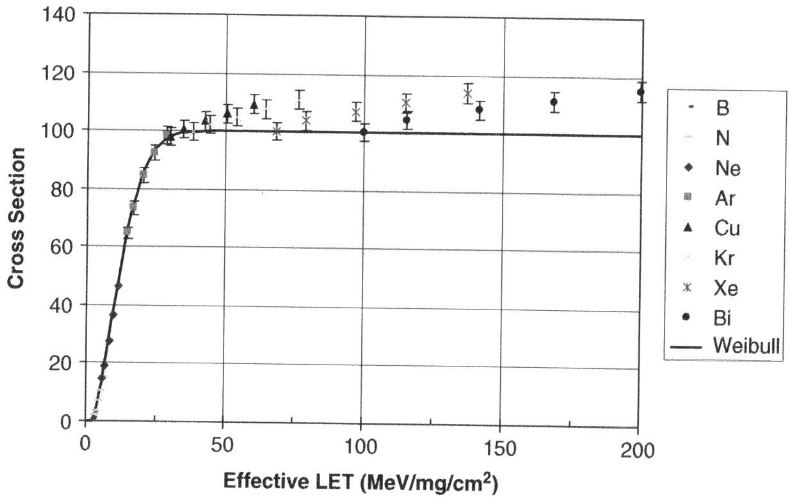

Figure 5-24 The expected measured cross-section curve for a 10 × 10 × 1 device. The effect of the depth is evident by the differences in the high LET points between 0 degree data and 60 degree data. The same relationship will hold if the device is scaled down by a factor of 2, or 5, or 10.

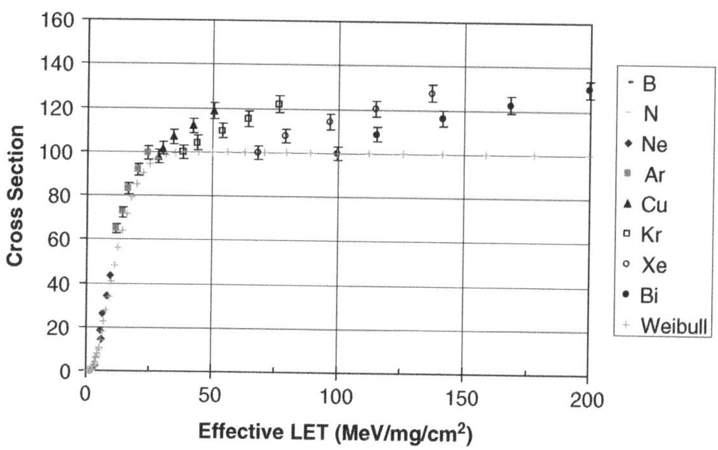

Figure 5-25 The expected measured cross section for a 10 × 10 × 2 device. The effects of the depth of the device are now much more notable. The effects stay the same as devices are scaled down.

the data. Again the same effect will be present for a $2 \times 2 \times 0.4\ \mu m^3$ device.

For many devices, the depth and funnel length can be estimated without the complete fitting procedure. If the basic Weibull curve has been obtained from points at normal incidence, then the high LET–high angle point can be approximately described as having a bridge value of one, so that its ratio to the expected Weibull curve is directly related to the depth to width ratio. If one now examines the middle of the curve, where the bridge term is near zero, the deviation away from the curve is primarily due to the funnel. The ratio of the funnel to the depth can be obtained by examining the deviation from the Weibull curve as a function of angle. The best set of data for this purpose is probably one that has the normal incidence LET at approximately one-quarter of the height, so that the 60° point is near the 50% portion of the curve and has primarily the funnel LET correction.

In principle, one can obtain the six SEU parameters from six measurements. The SEU parameters are the Weibull parameters (cross section, threshold, width, and shape) and the depth and funnel length. One then takes one point at normal incidence with the maximum LET beam available, three points at normal incidence along the face of the curve and at approximately factors of two in LET and adds a 60 degree point for the LET(0°) nearest the 25% portion of the curve and a 60 degree point for the highest LET beam available. The normal incidence points determine the Weibull parameters and the other two points determine the funnel and depth. In actual practice, if the funnel and depth are significant, then one probably wants several full sets of angular data and a complete fitting procedure.

5.3.4 Deep Device Structures

In Section 5.3.2 we discussed how the cross-section measurements are modified for structures that are no longer planar. When we first introduced the concept of device cross section for planar devices, we introduced a cosine factor to allow for the fact that the area being tested was larger than the cross-sectional area of the beam. Using effective LET lets us use a universal scale to include both normal and angular incidence testing. Then in our discussion of RPP shapes, we assumed that the quoted cross section included the cosine factor, and immediately removed it to get back to the number of upsets so that we could do the appropriate geometrical correction.

As we consider devices even deeper than those shown in Figure 5-20, it is no longer useful to report the cross section that includes the cosine factor, or to use effective LET in reporting the data. The fact that these two concepts are no longer useful does not invalidate the RPP formalism for calculating upset rates, as neither is basic to this approach. The lack of validity of the effective LET concept may introduce problems for the effective flux approach to upset rate calculation, but this has not been carefully examined.

In discussing deep device structure, we are considering structures such as shown in Figure 5-26. We show sample tests on tracks through the structure. If we test on the low end of the cross-section curve, so that long tracks are necessary for upset, we may observe the upset rate decreasing as we go to 60 degrees, perhaps even becoming zero. This behavior has been observed in Sage HBT structures [Marshall 2005, Pellish 2007]. The decrease at large angles is a property of the device sensitivity, the device geometry, and the test procedure. Therefore, it is an artifact of the testing and has nothing to do with the validity of the RPP rate calculation approach. It appears that the best display of

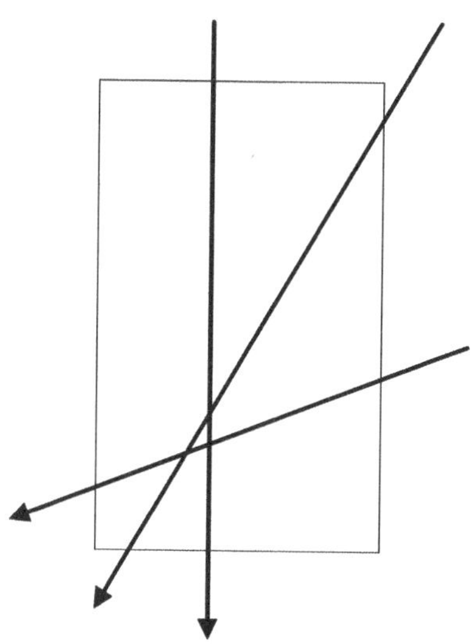

Figure 5-26 Deep RPP structure with possible ion tracks.

this type of data is to show each data series as a function of angle. We need to become more comfortable with this type of data.

Figures 5-27, 5-28, and 5-29 show the expected results for several devices using the Connell equations. Figure 5-27 shows the expected increase of cross section expected from geometry for a near planar device. However, one is making a mistake if he expects this behavior for deep devices. Figure 5-28 indicates that large angle data decreases for a $2 \times 2 \times 2$ device.

Figure 5-29 shows the behavior of the measured cross section for a very deep device. In this situation, all of the cross-section curves decrease at large angles. This is due to the combination of the geometry and the cross-section curve. Similar data has been acquired in deep SiGe HBT devices by Marshall and discussed by Pellish and Montes [Marshall 2005, Pellish 2007, Montes 2008]. They concluded that this type of data showed that the RPP model was invalid. On the contrary, this type of behavior is exactly what one expects from the IRPP model that combines the variation of upset probability described by the cross-section curve with the path length distribution described by the RPP model. Observation of decreasing cross section with increasing angle for deep devices is experimental confirmation of the IRPP approach for upset rate calculations.

There are two possible ways of obtaining good data for a cross-section curve that can be used in the IRPP upset rate calculations. The

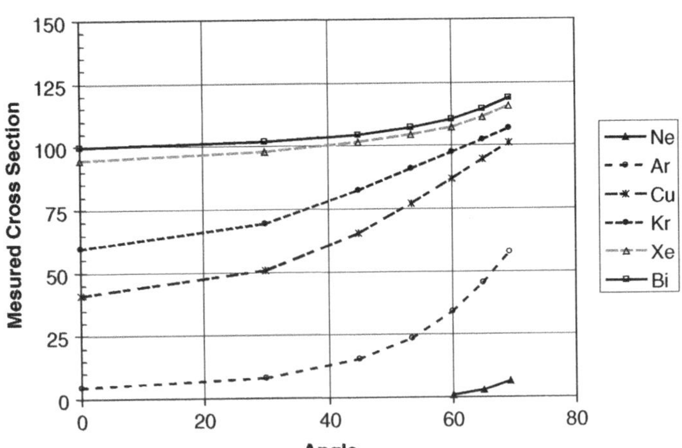

Figure 5-27 Measured cross section for a $10 \times 10 \times 1$ device with $L0 = 10$, $w = 30$, $s = 1.6$, and $cs = 100$.

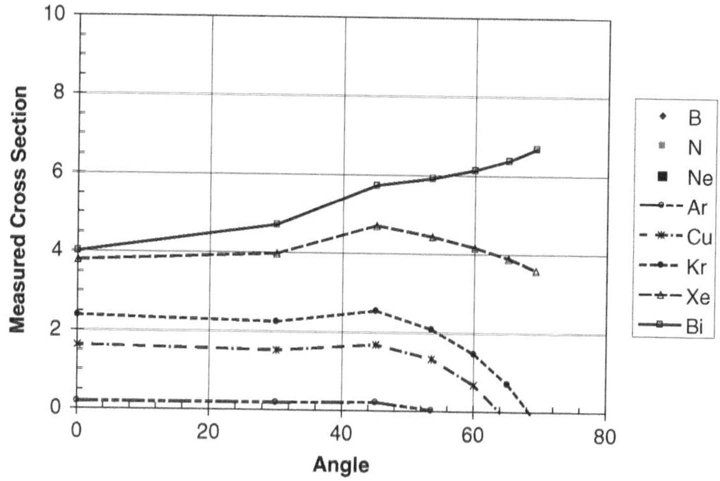

Figure 5-28 Measured cross section for a 2 × 2 × 2 device with L0 = 10, w = 30, s = 1.6, and cs = 4. With the threshold of 10, one would expect to measure a cross section with Ne at angles of 60 degrees and more. However, nothing will be observed there. The first observed upsets would be with Ar 0 degrees at 14.3. The peak at 45 degrees indicates that the device is nearly a cube.

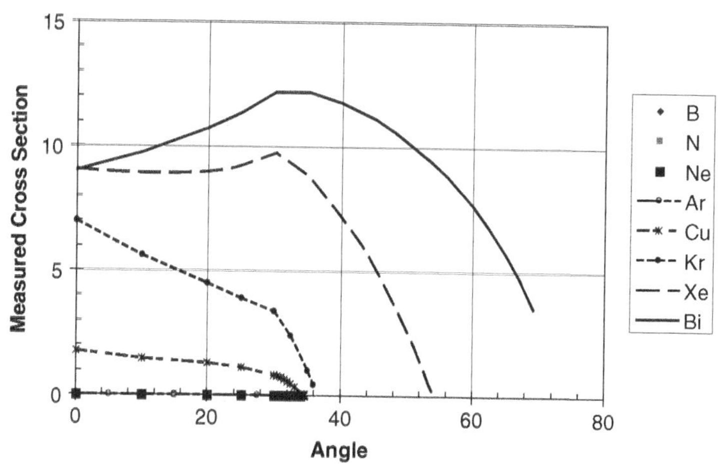

Figure 5-29 Measured cross section for a 3 × 3 × 5 device with L0 = 0, w = 30, s = 7.95, and cs = 9. All of the low LET beams measure zero cross section past 34 degrees. The peak in the cross section at 30 degrees indicates that the device is deeper than it is broad.

first approach is to employ a large number of ions at normal incidence. This entails a number of runs, but the results are immediately useful.

The second approach is to take the data at a series of angles, and then use the complete Connell type geometrical calculation, including the tracks of particles that do not pass through the top or bottom of the device. In this case you do a least square fit to the angular data as you would with the cross-section curve. The mathematics is the same, but you're using a different view of the data. Some people may be more comfortable looking at these figures than at Figure 5-23. In this case each data point can be identified with a point on the cross-section curve versus LET and so completely specify the curve as if you were using effective LET.

Pellish observed that the decreasing cross section past 60 degrees implied that the large number of incident particles near the plane of the device would lead to a harder device. While the decreasing cross section at high incident angles can appear to improve the device hardness, one also needs to consider that the abundance of particles within 60 degrees of the vertical plane might cancel this effect. The environment is isotropic, so the rate of upsets does not depend on the orientation of the RPP in the device.

5.3.5 Cross-Section Curves on Rotated RPP Structures

The study of structures such as that in Figure 5-26 brings up questions about the cross-section curve if we were to observe the same structure from different orientations. Do we expect the curve to be the same regardless of the observation axis? Of course, the answer is no. The total cross section depends on the area normal to the beam. The critical charge of the device does not depend on device orientation, but on circuitry, so it remains the same. However, as you change orientation the depth changes, so the critical LET changes. We would expect the variation of deposited charge for upset to remain basically the same, so the shape of the curve should change very little. Therefore, we expect a family of three different cross-section curves. The IRPP calculations examine the deposited charge for each point of the cross-section curve and then examine the integral over all possible paths to determine the contribution to the upset rate from that charge deposition. Therefore, the resulting upset rate will be the same for the three different test orientations.

The figure of merit calculation can vary by a factor of 2 as you change orientation; however, as it is used primarily for determining the order of magnitude of the upset rates, that is not a significant variation.

The previous paragraph assumed that the distribution of charge sensitivity does not change with device orientation. There may be a few cases for which this is not true. The device modeling of Pellish [Pellish 2007] seems to show that their structure has different charge collection mechanisms that depend on track orientation. However, the critical charge depends only on the circuit, and not on the track orientation. Therefore, the critical LET does not change. The calculated upset rate varies little with minor changes in the shape of the cross-section curve, so these effects will have little impact on upset rate.

5.3.6 Charge Gain Effects on Cross Section

Some devices appear to upset if an ion impinges on any part of them. One explanation of this is that there is an additional gain mechanism (such as parasitic bipolar gain) that is turned on. So, a large signal is generated after a small charge deposition [McMorrow 1994]. Figure 5-30 shows the geometry in this case. The lines at the top indicate the area that is expected. The lines at the bottom indicate the area that is observed. The effect shows up in experiments as if the cross section increases at large angles for both low and high LET ions.

$$\sigma_0 = \sigma_{\mathrm{m}} \times \left(\cos\Theta + \frac{z}{x}\sin\Theta\right)^{-1} \tag{5.15}$$

This leads to situations where data at one effective LET presents a much larger cross section if taken at large angles. Figure 5-31 shows

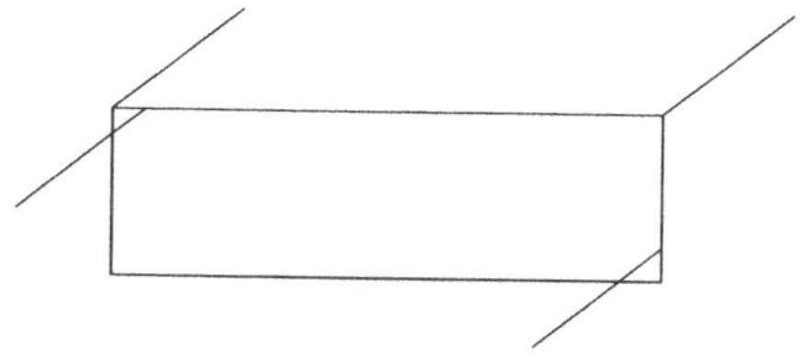

Figure 5-30 Apparent increase of area at large angles due to funnel.

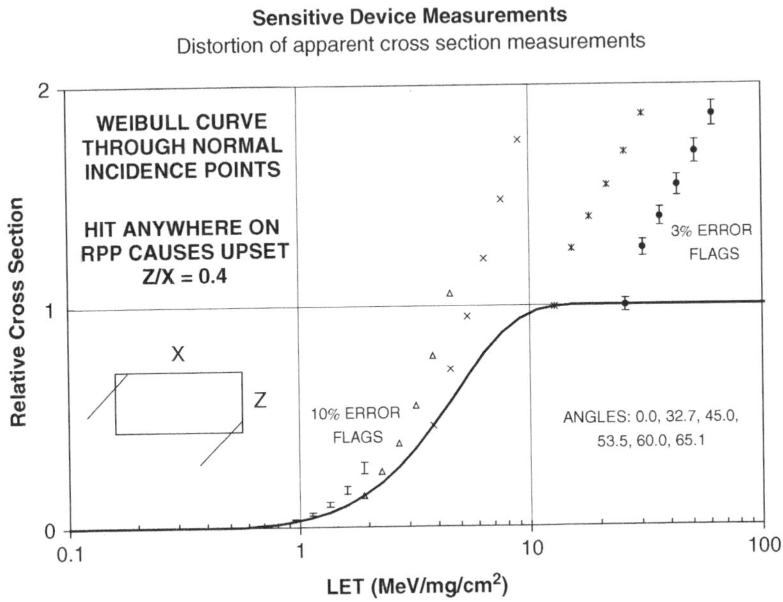

Figure 5-31 Angular data when a hit anywhere on the device will cause upset.

how the data might appear. The data in this situation appears extremely discontinuous, and there is no evidence of saturation at the limiting cross section. Although this behavior has been observed on a number of devices, we are not able to predict when it will occur.

5.4 POSSIBLE PROBLEMS WITH LEAST SQUARE FITTING USING THE WEIBULL FUNCTION

5.4.1 Multiple Good Fits

The previous section discussed how problems with data acquisition can make the interpretation of data difficult. It is also possible to have problems even if the data is good data. This especially arises in SEE analysis where data is fitted with the Weibull function. The function is so general that it will often fit data in ways that are not desirable. Normand has given examples in which possible Weibull

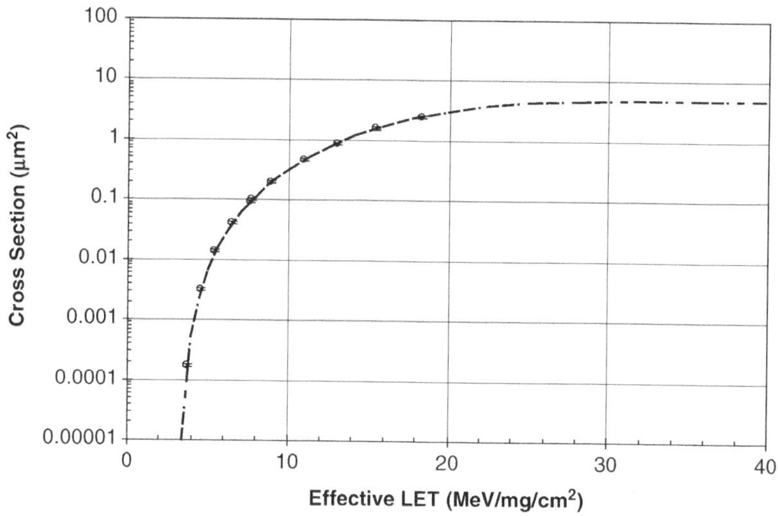

Figure 5-32 Cross-section data using 150 MeV F and 195 MeV Si that covers LETs from 3.2 to 18 MeV/mg/cm^2. The data appears to saturate at 4.8 μm^2, rather using a visual fit or using a least squares fit with an integral Weibull function. The threshold is 3.2, s = 2.9, and w = 16.8.

fits with different widths give different FOM values for the devices [Normand 2004]. This section will examine a set of data where we are using curve fitting to describe the data. In these examples we will avoid systematic errors by constructing the data [Petersen 2008a]. We choose a Weibull function for a plausible part, choose LET points consistent with those often used in accelerator testing, and generate data points based on the curve using a normal error distribution. The error distribution has its center at zero, and a width of one standard deviation. We then assign the one standard deviation value at 5% of the calculated value, and calculate the value of each data point.

Let us consider a measurement of single event upset cross section using low LET heavy ion beams. Figure 5-32 shows such a data set. The cross section appears to saturate at a cross section of 4.8 μm^2 using either a visual fit or using a least squares fit with the integral Weibull function. The least square fit has R^2 of 0.99999, which would seem to indicate an outstanding fit. The calculated upset rate is 7.0e−8 upsets per bit-day for an assumed depth of 0.22. The upset rates are calculated for geosynchronous orbit, solar minimum conditions, and 100 mils of shielding.

It is normal to extend the experiment with at least one more set of data points. We will add data taken with 105 MeV Cl, which adds data points from 15 to 35 MeV/mg/cm^2. The results are shown in Figure 5-33. The apparent saturation cross section has clearly changed.

We can complete the experiment by using 255 MeV Ni and 320 MeV I. This will add LET points from 27 to 142 MeV/mg/cm^2. The nickel data appear to move the saturation cross section to 23.6 and the iodine data moves it to 24.6 (Figure 5-34). The data was generated with the limiting cross section of 25, but the result is consistent with the 5% standard deviation on the points.

The results and the error bars are shown somewhat cleaner in Figure 5-35. The final fit has a threshold of $L0 = 3.24, W = 32.4, S = 2.9$, and $CS = 24.6$. The calculated upset rate is 1.1e−7 upsets per bit-day for an assumed depth of 0.5 micron. Note that the change in

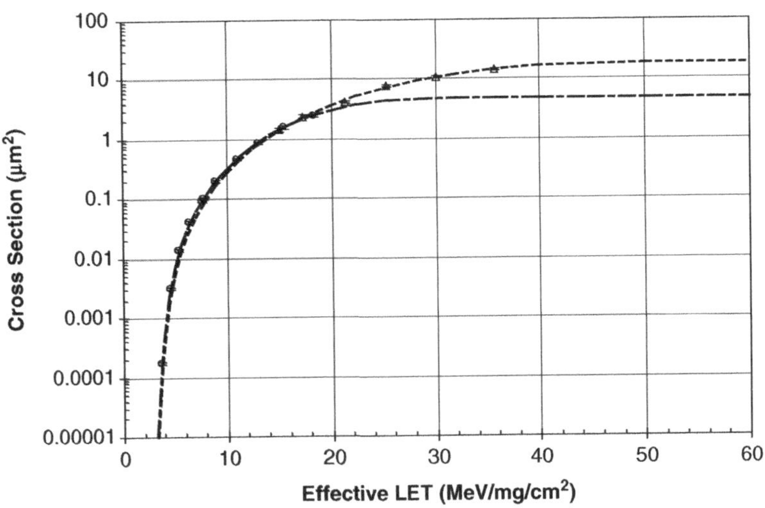

Figure 5-33 The data of Figure 5-32 with additional data from LETs of 15 to 35. The curve now appears to saturate at a cross section of 18 μm^2. Again there is agreement between the fit by eye and the least squares fit using the Weibull function. If we were to view this data on a liner cross-section scale, or on a logarithmic LET scale, it would be clear that we are not yet approaching saturation. Both the log-linear scale and the Weibull fit are misleading.

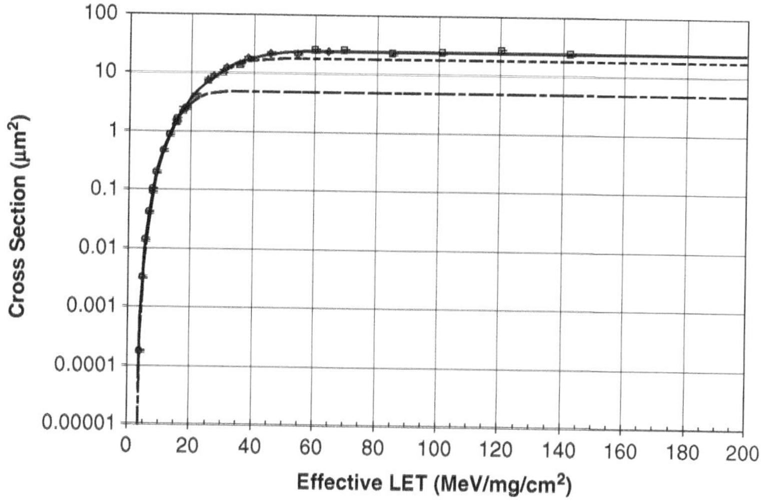

Figure 5-34 The complete data set now shows the saturation cross section at a cross section of 24.6. There is now adequate data in the saturation region so that both the eye and the Weibull fit can truly see that there is saturation.

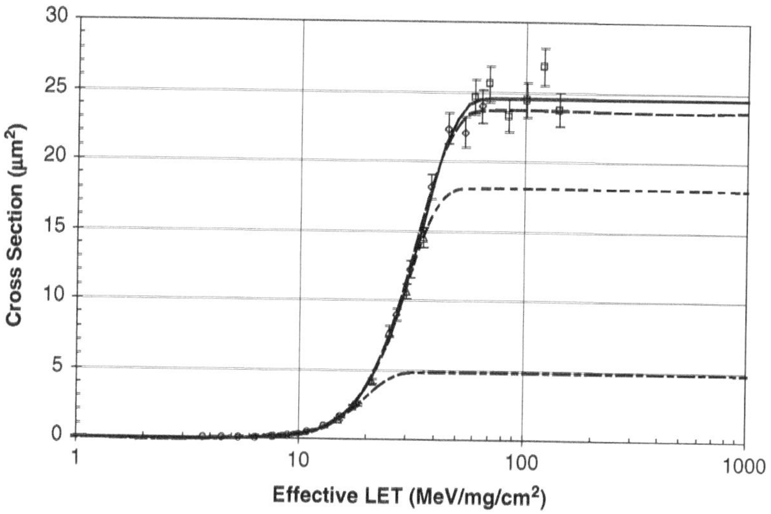

Figure 5-35 The data replotted on linear cross section, logarithmic LET scales. The four curves reflect the four different fits to the data as it was accumulated. Note that if this display had been used initially, there would not have been the temptation to assume that the first set of data reached saturation. The Figure of Merit for these fits varies from 2.4 E-8 to 4.3 E-8. This reflects the relative predicted upset rates.

calculated upset rate depends both on the changed fit and on a different assumed depth.

5.4.2 Reason for Inconsistent Weibull Fitting

We can see what the problem is by examining Figure 5-36. If one examines the fit as a function of the limiting cross section, we get very good fits for virtually any limiting cross section if we allow the width W to vary. We get good fits for limiting cross sections from 3 to 100,000 with R^2 exceeding 0.999 for the entire range. So although the mathematics chooses a cross section of 4.8, there basically is little difference between this solution and any other in a wide range. R^2 for a cross section of 25 is 0.9993.

We can get another view of the problem by examining Figure 5-37. The linear-linear display shows that you can achieve reasonable extrapolations to a wide range of limiting cross sections if the data does not actually extend to the saturation region.

A plot of the interplay of the width and the limiting cross section for the data in Figure 5-33 is shown in Figure 5-38. This demonstrates

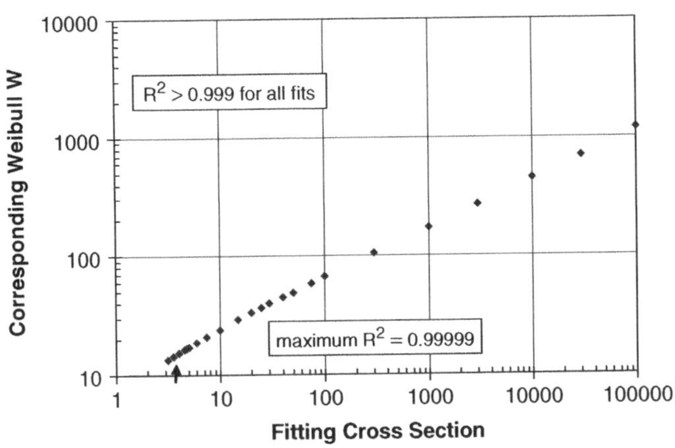

Figure 5-36 Plot of the Weibull width as a function of limiting cross section for the data in Figure 5-32. The resulting fits are basically equivalent over the entire range, so the best least squares fit using the Weibull function is misleading and untrustworthy when the data does not extend clearly into the saturation region.

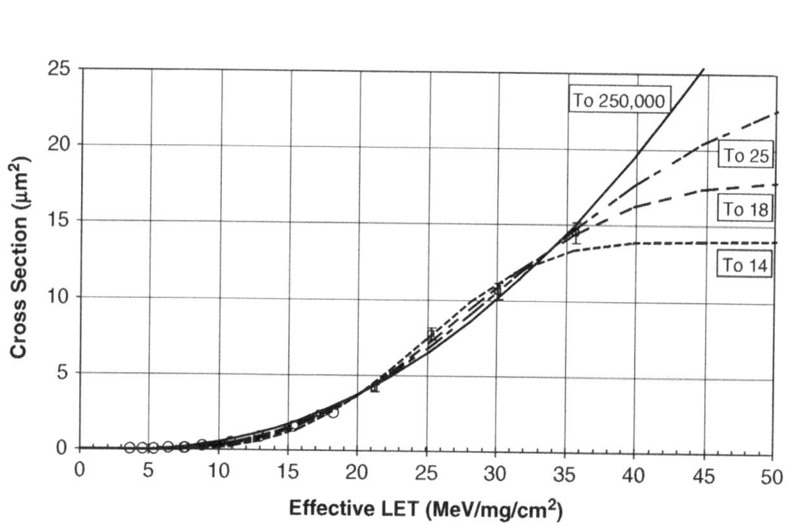

Figure 5-37 Figure 5-33 with various fitting curves showing possible extrapolations. The limiting cross section of 18 resulted from a least squares fit with the Weibull function. The data was generated with a limiting cross section of 25. The entire range from 14 to 250,000 gives reasonable R^2.

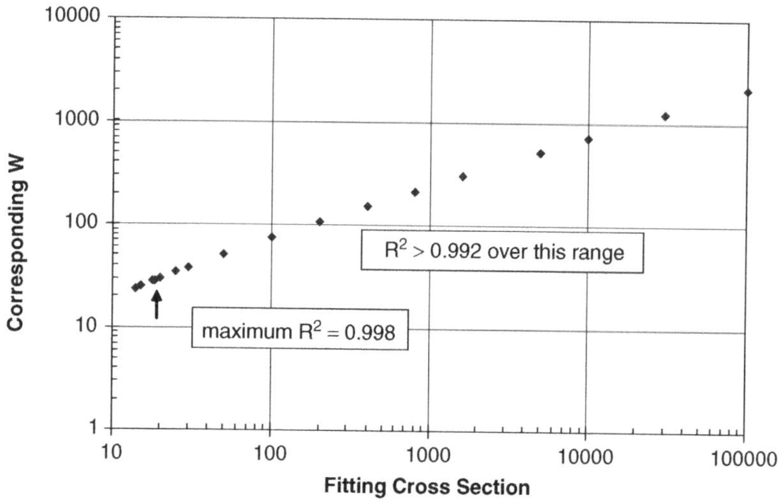

Figure 5-38 The relationship of the fitting width for a range of assumed limiting cross sections for the data of Figure 5-33. Least square fitting with the Weibull function does not work for this data set.

why least square fitting does not do a good job of discriminating between the various curves shown in Figure 5-33.

Some experimenters attempt to avoid the complexity and problems involved with least squares fits to the data by using the CREME option to insert data directly. This option inherently performs a less accurate integration and there is still the problem of determining the depth of the device.

CHAPTER 6

Analysis of Various Types of SEU Data

This chapter will examine the physical mechanisms responsible for single event upsets. The initial discussion covers microscopic and macroscopic characteristics of basic SEU mechanisms. Other authors have examined detailed SEU mechanisms in more depth and with more references [Massengill 1993c, Massengill 1996, Dodd 1999, Dodd 2003]. We will then concentrate on the interplay of experimental observations of single event phenomena and the developing analysis of the events.

6.1 CRITICAL CHARGE

Several groups examined the transients produced in diodes by ionizing particles. The pulses are very fast and stretched the available experimental techniques. Figure 6-1 shows alpha particle pulses with a rise time of 155 ps and full-width-at-half-maximum of 360 ps. The system response was 32 ps [Wagner 1986]. Similar fast pulses are observed with other ions [Wagner 1988].

The response of circuits to the very fast particle ionization pulse can be simulated in circuit models such as SPICE [Diehl 1982, Johnson 1985, Weatherford 1986, Kerns 1989]. The common model for the charge collection at a junction due to direct funneling or diffusion is

Single Event Effects in Aerospace, First Edition. By Edward Petersen.
© 2011 the Institute of Electrical and Electronics Engineers, Inc.
Published 2011 by John Wiley & Sons, Inc.

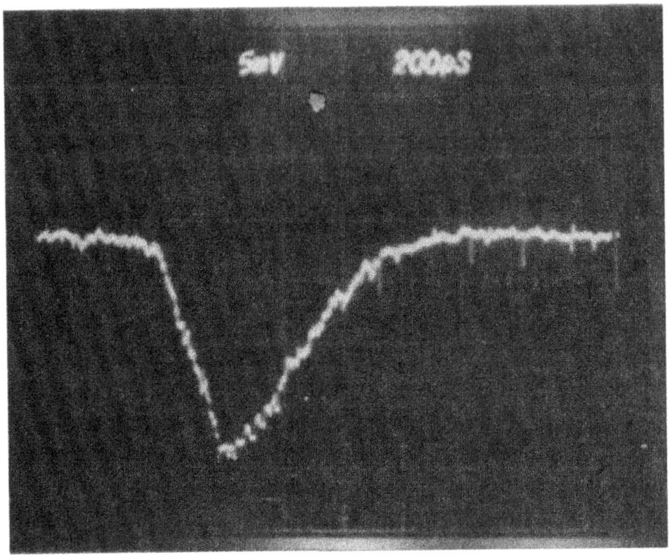

Figure 6-1 Sampling measurement of alpha particle single event current transient in a silicon diode. The rise time is 155 ps [Wagner 1986]. Copyright ©1986 IEEE.

a double-exponential time dependent current pulse [Messenger 1982]. The circuit models often have to be modified as the single events push them outside the normal operating parameters. In these papers the term "critical charge" is used to indicate the minimum integral of the current collected from direct ionization capable of initiating upset. In the case of some devices such as CMOS/SOI devices with their bipolar amplification, it is not the total charge that drives upset [Kerns 1989]. Modelers then can use several input wave forms with various time constants in SPICE modeling to determine which pulses lead to upset. It turns out that the amount of charge in the upsetting pulse is nearly independent of the pulse shape or time constants [Sexton 1991]. Therefore, to first order, the critical charge Q_C is only a function of the circuit and does not depend on the ion or the particle LET.

6.2 DEPTH AND CRITICAL CHARGE

The early authors all assumed that the critical charge for upset corresponded to the threshold LET. This concept came into question as there were more calculations of critical charge based on circuit modeling [Petersen 1992b]. The circuit calculation should lead to the critical charge of the median part of the cell if nominal circuit parameters are used. Several examples exist in the literature where a careful calculation has been performed, but when the results are compared with a charge derived from the measured onset threshold, it is necessary to assume an unexpected charge collection depth or an unexpected charge multiplication factor to reconcile the calculations with the experiment [Davis 1985, Shiona 1986, Rollins 1987, Hughlock 1990, Worley 1990, Sexton 1991]. These authors all discuss their expected charge collection depths based on the technology. They then compare this depth with a calculated depth based on the threshold LET and observe differences of a factor of 2. Figure 6-2 shows this comparison, and also the calculated depth using the median cross-section LET. The latter comparison gives much better agreement with the author's expected depths.

The 50% value of the cross-section curve corresponds to the center of the lognormal distribution. As it corresponds to the critical charge,

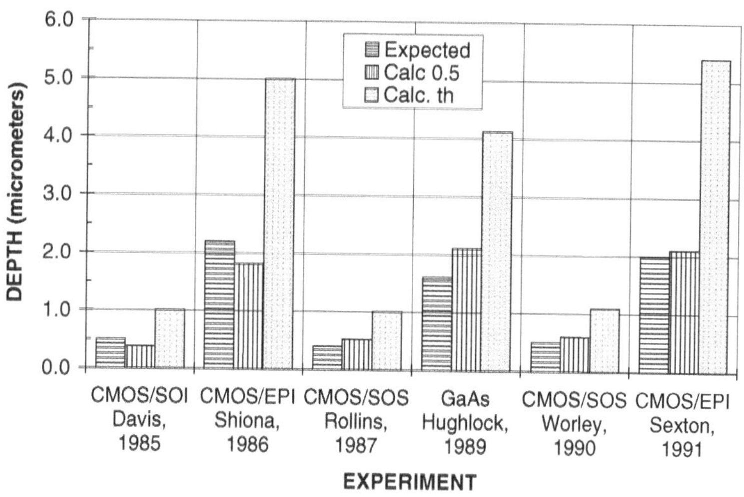

Figure 6-2 Comparison of expected and calculated charge collection depths in single event experiments. The expected depth is obtained from knowledge of the technology. The calculated depths are obtained by combining the critical charge, calculated using circuit analysis, with either the threshold LET or the LET at 50% of the saturation cross section. The agreement of the expected values with the 50% value indicates that the circuit analysis calculation of critical charge is for the median location in the cell.

we call that the critical LET. This is one of the three parameters that determine the lognormal distribution. We will see later that, for some very small feature size devices, the critical charge is not well defined. However, even in these cases the cross section is still described by the lognormal distribution, so the critical LET corresponding to the center of the cross-section distribution is still a very useful concept.

6.3 CHARGE COLLECTION MECHANISMS

There are several charge collection mechanisms that are important for single event effects. These are discussed in detail elsewhere and will be mentioned only briefly here [Dodd 2003, Reed 2008]. The different mechanisms are important for this discussion in that they have different time characteristics, and that they may necessitate the modeling of the entire transistor, the entire memory cell, or even of adjacent memory cells.

6.3.1 Drift Process and Funneling

IBM researchers used numerical device simulators to compute the response of reverse biased p/n junction to alpha particle strikes. They discovered a fast transient due to disturbance of the electrostatic potential, which they call the "field funnel." This funneling effect increases charge collection by extending the electric field deep into the substrate [Hsieh 1981a, Hsieh 1981b, McLean 1982, Hsieh 1983, Edmonds 1991, Dodd 1994, Dodd 1996a, Edmonds 1997, Edmonds 1998, Edmonds 2001, Oldham 2003]. The role of the funnel is less significant in the case of static circuits such as SRAMs, where reverse-biased transistor junctions are connected to active external circuitry. There can be a loss of bias at the struck node that tends to lessen the importance of drift collection (and hence the funnel) as the transient proceeds [Dodd 1995]. The fast transient may initiate the upset, but it is the later, slower diffusion process that ensures the bit flips.

6.3.2 Diffusion Process

In the absence of the strong electric fields that lead to fast drift of the charged particles, the slower diffusion process dominates the charge collection. This is relevant for hits near to but outside the drain. In the absence of well-defined boundaries, diffusion may extend charge collection outside the normal device dimensions [Smith 1995].

6.3.3 Plasma Wire Effect

It appears that the ion track can induce at the electron–hole pair a plasma wire shunt path for the current in complex structures for hits to the drain. This was initially shown experimentally by Knudson and co-workers [Knudson 1984] and verified by charge collection studies by Kreskovsky and circuit and numerical charge collection studies by Zoutendyk [Kreskovsky 1985, Zoutendyk 1987].

6.3.4 ALPHEN (Alpha-Particle–Source–Drain Penetration Effect)

For submicron devices, it is possible for the particle strike to pass through both the source and the drain. There is a short source–drain conduction current that mimics the on state of the transistor [Takeda 1989]. For 0.3 μm gate length devices, even normal incidence strikes can lead to charge multiplication [Velacheri 1994]. A similar phenomenon has been found to occur between adjacent trench storage capacitors in trench-type DRAMs [Chern 1986].

6.3.5 Bipolar Transistor Effect

It is possible to have parasitic structures in both bulk and SOI devices in which a source acts as the emitter, the channel as the base region, and the drain as the collector. This adds to the original particle induced current at the drain and causes an increased SEU sensitivity [Davis 1985, Fu 1985, Woodruff 1993, Dodd 1996b, Kerns 1989, Musseau 1996a, Massengill 1990, Musseau 2000, Schwank 2003].

6.3.6 Recombination Effects

Although not strictly a charge collection effect, the recombination of holes and electrons through direct or redirect processes can have a large effect on the charge available for collection [Dodd 1999].

6.4 CHARGE COLLECTION AND THE CROSS-SECTION CURVE

At one time it was believed that the shape of the cross-section curve was determined primarily by a spread of critical charges from cell to cell inside a device [Petersen 1992b]. In this interpretation the cross-section curve reflects the fraction of cells that upset at any given LET. This implied that a point on the low cross-section portion of the curve would correspond to only a few cells upsetting. This was shown to not be true by the experiments of Cutchin and of Buchner [Cutchins 1993, Buchner 1994, Petersen 1993]. Cutchins's experiment showed that a point on the low cross-section portion of the curve corresponded to all of the cells having the same low cross section, while Buchner's experiment showed that all cells had the same critical charge for upset. It now appears that there is no support for the concept of critical charge variation from cell to cell across the chip being primarily responsible for the shape of the upset cross-section curves.

A great deal of evidence has accumulated in favor of the concept that the variation of cross section with LET commonly observed is due to variations inside the cell. Much of this evidence was obtained while pursuing other questions and was not viewed in this light. We will discuss representative results for a number of technologies [Petersen 1996a, Petersen 1996c, Petersen 1997b].

6.4.1 CMOS

There is a great deal of information showing the variation across the device for CMOS technology. The Sandia group has shown variations in upset sensitivity as they scan a microbeam across transistors in a device [Horn 1992, Sexton 1993]. They showed the largest charge collection in the center of struck n^+ and p^+ junctions. Buchner and co-workers have examined the upset sensitivity of CMOS RAMs using a scanned laser beam. The 6504RH was examined in two different papers [Buchner 1990, Buchner 1992]. They obtained a wide spread in response as they moved around the transistor, but a very narrow distribution of sensitivities when the same location is examined at various transistors across the device. They also obtained similar results on a 64K CMOS RAM [Buchner 1992]. Similarly, McMorrow and co-workers have examined SEU in a range of hardened and unhardened

CMOS parts with laser excitation; all parts tested exhibit significant intracell variations in the SEU threshold [McMorrow 2000].

Upset modeling studies also support the variations of upset sensitivity across the chip and with angle [Woodruff 1993]. Woodruff used three-dimensional modeling to examine sensitivity with angle and position. They only studied two positions, but there had to clearly be a continuous variation of sensitivity between the two positions they reported.

One study contained both microbeam studies and three-dimensional simulations [Metzger 1994]. Metzger and co-workers examined the MH65162 16 kbit CMOS RAM. The computer simulations show the contours of collected charge essentially following the outline of the assumed device dimensions, and increased charge collection at the center of the device.

Detcheverry and co-workers modeled across one direction of a cell. They approximately fit a cross-section curve by using physical geometry in the unmodeled direction [Detcheverry 1997].

6.4.2 Hardened CMOS

Buchner also studied a hardened CMOS device, the HM6504RHH, which incorporates decoupling polysilicon resistors [Buchner 1994]. They again observed a narrow spread of thresholds across the device, but a wide spread as they examined different locations on a single transistor. Kohler examined hardened CMOS RAMs [Kohler 1989]. He pointed out that the statistical distribution of hardening resistor values will introduce a spread of critical charges, in addition to the normal width factor of unhardened RAMs.

6.4.3 Bipolar Devices

Zoutendyk performed several experiments on bipolar devices in which he observed the variation of upset cross section with particle energy, not particle LET [Zoutendyk 1984, Zoutendyk 1985]. These are essentially examining the variation with charge deposition versus depth. Smith has reanalyzed the data in terms of the cross section versus LET and the results show a standard smooth curve of cross section versus LET [E. C. Smith, private communication].

The sensitivity of the 93L422 as a function of position was examined directly using laser pulses on various regions of the device [Buchner 1994]. There is a wide distribution of sensitivities as the laser light is moved to different positions around the cell. If the light is moved from cell to cell, exposing the same position, there is a very tight distribution of sensitivities.

6.4.4 CMOS-SOI

Massengill discusses the distributed bipolar gain effect that is present in an SOI device [Massengill 1993a]. The gain varies with location across the transistor, so the signal generated by an ion hit will depend on the location of the hit. Massengill also examined the combined effects of bipolar gain variation in a SOI transistor and parameter variation from cell to cell across the device [Massengill 1993a, Massengill 1993b]. The primary effect is the distributed bipolar gain, which varies with position across a single transistor. In both situations, the better process controls on later chips will reduce the contribution of the variation across the chip [Petersen 1996c, Jones 2000].

Kerns and co-workers first described critical charge sensitivity (due to bipolar gain) along the SOI gate width relative to the body tie location [Kerns 1989]. Warren and co-workers demonstrated the hit-location variability through two-dimensional device simulations and Musseau later experimentally verified the effect through laser probing [Warren 1999, Musseau 2000].

6.4.5 NMOS–Depletion Load

Zoutendyk examined upsets in a NMOS device as a function of ion energy [Zoutendyk 1986]. As he varied the ion energy, cross sections corresponding to various areas of different sensitivity were shown.

6.4.6 NMOS–Resistive Load

NMOS SRAMs that use a resistive load have very long time constants for charge collection. These are sometimes called RMOS devices. Smith has examined the effects of diffusion in this case [Smith 1995].

He found that the diffusion was very important in this type of device, and that it leads to a cross-section curve that increases gradually as the LET is increased. The shape of the cross-section curve was well approximated by the Weibull function.

6.4.7 GaAs HFETs

Remke, Flesner, and co-workers have used a focused electron beam to study the upset sensitivity of AlGaAs/GaAs heterostructure field effect transistors (HFETs) and HFET SRAMs [Remke 1989]. They measured a variation of sensitivity in the region between the source and drain. Pulsed laser charge-collection measurements on GaAs MESFETs and HFETs exhibit a strong dependence on the location of the injected charge within single transistors [McMorrow 1994].

6.4.8 GaAs C-Higfet

Cutchins performed an upset equilibrium experiment to examine the hypothesis that the shape of the upset cross-section curve was due to cell to cell differences [Cutchin 1993]. He accumulated upset results at a LET that corresponded to 5% of the limiting cross section. He observed an equilibrium corresponding to 50% of the cells upsetting. This means that all of the cells are susceptible to upset. Therefore, the small cross section must correspond to the small probability of an individual cell upsetting.

6.4.9 VLSI Process Variation

A study at IBM has examined in detail the variations of critical charge that can arise from VLSI process variations [Freeman 1996]. They obtained a basic variation of 5% from the process variations. In their technology, the spread became 14% when it included the effect of position along a word line. These variations are significantly less than the variation implied by the widths of heavy ion cross-section curves. The IBM groups handle the charge collection processes across a cell by using Monte Carlo calculations [Srinivasan 1994, Murley 1996].

6.5 EFFICACY (VARIATION OF SEU SENSITIVITY WITHIN A CELL)

We now want to address the implications of variations in the SEU sensitivity within a cell. This variation can be described as a position-dependent internal SEU gain, or efficiency. However, because these terms are used elsewhere in circuit design, and because the terminology "charge-collection efficiency" has specific historical and mechanistic implications, we introduce the term "efficacy." The intent here is to use a term that is applicable across all technologies, irrespective if the "gain" is less than or greater than unity.

In what follows we introduce the concept of efficacy by examining how it can be determined from cross-section curves. This leads to efficacy probability curves and shows how the efficacy contribution can be plotted as a function of fractional sensitive area. The efficacy curve can also be determined from pulsed laser or ion microbeam measurements, or from computer simulation. We present a set of laser measurements and illustrate how to derive the cross-section versus linear energy transfer (LET) and efficacy curves from this data. These curves in turn provide the information necessary to calculate the SEU sensitivity of the device as given by the figure of merit (FOM).

6.5.1 Cross-Section and Efficacy Curves

Efficacy is the correlation between single event (SE) strike locations and actual upset sensitivity—that is, it relates physical device geometry to the circuit response. We now attempt to quantify this relationship. It is assumed that there is a unique critical charge for upset. This is the charge needed at the next circuit node for reset. The critical charge corresponds to the product of the LET at 50% of the limiting cross section and the device nominal depth [Petersen 1992b] (using the conversion factor of 1 pC/μm = 98 MeV/mg/cm^2). The limiting cross section is the saturation value as indicated by a lognormal or Weibull fit to the data. Assume that the SEU efficacy is one when the cross section is 50% of the limiting cross section (L0.5). We define the SEU efficacy value as L0.50/LET and assume constant depth. Therefore, it varies as 1/LET.

The definition is not circular. There are three different quantities involved: (1) the charge deposited by the ion, (2) the charge that is produced and presented to the next circuit node, and (3) the critical charge that is necessary at the next circuit node in order to produce an upset.

If an upset is produced at a LET that is one-half of L0.50, then the critical charge is being developed from charge deposition that is one-half of the circuit critical charge (one-half LET, same depth). Therefore, the efficacy for this hit is two. If an upset is produced at a LET that is twice L0.50, which means that twice the critical charge must be deposited to produce an upset (twice the LET, same depth), then the efficacy for this hit is one-half. Hits at this location by ions of lower LET will not produce an upset.

The efficacy and cross-section curves are perhaps better represented by the cumulative lognormal distribution than by the Weibull distribution because the lognormal distribution can be related to the device physics [Massengill 1993c].* The lognormal function is the normal distribution with the variable being LET. Let

$$z = \frac{\ln(LET) - m}{\sqrt{2}\sigma} \tag{6.1}$$

where m = mean of the function in terms of ln (LET) and σ is the standard deviation of the function in terms of ln(LET). The lognormal distribution is

$$f(z) = \frac{1}{\sqrt{2\pi}\sigma} e^{-z^2} \tag{6.2}$$

The cumulative lognormal distribution is

$$F(z) = 0.5\text{erfc}(-z) : z \leq 0$$
$$F(z) = 0.5 + 0.5\text{erf}(z) : z \geq 0 \tag{6.3}$$

*Note that the lognormal function contains one less parameter than the Weibull curve. When used to describe cross-section curves, the lognormal is sometimes poorer at low LET. Some devices act as if there is a cutoff at high efficacy; others show a large (relative) contribution at high efficacy. The calculated upset rate will be 1% to 2% larger or smaller if the Weibull distribution is used to describe the cross-section curve, instead of the lognormal function. This is discussed in Chapter 11.

6.5.2 SEU Efficacy as a Function of Area

The efficacy behavior of a device can be examined starting with the cross-section versus LET curve. One can convert the cross-section data as a function of LET to cross section as a function of the efficacy. This is a cumulative lognormal function (or integral Weibull function). We can invert the plot to obtain the efficacy as a function of area. Figure 6-3 shows the basic measured cross-section curve for the RMOS IDT 71256 SRAM [Koga 1988, Petersen 1996a], together with its differential form. The older device used here provides a clear example of this approach. The approach is equally valid for modern devices, although the results can be more complex. For the purposes of this development, the efficacy is defined to be unity at the 50% point of the cross-section curve (designated L0.50) and to have its value elsewhere determined by the ratio of the LET to L0.50. This is shown in Figure 6-3.

Figure 6-4 shows the efficacy probability curve obtained by plotting the differential cross-section curve versus the calculated efficacy for the lognormal distribution. The width of the curve can vary greatly for different technologies. Note that one approach to hardening devices is to narrow the efficacy probability curve so that there is less contribution from the high efficacy side. The relative areas that contribute

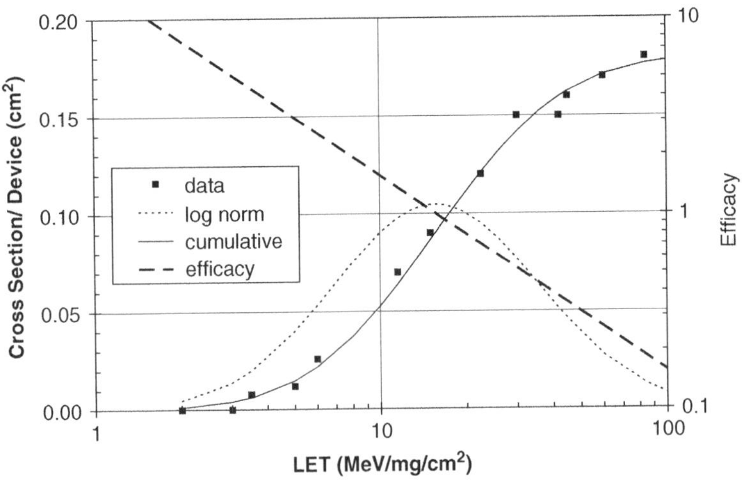

Figure 6-3 IDT 71256 (RMOS) cross-section data with the lognormal curves that describe it. The plot also shows the efficacy as a function of the LET.

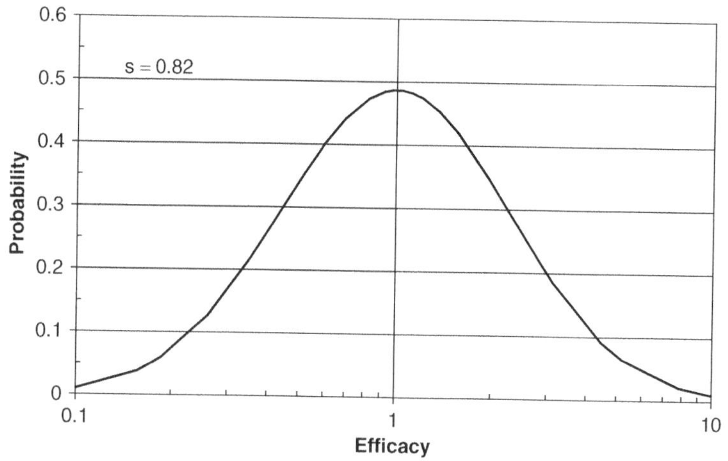

Figure 6-4 IDT 71256 efficacy probability curve. This is the differential form of the cross-section curve plotted against efficacy rather than LET.

to the efficacy can be determined from the normalized cross section as a function of efficacy, as is shown in Figures 6-5 and 6-6. Figure 6-5 shows the plot on a log scale and demonstrates that the high efficacy contribution comes from a very small fraction of the device area. The efficacy is greater than 5 for only 2% of the total area.

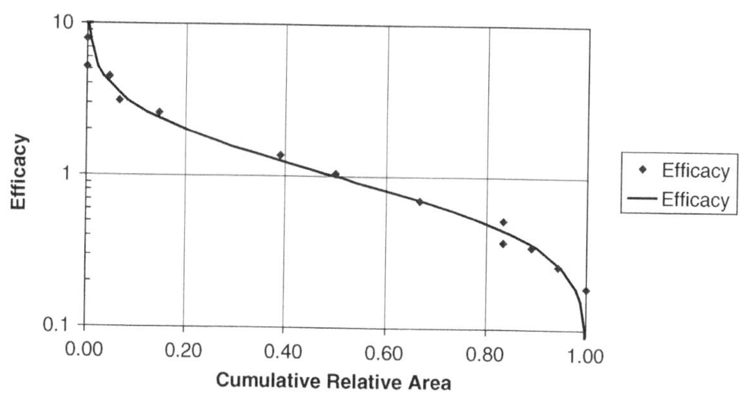

Figure 6-5 IDT 71256 efficacy as a function of the relative area that contributes to the efficacy value. If this curve is measured or calculated, it can be transposed to show the cross section as a function of LET. The LET at 25% of the limiting cross section can be determined from this curve and used directly to determine upset rates using the figure of merit approach.

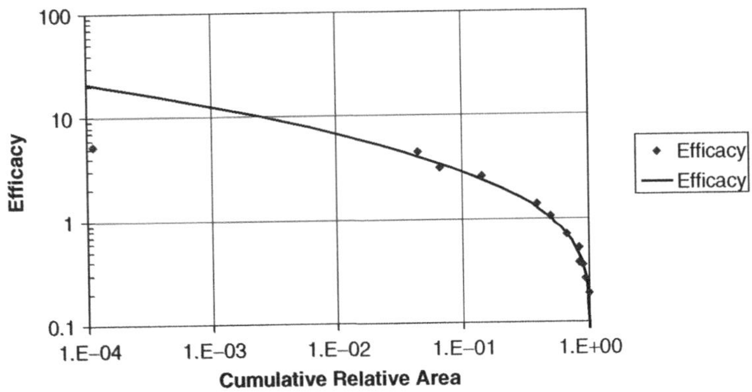

Figure 6-6 IDT 71256 data depicted on a log plot to show the relatively small areas that contribute to high efficacy values.

Each given technology generally has curves that group together. There is a large variation between technologies. The efficacy in this example approaches 10. Even larger efficacy occurs in devices in which diffusion is likely to be present [Petersen 1996b]. Bulk and SOI CMOS devices generally have maximum efficacies less than 10.

The concept of efficacy can be used for any technology. If a circuit designer can determine the efficacy distribution for her device, she can calculate the SEU rate for the device. Methods of determining the efficacy distribution are discussed in the next two sections.

6.5.3 Efficacy and SEU Sensitivity Derived from a Pulsed Laser SEU Experiment

Section 6.4 mentioned a number of experiments that showed different response across a device. We give as an example a pulsed laser experiment that mapped the SEU sensitivity completely across one device [Petersen 2005]. We obtain a cross section and, from that, determine the efficacy and information necessary for FOM calculations. Because of the large feature size of this device, it provides an excellent example of the approach.

6.5.3.1 Laser Experiment

Pulsed laser SEU measurements were performed on a 5 V HM-6504, 4K CMOS RAM at the IXL laser facility using 1 ps, 800 nm optical pulses with a spot size of 1.1 μm at the surface of the device [Pouget 2004, Petersen 2005]. The HM-6504 was first tested for single event upsets in 1979, before systematic measurements of cross-section curves were common [Kolasinski 1979].

A single memory cell, the "target" cell, is visually selected in the middle of the array. Its logical address is read from the memory tester by inducing a SEU with the laser. The addresses of the surrounding cells, the "neighbors," are obtained in the same way. In order to increase the test speed, the tester checks only the target cell and its eight neighbors.

A rectangular scan area is defined around the target cell so that it includes all the SEU sensitive regions for this address. Since this area slightly overlaps the neighbors, the laser may also flip them. During the laser scan, after each laser strike, (1) only upsets detected in the target cell are used to build the sensitivity mapping; and (2) the state of neighboring cells is monitored to ensure that the electrical environment of the target cell remains the same.

The two-dimensional SEU map is obtained by scanning the device in an x-y grid with 1 μm resolution. A single laser pulse is fired at each grid point, with each detected SEU recorded as a black spot at the corresponding location. The same area is scanned with pulses of increasing energy from 3 pJ to 60 pJ for both all-to-0 and all-to-1 test patterns. Figures 6-7 through 6-10 show representative results. At 4 pJ, there is only one pixel with sufficiently high efficacy to cause a 1 to 0 upset; no upsets are observed for 0 to 1. For the 0 to 1 pattern, the first transitions occur at 7.2 pJ (Figure 6-8). As the energy increases, more

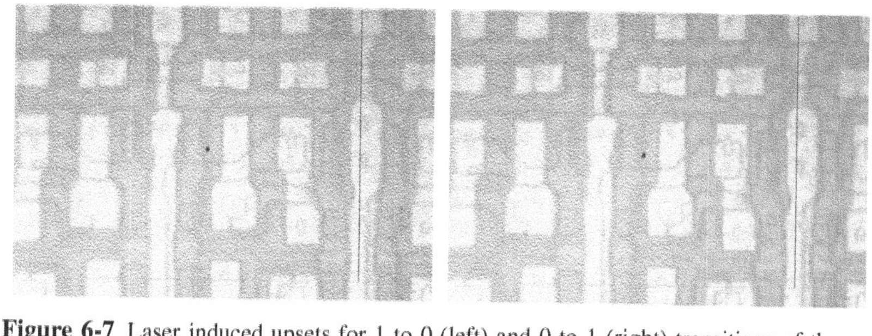

Figure 6-7 Laser induced upsets for 1 to 0 (left) and 0 to 1 (right) transitions of the HM-6504 with 4 pJ, 800 nm optical pulses. At this pulse energy there is only one pixel with high efficacy so that the hit causes an upset. No upsets are observed for 0 to 1 transitions [Petersen 2005].

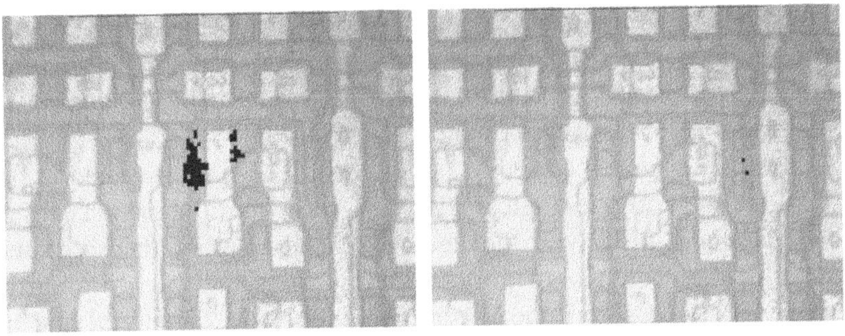

Figure 6-8 Laser induced upsets for 1 to 0 (left) and 0 to 1 (right) transitions of the HM-6504 with 7.2 pJ, 800 nm optical pulses [Petersen 2005].

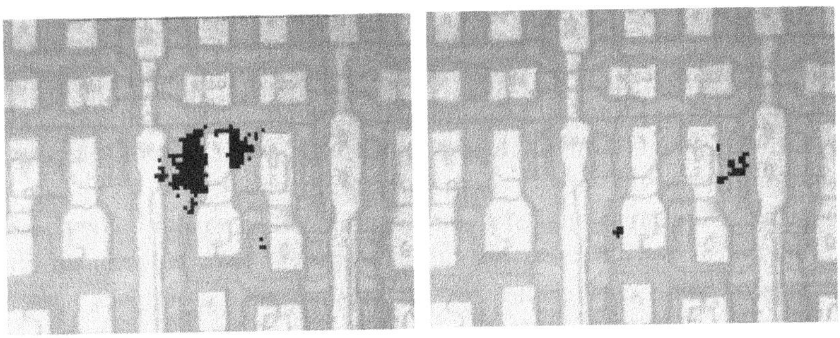

Figure 6-9 Laser induced upsets for 1 to 0 (left) and 0 to 1 (right) transitions of the HM-6504 with 10.4 pJ, 800 nm optical pulses [Petersen 2005].

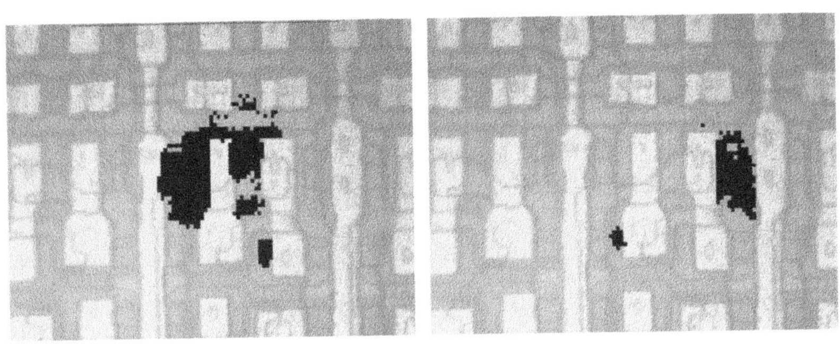

Figure 6-10 Laser induced upsets for 1 to 0 (left) and 0 to 1 (right) transitions of the HM-6504 with 20 pJ, 800 nm optical pulses [Petersen 2005].

and more locations have sufficient efficacy to produce upsets. These figures clearly illustrate that the sensitive area inside the cell increases with the quantity of deposited charge [Buchner 1994]. The experiment was repeated for several different target cells with analogous results.

No latchup is observed for incident laser pulse energies below 150 pJ. A similar series of plots have been produced using a microbeam by Dodd and co-workers [Dodd 2001]. They did not carry their results on to a cross-section curve.

In order to calculate the FOM, laser data have to be corrected for several effects, and the laser energy has to be calibrated into an equivalent LET.

6.5.3.2 Laser Cross Section

The total number of hits n is recorded for each energy level and each pattern. A single hit corresponds to the smallest sensitive area s that can be measured, which is determined by the scanning step size. In this test, a sensitive area of $1\,\mu m^2$ is associated with each hit. This elementary area gives the resolution of the measurement and can be seen as the inverse of the laser pulse fluence on the scanned area [Jones 2000]. The measured laser cross section is thus given by

$$\sigma_L = ns$$

Figure 6-11 shows the results as cross section versus laser pulse energy. It is necessary to make several corrections to the raw data. The

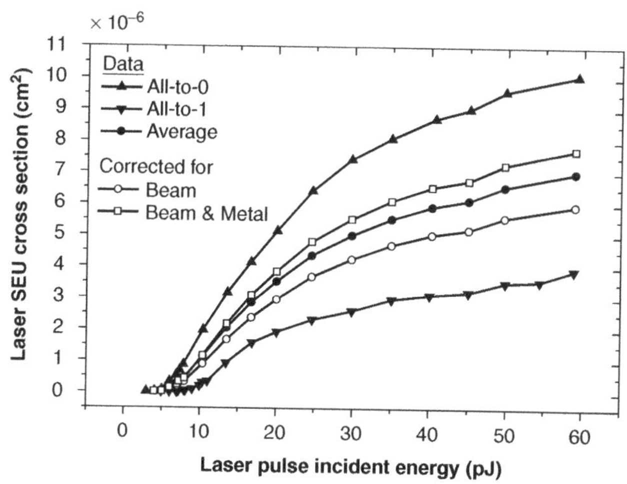

Figure 6-11 Laser cross-section results calculated from data of Figures 6-7–6-10 and Figure 6-12. The average values correspond to the heavy ion cross-section measurements made using a checkerboard pattern.

first is associated with the increase in the "effective" spot size with the pulse energy. An increase of the pulse energy does not change the radial width parameter of the Gaussian distribution of the beam intensity, but it does change the height of this distribution. Thus, the range over which the laser intensity, or equivalently the generated carrier density, is above a given threshold increases with the pulse energy. This effect leads to an overestimation of the cross section. As a first order approximation, if we consider that the sensitivity can be described by a threshold carrier density, this effect can be corrected using [Pouget 2000]

$$\sigma(E) = \left[\sqrt{\sigma_{\text{meas}}(E)} - w_0\sqrt{\frac{\pi}{2}\ln\left(\frac{E}{E_{\text{th}}}\right)}\right]^2 \tag{6.4}$$

where E is the laser pulse energy, E_{th} is the SEU threshold energy, w_0 is the beam waist, and σ_{meas} is the measured cross section. The resulting corrected average cross-section curve is shown in Figure 6-11 labeled "beam."

Another effect that has to be taken into account is the metal interference that prevents the laser light from reaching the silicon. Since some of these areas may be sensitive to upset but will not be counted as such, the metal interference effect leads to an underestimation of the cross section. This effect is clearly visible on Figures 6-9 and 6-10, where black pixels surround metal lines (clear structures). For VLSI devices, this effect is problematic and can require backside testing [Lewis 2001, Darracq 2002]. In the present case, image processing of a microphotograph of a memory cell reveals that approximately 30% of the cell area is covered by metal. Again using a first order correction, we assume that the same ratio applies to the sensitive areas, meaning that 30% of the sensitive areas are not detected by the laser test. The resulting corrected average cross-section curve is plotted on Figure 6-11 (beam and metal). This curve shows connected points. Figure 6-15 (vide infra) shows the data fitted with a lognormal distribution.

The laser cross-section curves of Figure 6-11 do not show a clear saturation, even when corrected for the effective beam size effect. However, from the 60 pJ distribution shown in Figure 6-12 a geometrical estimation of the limiting cross section required for the FOM

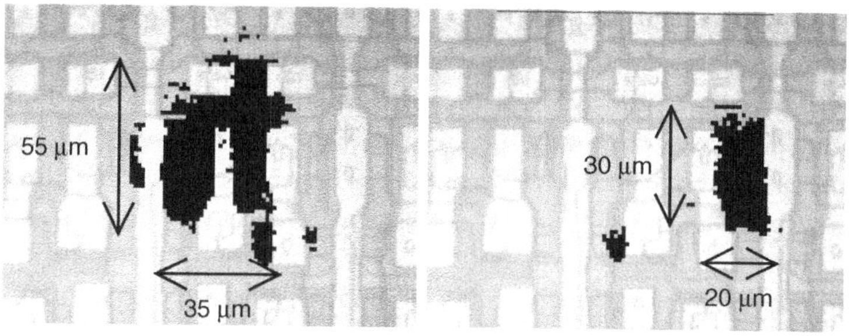

Figure 6-12 Laser induced upsets for 1 to 0 (left) and 0 to 1 (right) transitions of the HM-6504 with 60 pJ, 800 nm optical pulses. The estimation of the limiting cross section is indicated.

calculation is obtained. This gives $1.26 \times 10^{-5} cm^2$ for the average between both tested patterns.

6.5.3.3 Laser Efficacy and FOM

From Figure 6-11 a laser pulse energy of 20 pJ corresponds to the cross-section at 50% of the maximum. The efficacy is then given as the ratio of that value to the energy deposition. Figure 6-13 shows the efficacy as a function of the relative area. We compare these results with the efficacy curve for the IMS 1601, a MOS device, and with the 6504RH. We observe qualitatively very similar curves.

The information shown in Figures 6-11 and 6-13 can be used to plot the relative cross section as a function of relative energy. The relative energy when the cross section is 25% of the limiting cross section corresponds to L0.25 in the FOM calculation. This corresponds

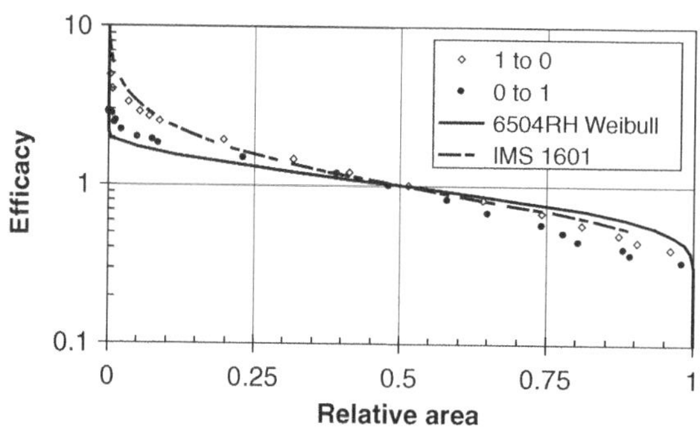

Figure 6-13 The efficacy curve derived from the results in Figure 6-11.

to 60% of the value at the relative energy of 1: that is, 60% of 20 pJ. We want to convert this energy to an equivalent LET.

To calculate the FOM for the HM-6504 we need to know both the saturated cross section and the LET at which the cross section is 25% of the saturated value. From Figure 6-11 we determine that the saturated cross section is 7.8×10^{-6} cm^2 for the device. The pulse energy at which the cross section is 25% of the maximum is also obtained from that curve and has a value of 12.5 pJ, which is equivalent (see above) to a LET of 18 MeV-cm^2/mg [Moss 1995, Melinger 1998, Petersen 2005]. The resulting FOM is 2.4×10^{-8}, which is a reasonable value for an early generation unhardened device (cf., Figure 3.3, vide infra rate vs. FOM). Figure 6-14 shows the relative cross section as a function of relative energy. Figure 6-15 shows the cross-section curves fitted with lognormal distributions.

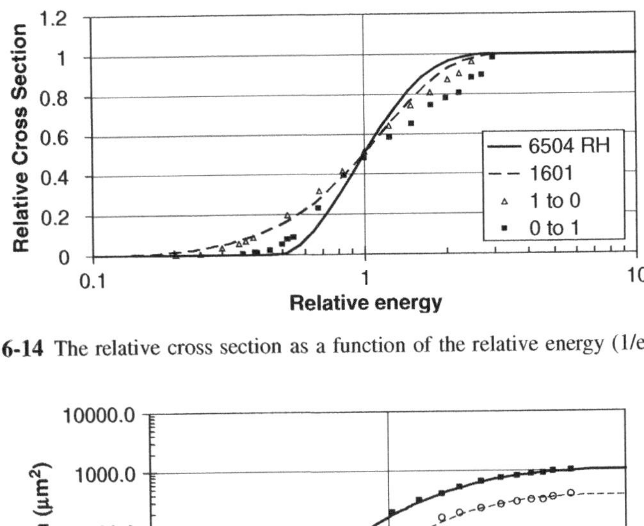

Figure 6-14 The relative cross section as a function of the relative energy (1/efficacy).

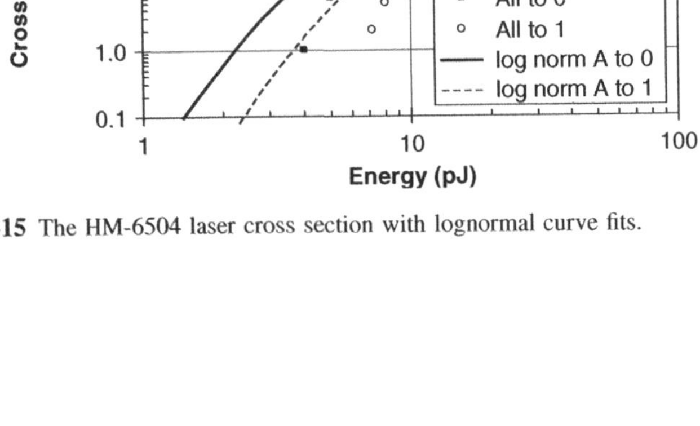

Figure 6-15 The HM-6504 laser cross section with lognormal curve fits.

6.6 MIXED-MODE SIMULATIONS

Mixed-mode simulations combine study of the charge-collection mechanisms in the device with the circuit analysis of the response to the pulse. This approach was initially developed by Rollins and Choma [Rollins 1987]. Similar modeling was also developed by Zoutendyk [Zoutendyk 1987]. Figure 6-16 shows the approach for one of the early examples of mixed-mode analysis [Dodd 1995]. They combined a DAVINCI three-dimensional device simulator with SPICE circuit models and examined the interaction of the circuit with the charge collection. Upset occurs in a SRAM cell when the recovery time of the cell exceeds the feedback time. Therefore, Q_{crit} is the charge collected up to the feedback time. Q is integral $I\,dt$.

Results similar to that of Figures 6-11 to 6-15 can be produced by calculations (computer simulation) based on knowledge of the device structure. In 1993, Massengill and co-workers presented a closed-form

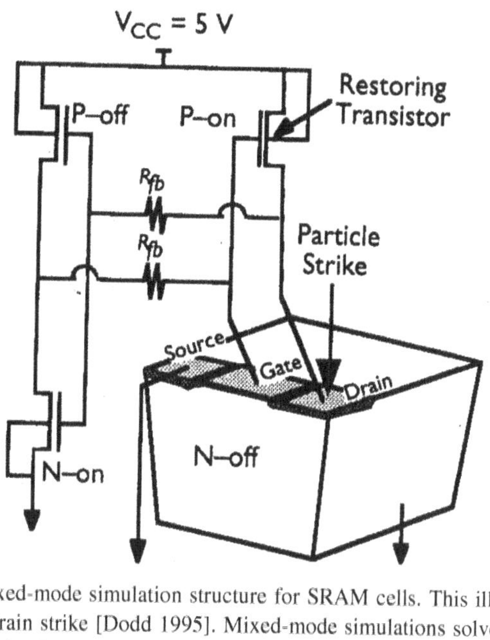

Figure 6-16 Mixed-mode simulation structure for SRAM cells. This illustration is of an n-channel "off" drain strike [Dodd 1995]. Mixed-mode simulations solve the device and circuit equations simultaneously, taking advantage of the localization of the ion strike. Only the struck device is modeled in the multidimensional device domain, reducing the computational burden [Dodd 1998]. Copyright ©1999 IEEE.

analytical relationship between parametric spreads affecting single event upset (either intracell or intercell) and measurable upset cross-section curves [Massengill 1993a]. This method can be applied to efficacy values in order to predict on-orbit upset rates. With knowledge of the critical charge and the device depth, the LET corresponding to L0.5 can be determined. Then a figure such as Figure 6-14 can be used to determine L0.25 and therefore the upset sensitivity of the device. There have been a very large number of mixed-mode simulations and we will attempt to discuss a representative sample.

6.6.1 Warren Approach

In 1999, Warren and co-workers used three-dimensional device and mixed-mode simulations to map single event sensitive areas in SOI devices [Warren 1999] and were able to generate cross-section values from the charge-collection parameters. The mixed-mode simulations mapped the single event sensitive areas in SOI devices, as shown in Figures 6-17 and 6-18.

Figure 6-19 shows the cross-section values obtained from the charge-collection parameters. The information in Figure 6-19 is

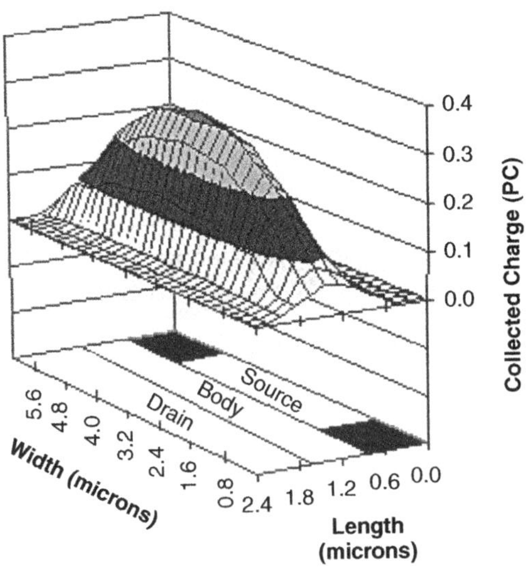

Figure 6-17 Charge collection profile map of a 0.8 μm SOI device for a LET value of 50 MeV/mg/cm² [Warren 1999]. Copyright © 1999 IEEE.

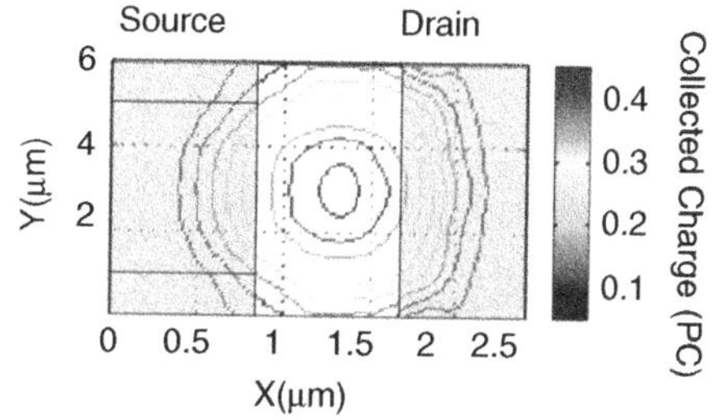

Figure 6-18 Another view of the charge collection at a LET = 50 MeV/mg/cm^2 [Warren 1999]. Copyright © 1999 IEEE.

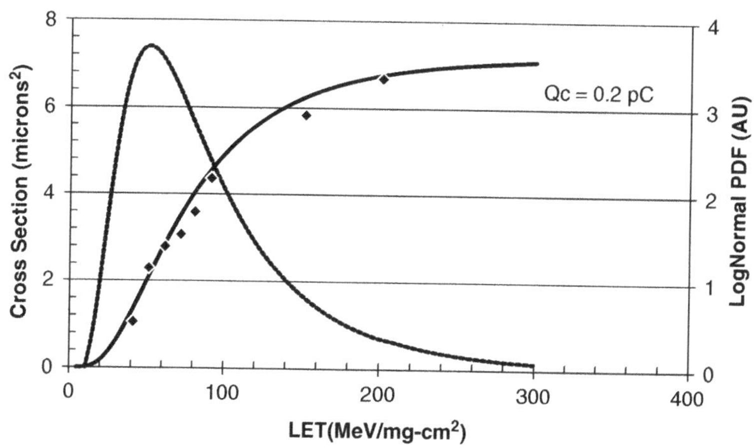

Figure 6-19 Cross-section values (data points) derived directly from two-dimensional device simulations. The curves show corresponding differential and cumulative lognormal curves [Warren 1999]. Copyright © 1999 IEEE.

adequate for determining the parameters for FOM calculations. Therefore, any designer can, in principle, determine the SEU sensitivity of his parts. Figure 6-20 shows the efficacy curve (not called that at the time) developed in this study which is parallel to those shown earlier for heavy ion and laser studies.

In circuits with simple 1-to-1 relationships between direct charge collection and upsets, this information can lead to the upset cross section and the efficacy values by simply accumulating physical cross-sectional areas defined by charge contours greater than the critical charge for the circuit at each simulated LET values, as shown in Figure 6-20.

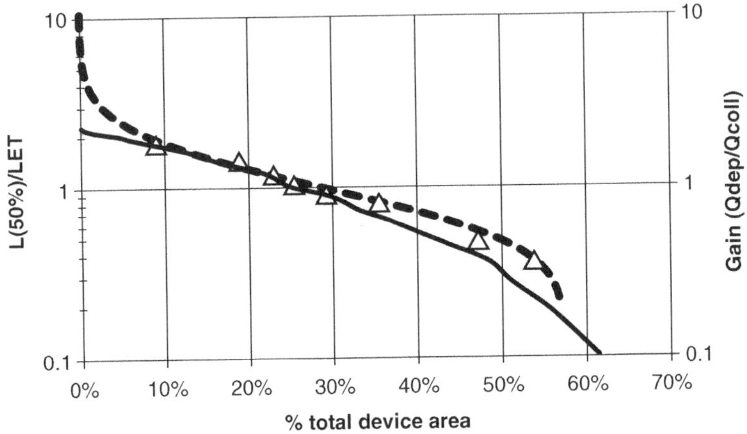

Figure 6-20 L0.5/LET (efficacy) derived from the two-dimensional device simulations [Warren 1999]. Copyright © 1999 IEEE.

6.6.2 Dodd Approach

Dodd and co-workers have performed similar simulations on bulk CMOS circuits [Dodd 2001], mapping the sensitive regions within and proximal to the sensitive devices. For ions with a LET of 11.5 MeV-cm^2/mg (just above the upset threshold), the SEU sensitive area is the center portion of the reverse-biased NMOS drain. This is shown in Figure 6-21. This is the expected most-sensitive strike location [Dodd 1996b]. Note that the sensitive area of the reverse-biased NMOS drain

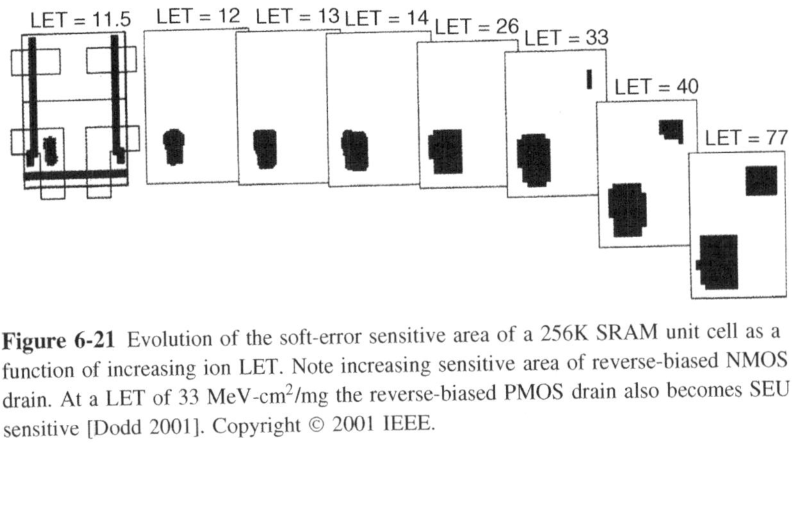

Figure 6-21 Evolution of the soft-error sensitive area of a 256K SRAM unit cell as a function of increasing ion LET. Note increasing sensitive area of reverse-biased NMOS drain. At a LET of 33 MeV-cm^2/mg the reverse-biased PMOS drain also becomes SEU sensitive [Dodd 2001]. Copyright © 2001 IEEE.

increases gradually with LET. For the highest LET values, the sensitive area around the NMOS drain is larger than the drain itself, because diffusion of charge from near misses is sufficient to cause upsets [Smith 1995]. At a LET of 33 MeV-cm^2/mg the reverse-biased PMOS drain also becomes susceptible. Figure 6-22 shows the measured and simulated upset cross sections.

6.6.3 Hirose Approach

Dodd's calculations used extremely large computer resources and are not likely to be repeated. However, later authors have developed approaches that have lower computer requirements. Similar simulations on SOI CMOS circuits have been used to generate cross-section curves that exhibit good agreement with experimental measurements [Hirose 2002]. This is shown in Figure 6-23. Hirose and co-workers produced cross-section curves using a commercial mixed-mode three-dimensional simulation using ATLAS (Silvaco International).

6.6.4 Simplified Approach of Fulkerson

The three-dimensional simulations noted above take large amounts of computer time. Fulkerson and co-workers have presented a simpler approach that may be more appropriate for many circuit designers [Fulkerson 2005, Fulkerson 2006, Fulkerson 2007]. They use simple

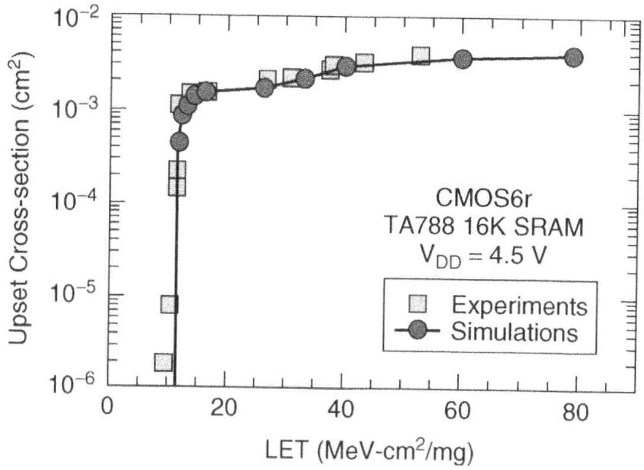

Figure 6-22 Measured and simulated upset cross sections in a CMOS bulk 16K SRAM [Dodd 2001]. Copyright © 2001 IEEE.

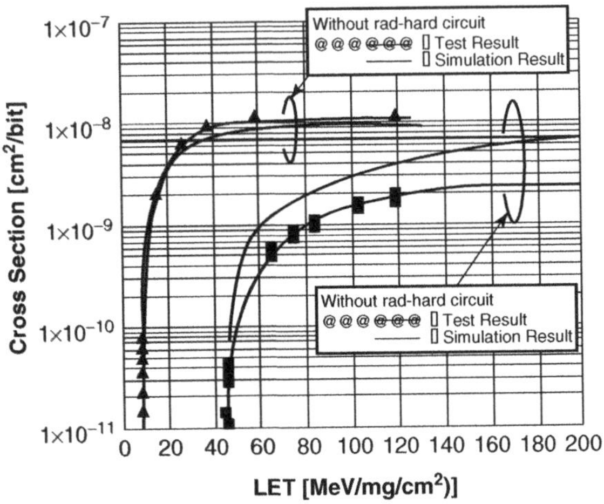

Figure 6-23 Measured and simulated full-SEU cross-section curves of 128 kbit SOI SRAM with and without RC hardening [Hirose 2002]. Copyright © 2002 IEEE. The simulation uses a commercial three-dimensional mixed-mode code.

closed-form one-dimensional solutions to the transport equations, combined with existing SPICE models, to obtain a cross-section versus LET curve that matches very well with experimental data for their device. They analyze the response of separate "boxes" of each transistor, Figure 6-24, and then sum the cross sections as a function of sensitivity, Figure 6-25. They also apply their "boxes" approach to situations where the ion has to hit two separated "boxes" to produce an upset [Fulkerson 2006].

6.6.5 The Imax, F (Tmax) Approach

Palau and co-workers have pointed out that we can think of two different cases in the CMOS/bulk situation [Hubert 2001, Wrobel 2006]. In the principal case, the ion crosses the drain and produces a fast pulse. The second case corresponds to the track passing close to the drain but not crossing the junction or the space charge region. This case can also induce a SEU due to the collection of charges arriving after their diffusion from the track. The current pulse is delayed and enlarged. In this case, the resulting slow pulse is not consistent with the fast pulses associated with the normal critical charge. The contributions may come from several microns from the drain in submicron technology. They

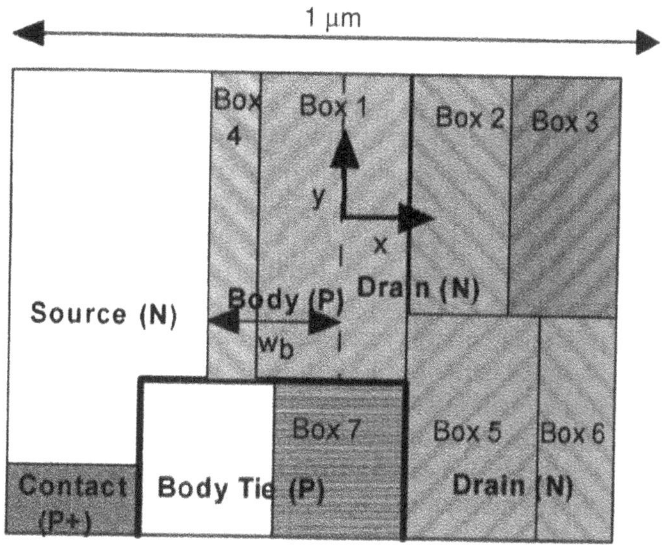

Figure 6-24 A single transistor broken into "boxes" of differing SEU sensitivities. This is done for all of the appropriate transistors in the circuit [Fulkerson 2006]. Copyright © 2006 IEEE.

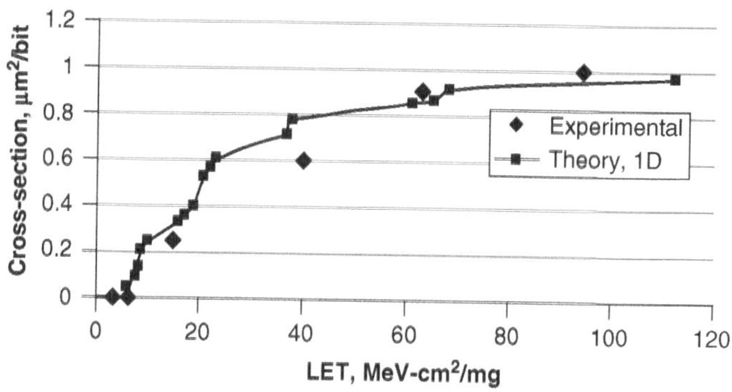

Figure 6-25 Cross section versus LET for a SRAM cell. The experimental points are from vertical strikes for a series of ions [Fulkerson 2005]. Copyright © 2005 IEEE.

examined this situation by using several shape pulses and a variation of pulse widths [Merelle 2005]. Initially this is done using SPICE type analysis and obtaining a plot of SEU upsets versus an Imax, Tmax locus. They can then compare charge-collection calculations with this locus. A great deal of time and effort are saved by separating the two types of calculation. Figure 6-26 shows a sample of their results. We

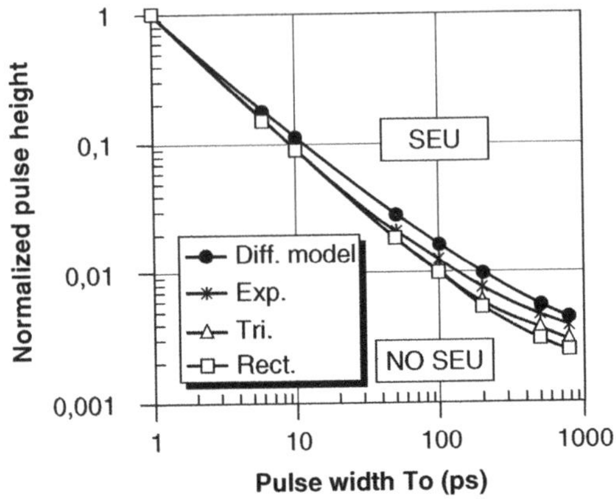

Figure 6-26 Results of SPICE simulation of a 0.6 μm feature size SRAM technology. Comparison of threshold input pulse height versus duration parameter for different current pulses [Merelle 2005]. Copyright © 2005 IEEE.

observe that the upset criteria can be represented by a single critical charge value for pulses up to a length of approximately 200 ps. Pulses longer than that require more charge. The authors compare their calculated Imax, T0, rather than the charge generated.

The authors model the charge collection using an ambipolar diffusion model. It computes the quantity of carriers arriving at the sensitive surface. For ion tracks that cross the sensitive surface, they use a threshold LET value to determine if the cell flips: that is, the fast pulses in the less than 10 ps regime. The diffusion current pulse is then used for the slow pulses. The model is illustrated in Figure 6-27.

This model has been extended to obtain a Monte Carlo calculation of the cross section [Correas 2007, Correas 2008]. The results are shown in Figure 6-28 for a 130 nm CMOS bulk SRAM. The cross-section scale is in arbitrary units and has an undisclosed normalization between experimental and simulated results. They considered ion strikes for paired NMOS-off and a PMOS-off transistors and their Figure 6-13 shows a cross section of 10 μm^2 for a 23.46 MeV/mg/cm^2 ion.

Their results can be compared with cross-section results for a 160 nm CMOS bulk SRAM. That device has a limiting cross section of 8 μm^2 and the area of the entire memory cell was 6.0 μm^2 [Hafer 2006b]. In this case the entire memory cell is upsetting. Modeling by Morris and co-workers indicated that all parts of the cell contribute

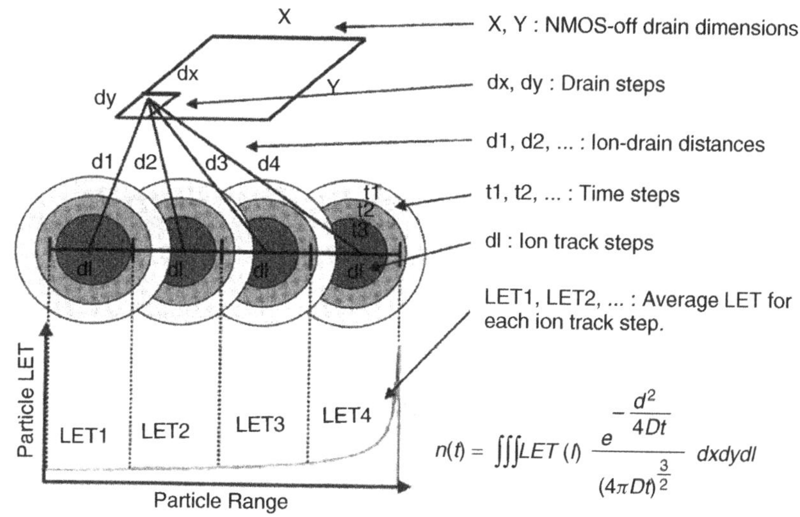

Figure 6-27 Ambipolar diffusion model. It computes the quantity of carriers arriving at a given time to the sensitive surface. D is defined as the ambipolar diffusion constant [Merelle 2005]. Copyright © 2005 IEEE.

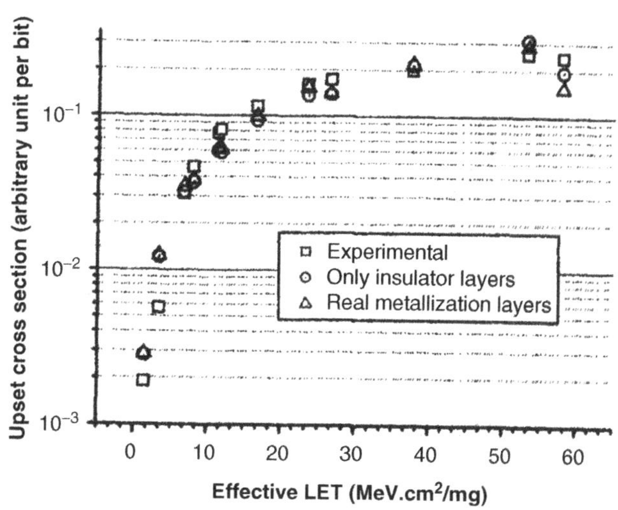

Figure 6-28 Experimental and simulated SEU cross sections for two passivation layers [Correas 2007]. Copyright © 2007 IEEE.

to the upset due to parasitic bipolar actions [Morris 2006]. It would appear from these results that it is not adequate to model only a pair of transistors as per Correas.

6.6.6 Circuit Level Simulation to Upset Rate Calculations

Warren and co-workers have produced mixed-mode calculations combining the HSPICE circuit simulator and the MRED Monte Carlo energy deposition code [Warren 2008]. They combine both radiation transport and SPICE simulation tools into a single automated simulation flow. It is possible to identify single event vulnerabilities from both a layout and circuit perspective with no user intervention during run time. It utilizes a SEU model based on the physics of energy deposition, the charge collection at each transistor, and circuit response to the collected charge or charges. They were analyzing a DICE cell and so could examine coincidence pulses and could carry the calculation on to the upset rates in space.

6.6.7 Multiple Upset Regions

In Figure 6-29, areas outside a single "struck" device are sensitive to upset due to charge collection at multiple nodes. The efficacy of this mechanism would not be captured by simply the charge-collection efficiency of a node as compared to critical charge for upset. Here, the efficacy must be derived by an analysis of the sensitive area across the entire region (e.g., across an entire RAM cell). From these contours, the efficacy is derived by a straightforward accumulation of upset areas normalized by the 50% cross-sectional area at an efficacy of unity.

For each of these examples the cross-section or efficacy curve supplies the parameters required for FOM calculations. Therefore, a designer can, at least in principle, calculate the SEU sensitivity of devices.

In circuits with simple 1-to-1 relationships between direct charge collection and upsets, the upset cross section and the efficacy values can be determined by simply summing physical cross-sectional areas. The areas are defined by charge contours greater than the critical charge for the circuit at each simulated LET value. However, as

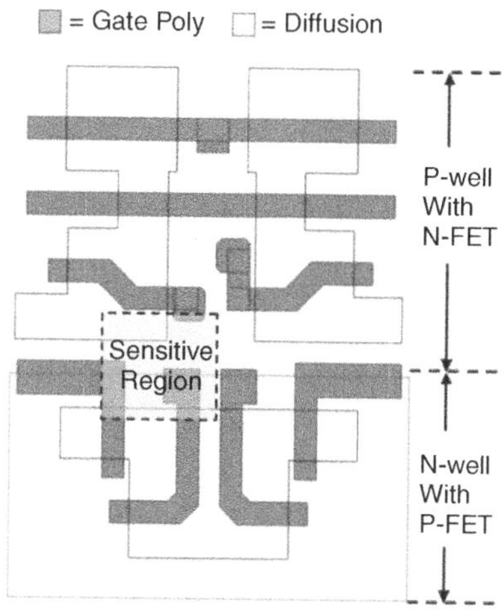

Figure 6-29 Top layout view of a 6-T 250 nm bulk CMOS SRAM cell showing the sensitive region that was mapped via three-dimensional TCAD simulations for a particular LET value [Olsen 2005]. Copyright © 2005 IEEE. Note that the sensitive area is not contained within one device, but spans devices and even a well boundary.

described by Olson and co-workers [Olsen 2005] and Warren and co-workers [Warren 2005] and supported by experiments of McMorrow and co-workers [McMorrow 2005], modern technologies rarely exhibit upset patterns that map directly to a single-device charge-collection profile. They instead exhibit complicated upset patterns based on charge sharing among several nodes, long diffusion processes, and parasitic effects impacting nodes located some distance from the strike location. In order to capture these effects, detailed three-dimensional mixed-mode simulations of geometric structures larger than a single device most likely are necessary. Note also that even though the feature sizes may be approaching track sizes, the dimensions of the sensitive region are remaining significantly larger than either.

Although this section has used combined charge-collection and circuit modeling, other approaches could be used. For example, a subset of the IBM Monte Carlo approach would lead to the efficacy and cross-section distributions [Srinivasan 1994, Murley 1996].

6.6.8 Efficacy and SEU Threshold

The ability to find upsets with very small cross sections at very small LETs corresponds to finding very small regions with high efficacy.

The curve in Figure 6-6 gives one example. There are regions in which the electric fields are more intense, allowing for large values of charge multiplication. The high efficacy regions can also be explored in charge-collection and circuit simulations by using finer and finer grids in the regions of high efficacy. This is one of the advantages of the simulation approach.

The concept of onset threshold generally is not valuable in upset rate calculations [Pickel 1996, Petersen 1997b]. However, threshold behavior becomes important when examining very low LET cross-section and efficacy data. We recommend fitting heavy ion or laser data with the lognormal distribution. However, in nearly all cases, the lowest LET (highest efficacy) points do not fall on these curves. The IDT 71256 data discussed above has a single low LET point that falls at higher LET (lower efficacy) than predicted by the curves that fit the rest of the data. This effect is evident again in the laser data of Figure 6-3. Warren and co-workers discuss lognormal and Weibull fitting of their data and note that the lognormal distribution does a better job of fitting the data near saturation, but deviations are observed at low LET [Warren 1999]. Petersen notes that this is true for most data. It is estimated that introducing a low LET threshold on the lognormal curve would improve rate predictions by 1% to 2% (see below) [Petersen 1996b].

We have introduced efficacy to describe various internal processes (gains) that multiply the charge generated in the initial ionization. The curves through the data predict that these gains will keep increasing as the area involved decreases. The limitations on sharp corners and boundaries during device fabrication will limit these effects for real devices. Therefore, all devices are expected to show lower efficacy at very small dimensions than predicted by the extrapolation of the cross-section data at larger dimensions.

Some recent data on submicron parts show unexpected contributions at very low LETs [Ecoffet 1992, Liu 2004, Warren 2005, Liu 2006a]. It is a temptation to claim that the manufacturers have found new ways of producing high efficacy in their devices. This is probably not realistic. The most likely explanation is heavy ion nuclear reactions (hadronic reactions) in the materials above the sensitive volume [Warren 2005]. Other explanations are possible [Liu 2004, Liu 2006a].

The contributions to the upset rate as a function of LET in the IRPP calculation are illustrated in [Petersen 1997b, Section 3.3]. As one moves from the maximum cross section to the threshold, the probability for upset decreases by four orders of magnitude. Meanwhile, the RPP upset rate that is being sampled increases by approximately two orders of magnitude. Therefore, the threshold region contributes only a few percent to the upset rate. Changing the description of the threshold behavior using a Weibull distribution, a lognormal distribution, or a lognormal distribution with cutoff, only changes the total upset rate by a few percent. The exception to this conclusion is for very hard devices for which the maximum cross section occurs at LETs above the iron shoulder in the upset rate versus LET curve. In this case the threshold change can contribute significantly to the total upset rate (10% to 20%).

An additional factor is important in the low LET hadronic contribution to the space upset rate. If the low LET cross section is ascribed to hadronic interactions in the overlayers, the cross-section curve has a nature entirely different from the cross-section curve due to direct ionization. The normal cross section indicates the probability of upset due to ions of that LET coming from 4π steradians. The overlayer hadronic reaction upset cross section, on the other hand, corresponds to the probability of indirect upset by ions of that LET coming in from a fraction of 4π steradians.

6.6.9 From Efficacy to Upset Rates

As described in Section 6.3.3, efficacy is related to the correlation between single event strike locations and actual upset sensitivity; that is, it relates physical device geometry to the circuit response. As such, efficacy measures could include geometric-dependent parasitic effects (e.g., the intracell variability of Chapter 2) as well as other, indirect nonidealities that affect circuit upset as a result of strike location. An example of indirect modalities is the parasitic multiple-bit charge sharing effect described by Olson and co-workers [Olsen 2005]. In fact, any single event geometric spread that can be mapped to upsets can be captured in the efficacy measure. However, in this discussion, we concentrate our attention on strike location dependencies within a single device.

There are three diagnostic methods for determining efficacy distributions: (1) analysis of measured cross-section curves, as described in Section 6.5.1, (2) sensitivity mapping via experimental laser/microbeam scanning, as described in Section 6.5.3, or (3) upset contour analysis from calibrated mixed-mode simulations, as described in Section 6.6. Regardless of the source, efficacy curves can be used to predict error rates that include the statistics of efficacy modalities. The quick way is to estimate the LET where the cross section is 25% of the limiting cross section and combine that directly with the limiting cross section in the FOM calculation.

The more complete method is to fit the cross-section curve and determine the FOM parameters from it. The median LET L0.5 (or mean parameter, m) is extracted from analysis of the circuit, the classical critical charge for upset, along with detailed knowledge of the expected process technology charge-collection efficiency. From an analytical lognormal curve fit to the efficacy distribution (as shown in Figures 6-19 and 6-20), the standard deviation (or shape parameter, s) can be extracted.

From these two parameters, the designer can calculate the FOM:

$$FOM = \frac{\sigma_{\mathrm{HI}}}{L_{0.25}^2} \left[\frac{(\mathrm{MeV/mg/cm^2})^2}{\mathrm{cm}^2} \right] \tag{6.5}$$

where σ_{HI} is the limiting heavy ion cross section per bit at large LET and $L_{0.25}$ is the LET at 25% of the limiting cross section [Petersen 1983a, Petersen 1993, Petersen 1995, Petersen 1998]. For the lognormal distribution $\ln(L_{0.25}) = m - 0.674\,s$.

6.7 PARAMETRIC STUDIES OF DEVICE SENSITIVITY

There have been a number of papers that examine families of single event effect (SEE) cross-section curves [Petersen 2007, Petersen 2008a]. These families result as one or another parameter of the device or circuit is modified.

6.7.1 Data Display and Fitting

Figure 6-30 shows three sets of data that apparently saturate at different limiting cross sections [Hafer 2007]. Both the fit by eye and the fit with the Weibull function suggest that the limiting cross section is

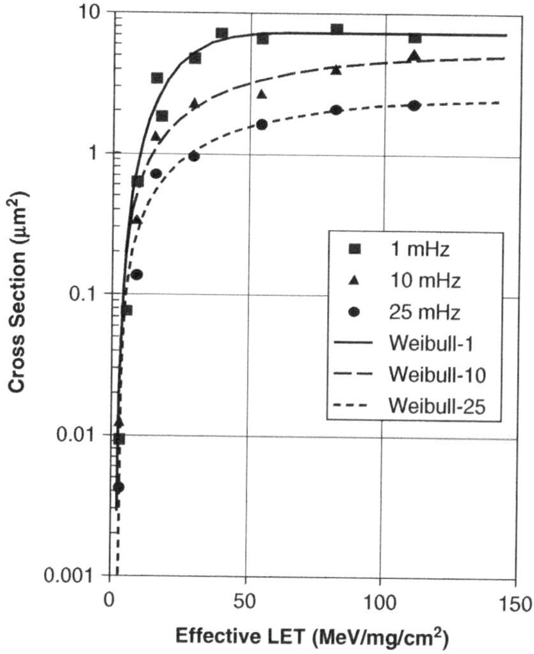

Figure 6-30 Cross-section curves measured at three different frequencies [Hafer 2007]. The apparent saturation in the lower curves is due to the log-linear scales used on the plot and is an artifact of those scales, not a property of the data.

changing with frequency. However, we know that the apparent saturation of the cross section when using a linear LET scale is an artifact of the display, not a characteristic of the data. We have discussed in Section 5.4 how the Weibull fit cannot be trusted. Furthermore, we expect the limiting cross section to be a function of geometry, so it should not change with parameters such as frequency. For the next step, we plot the three curves assuming that they have the same limiting cross section. This is shown in Figure 6-31.

Figure 6-31 shows the three curves plotted on a linear scale and fit with the Weibull function using the same limiting cross section. It is clear that the lower curves are not saturating and the fits are still very good. However, we can learn more if we fit with the lognormal function. We do this in two steps. First we fit the three curves separately. We then plot the coefficients of critical LET L_c as a function of frequency. In this case, they appear to be nearly a quadratic. We then fit the three sets of data simultaneously, varying the cross section,

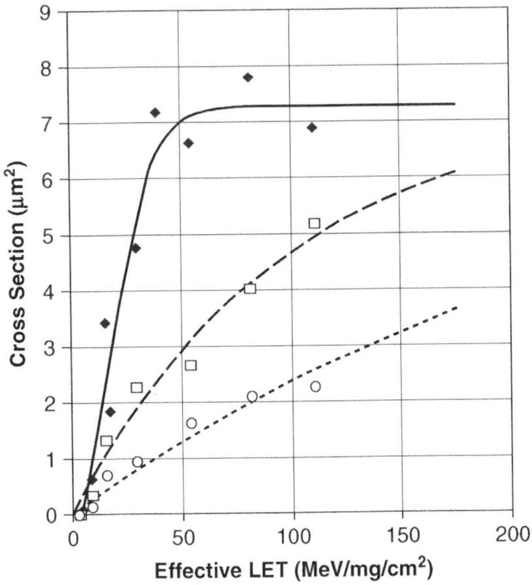

Figure 6-31 The data of Figure 6.30 fit with the same limiting cross section. The three curves are described by 12 parameters.

the shape parameter, and the three quadratic coefficients. In this case, the fit is a little better if the shape parameter also varies with the frequency, but that is very unusual. Figure 6-32 show the combined fits.

The reason that the logarithmic display is much more informative than the linear display is due to the logarithmic nature of the cross-section curve. It has become common to use the integral Weibull curve with its comfortable four parameters. The details of the Weibull distribution and other mathematical functions used in this chapter are placed in an appendix. However, no one has been able to relate the curve parameters to the single event characteristics of the devices. Furthermore, as Normand has demonstrated, one has to be very careful in the curve fitting to avoid false minima in the least squares fit [Normand 2004]. The use of the lognormal distribution for describing SEU cross section was introduced and advocated by Massengill and co-workers [Massengill 1993a, Warren 1999] and recommended by Petersen [Petersen 1996c, Petersen 1997b, Petersen 2005]. Those references did not recognize the full utility of this approach. We use the lognormal distribution where ln(LET) is normally distributed with

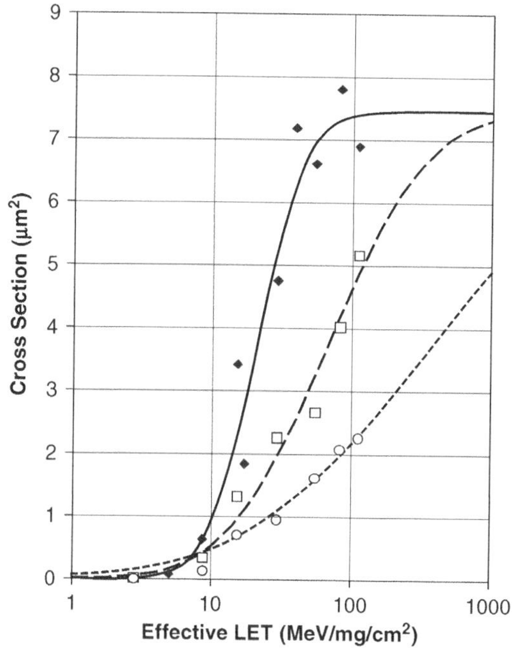

Figure 6-32 The six parameter fit to the data. $CS = 7.4\ \mu m^2$, $s = 0.55 + 0.07$ F, $L_c = 22 - 1.9\,R + 0.64\,F^2$.

parameters mean (m) and standard deviation (s) of ln(LET). The mean is the logarithm of the center of the distribution, which occurs at 50% of the integral distribution (L0.5). We have previously demonstrated that L0.5 corresponds to the critical charge [Petersen 1992b]. McNulty and co-workers have also recommended the 50% cross section as the threshold value for comparing devices [Roth 1993]. We have also demonstrated that the standard deviation of the distribution is a measure of the spread of the charge efficacy across the device; that is, the variation of charge deposition that is necessary to produce the critical charge at the sensitive node [Petersen 2005]. The limiting cross section (CS) of the distribution corresponds, to first order, to the effective charge collection area of the device [Petersen 2005]. Therefore, the three parameters in the lognormal distribution are basically related to three independent single event characteristics of the device. We can use the relationship, for silicon, that 1 pC/μm = 98 MeV/mg/cm^2; and determine the critical charge equivalent to L0.5 for a known thickness device. As a reminder that L0.5 is a critical LET related to the critical charge, we will instead use L_c for critical LET in this paper.

We are not claiming that the lognormal distribution gives better fits to the experimental data than the Weibull distribution. We are claiming that the fits are much more meaningful. We show that the two curves are nearly equivalent and that it would take very high precision data to differentiate the two.

Curve fitting of the lognormal distribution to the data is easily done using the least squares approach, weighting each point with its fractional standard deviation. It is often desirable to use both a linear fit and logarithmic fit, which compares the logarithm of the data point with the logarithm of the calculated point. This is because the linear fit is determined primarily by the top decade of points, fitting the limiting cross section, while the logarithmic fit equally weighs the data over all decades, fitting the shape and threshold region. This is similar to the common procedure when fitting the Weibull function, where the threshold is determined from the logarithmic fit and the limiting cross section from the linear fit.

Most hardening approaches change only one of the device parameters: the limiting cross section, the width of the distribution, or the critical LET. The limiting cross section depends on the device area, and the width of the distribution is determined by the device technology. The critical LET changes with depth or with changes of the device parameters. We will observe the changes of critical LET with changes in circuit behavior in our examples. This book will emphasize the variation of critical charge derived from the data. Details of motivation, devices, and measurement techniques can be found in the original papers.

6.7.2 Device Parameters and SEU Sensitivity

This section examines a number of papers that have measured device sensitivity. The majority of them have plotted the results on a linear LET scale and mistakenly concluded that they are seeing variations of limiting cross section. There is sometimes a great deal of scatter in the data, but the authors have not included error bars, so it is not possible to tell if the scatter is statistically reasonable, or the result of occasional dosimetry blunders.

We use least squares fitting using the Excel tool Solver. In principle, the least squares fit should use the data errors (fractional standard deviation) to weight the data points. The authors do not show their error bars, so this is not possible in our case. However, when authors fit their own data, they should use the proper error weighting. Mathematical fitting is very straightforward and should always be used in preference to fitting by eye.

We calculate the coefficient of determination R^2, which evaluates the explained sum of the squares. It is derived by examining the ratios of least squares deviation from the fitting line to the least squares value of the data points about their average. The fit is very good if R^2 approaches one. An R^2 of 0.80 says that we are describing 80% of the data variation from its average value, while an R^2 of 0.25 says that we are explaining only 25% of the variation.

This section includes a large number of examples so as to demonstrate the many types of parametric studies and to show how the analysis presented here expands what can be learned from such experiments. In all cases we started fitting with several more variables than the final fit. We attempted to reach a minimum number of fitting parameters while retaining a good R^2.

6.7.2.1 Applied Voltage Variation in Memories

Kolasinski and co-workers measured the sensitivity of the HM-6508RH to variation of supply voltage [Kolasinski 1981] Barak and co-workers measured the voltage variation of the HM-65162 [Barak 1999a]. Kolasinski used six voltages, while Barak used three voltages. Barak's results are shown in Figure 6-33. Both experiments showed the critical charge increasing linearly with the applied voltage. The plot shows a linear fit and the resulting R^2 is 0.994. The large value of R^2 supports the assumption that the curves have the same limiting cross section. Note that one can perform tests at low voltage and determine behavior at higher operating voltages and LETs where you cannot test.

6.7.2.2 Resistor and Temperature Variation in Latches

The change of critical charge for resistor-hardened d-latches was examined by Sexton and co-workers [Sexton 1991]. The three

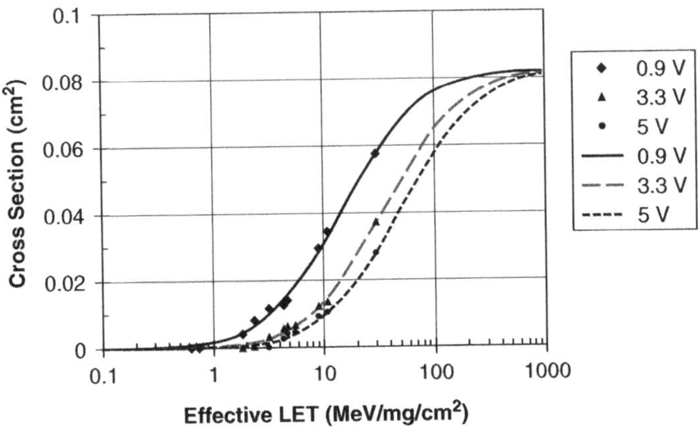

Figure 6-33 Voltage variation of critical charge. Four parameters describe the data: $CS = 0.082\,\text{cm}^2, s = 1.3, L_c = 7.1 + 8.5 \times V$. Despite their appearance on a linear LET scale, the three curves have the same limiting cross section. The space upset rate will vary with the square of the voltage.

families of curves are described by a single lognormal function with five parameters and are shown in Figure 6-34. This is a logarithmic fit and R^2 is 0.933.

6.7.2.3 Single Event Transients Versus Differential Input Voltage in a Comparator

An interesting set of experiments was carried out for single event transients in a comparator by Johnston and co-workers [Johnston 2000]. We can describe their five curves using four variables. In this case, the critical charge varies as the square root of the input voltage. This is a linear fit and R^2 is 0.866. The results are shown in Figure 6-35.

Johnston also looked at the dependence on the output triggering level for his circuit [Johnston 2000]. In this case, the device becomes more sensitive linearly at lower voltages.

6.7.2.4 Comparison of Ones and Zeros

The type of analysis that we are performing lends itself to many applications. Swift has looked at registers in the Motorola PowerPC750 [Swift 2001]. When we plot his results in Figure 6-36, we observe that the critical LET (L_c) has changed from 10 to 18 for the two types of upsets. This is a log fit and R^2 is 0.927.

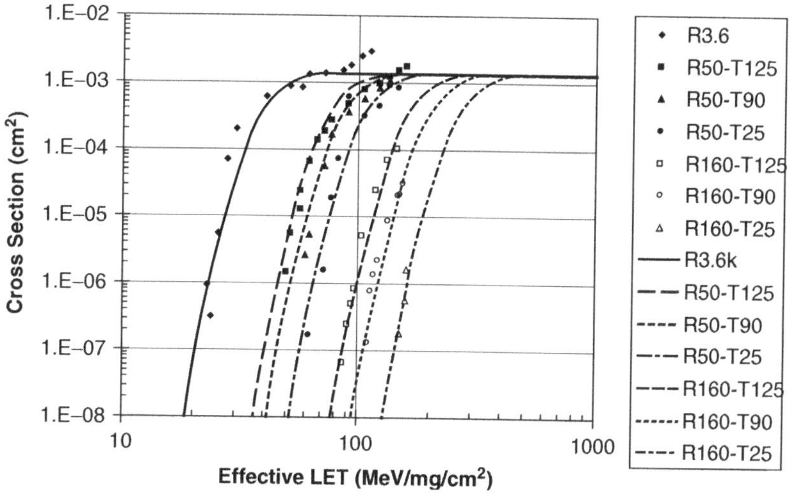

Figure 6-34 The effect of resistor and temperature variation. The five lognormal parameters that describe the three families of curves are: $cs = 1.32 \times 10^{-3}, s = 0.2, L_c = 39.8 + (1.8 - 0.0073T) \times R$. The critical charge increases linearly with the resistance and decreases linearly with the temperature. R^2 for the combined fits is 0.933. The space upset rate will vary with the square of the resistance and temperature.

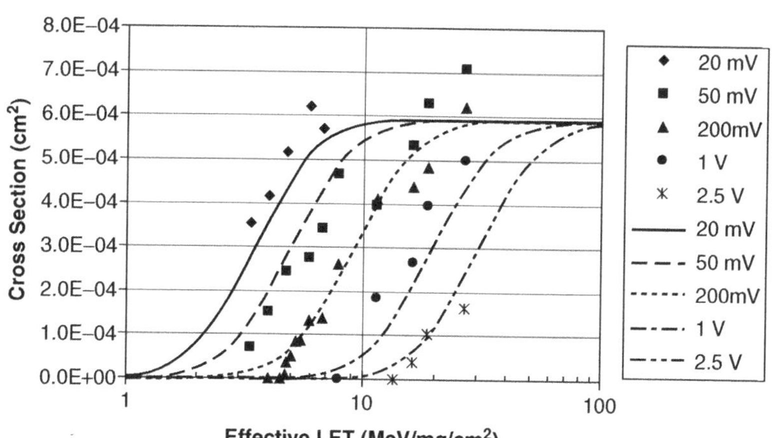

Figure 6-35 Variation of critical charge with differential input voltage for the PM139. Four parameters describe the data: $CS = 5.9 \times 10^{-4}, s = 0.49, L_c = 0.78 + 0.58\text{sqrt}(V)$. The critical charge varies linearly with the square root of the voltage. The space upset rate will vary linearly with the voltage.

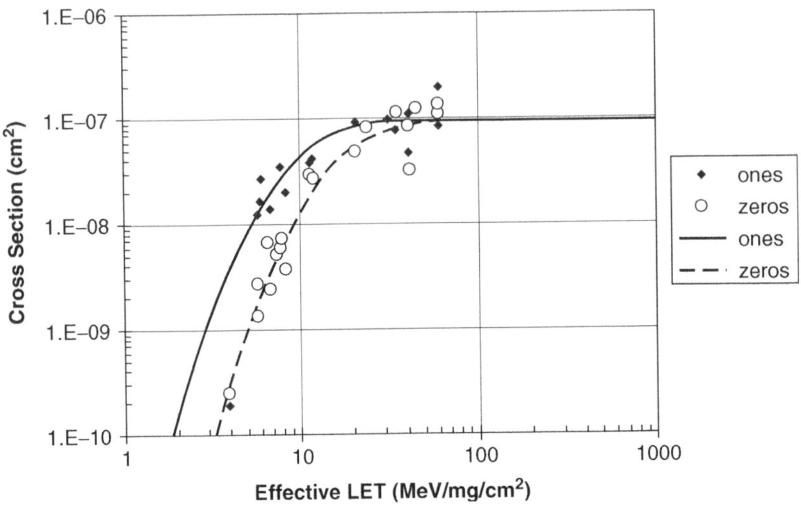

Figure 6-36 The difference of critical LET for two types of circuit element in a microprocessor can be determined by this type of analysis. The critical LET has increased from 10 to 18 MeV/mg/cm^2.

6.7.2.5 Shift Register Circuit Modules and Voltage

Crain and co-workers measured cross sections of shift registers made with different circuit modules [Crain 2001]. Although the results seem to indicate different cross sections, the data can be fit with a single limiting cross section and width of curve for several devices. The calculated limiting cross section is 7.2×10^{-6} cm^2. The 3% changes in the limiting cross section produce a 10% increase in the sum of the squares. We used a linear fit and R^2 for the combined fits is 0.975. The results are shown in Figure 6-37. The only change between the curves is the critical LET.

6.7.2.6 Variation with LET at Several Frequencies

Marshall and co-workers have investigated the effects of frequency on the upset cross section in very fast circuitry [Marshall 2004]. Figure 6-38 shows their results. The critical charge decreases linearly with frequency. This is a linear fit and R^2 is 0.945.

Heidergott and co-workers have measured the event cross sections in high speed ADCs [Heidergott 2001] (not shown). In this case, the critical charge decreases with increasing voltage and frequency. R^2 is

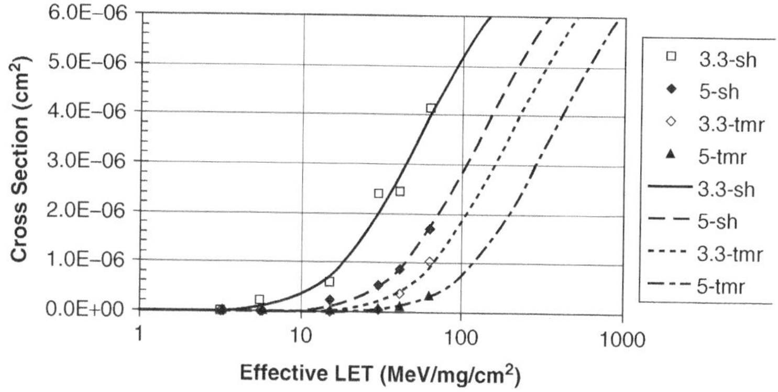

Figure 6-37 Two different types of shift register are fit with the same limiting cross section (7.2×10^{-6} cm^2) and shape parameter (1.0). The only shifts are in the critical LET L_c. The critical LET again increases with voltage. On a linear LET scale these curves would appear to have different limiting cross sections.

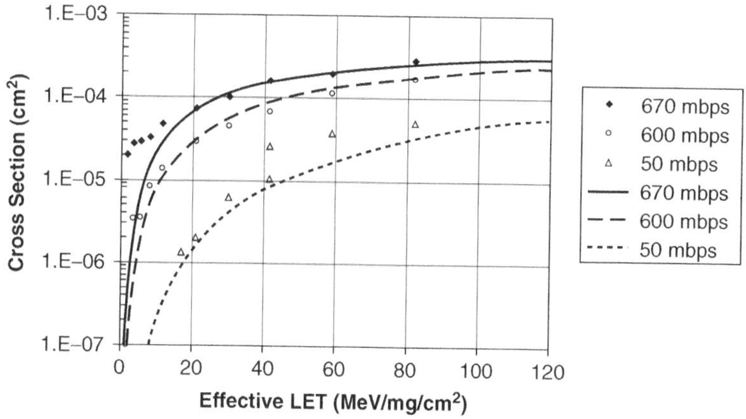

Figure 6-38 Effects of frequency. The data is described by the lognormal function with $cs = 4.2 \times 10^{-4}, s = 1.1, L_c = 449 - 0.58F$. The critical charge decreases linearly with frequency.

0.986. (For there to be a smooth variation with voltage, one pair of voltage labels must have been interchanged in their figure.)

Note that our fits with good R^2 are describing the data. They are not explaining it. The results of these last two cases may have been explained by Buchner [Buchner 1996]. As circuit clocks become faster, the circuit's SEU cross section will increase until the time of one clock cycle becomes comparable to the charge-collection time, at which point the cross section should saturate because the nodes will be sensitive all of the time. The saturation cross section will be determined by the geometry of the device.

This effect is shown in Figure 6-39 for the fits of Figure 6-38, where we plot the cross section directly as a function of frequency. Data that follows this type of curve has been shown previously in an experiment by Reed, Buchner, and co-workers [Reed 1996]. The cross section starts increasing nearly linearly with frequency, but then rapidly rises to saturation. Alternatively, we could say that the critical LET is decreasing or that the duty cycle is increasing.

This explanation does not explain the data in Section 6.7.1, which has the cross section decreasing with increasing frequency. In general, the analysis that we are performing needs to be used in conjunction with other tools for a complete explanation of the data.

The examples of frequency dependence that we have studied are consistent with constant limiting cross sections as shown by good R^2. In some types of combinational logic, the number of strikes that can contribute transients that lead to upset in a latch may increase with frequency. In this case the apparent cross section may become larger than the area of the latch. There are a plethora of different devices

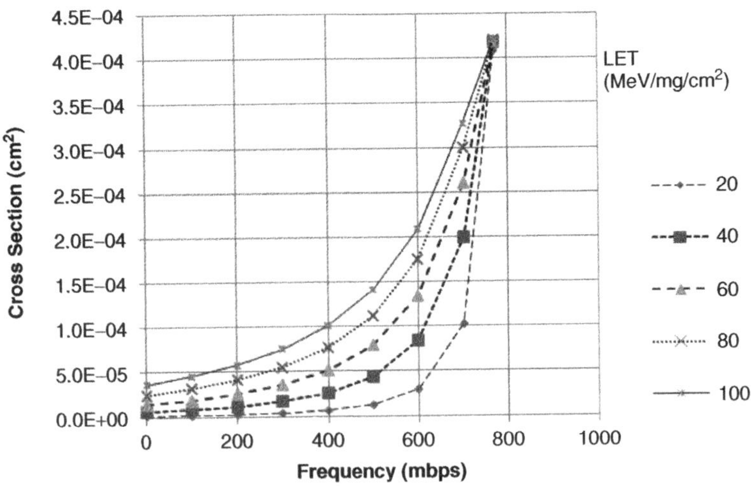

Figure 6-39 The fits of Figure 6.38 plotted as a function of frequency for various LETs. The curves start out nearly linear, but then rise to the saturation cross section at highest frequency.

that might show frequency dependence and we are presenting one tool that can be used to help understand the effect in many devices.

6.7.2.7 Variation with Frequency at Several LETs

Reed, Buchner, and co-workers have investigated the variation of cross section with frequency at several LETs [Reed 1996]. This data set is again fit with five parameters and $R^2 = 0.85$. The fit has an inverse frequency term that helps fit the high LET, low frequency portion of the data and which makes approximately a 5% improvement in R^2. These fits are shown in Figure 6-40.

6.7.2.8 Influence of Single Event Transient Pulse Length

Benedetto and co-workers measured the single event transient (SET) cross section as a function of transient pulse length using a variable temporal latch [Benedetto 2005]. They obtained the family of curves shown in Figure 6-41, which they fitted using six different Weibull functions (24 parameters)! We describe the family of six curves using a single

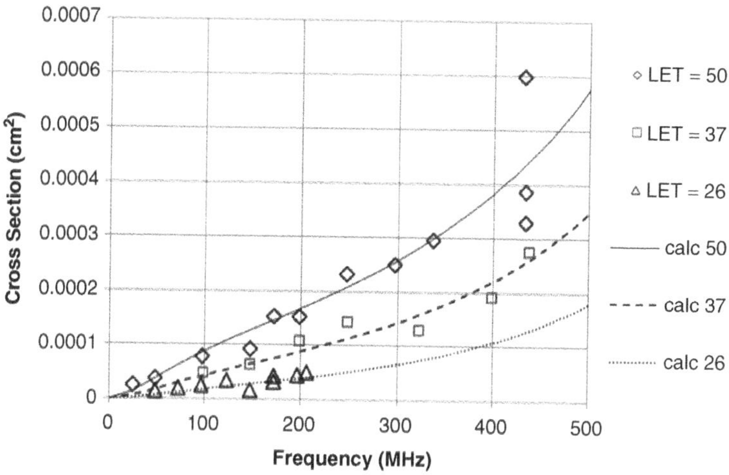

Figure 6-40 Effects of frequency at various LETs. The data are described by the integral lognormal function with $cs = 6.6\text{e}-3, s = 1.16, L_c = 466 - 0.56F + 2.3\text{e}+4F^{-1}$. The critical charge decreases linearly with frequency, but has an inverse frequency term that is effective at low frequencies.

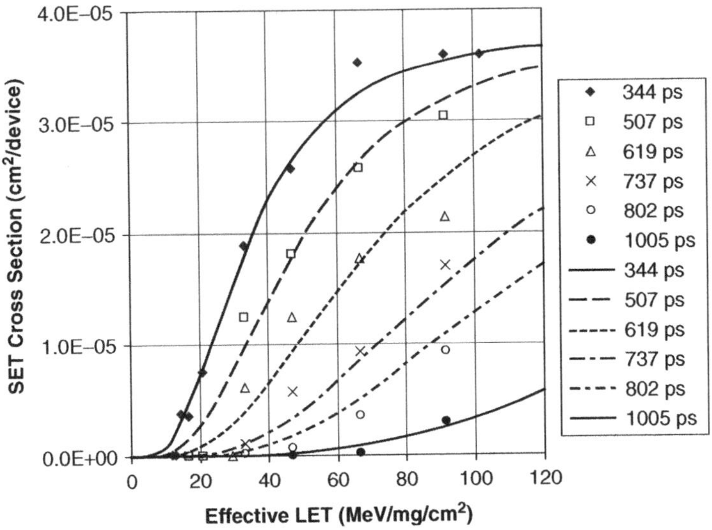

Figure 6-41 Heavy ion SET cross section as a function of internal delay. The data is described by a lognormal distribution with parameters $cs = 3.72 \times 10^{-5}$, $s = 0.8$, $L_c = 71.4 - 0.24D + 3.93 \times 10^{-4}D^2$. R^2 is 0.975. The space SET rates will vary as the fourth power of the delay.

lognormal distribution and five parameters. The fit is linear and R^2 is 0.975. The critical charge varies quadratically with the delay time.

6.7.2.9 Effect of Total Dose Irradiation on Single Event Hardness

J. R. Schwank and co-workers have investigated the effect of total dose on the SEU cross section [Schwank 2006]. They found opposite effects for checkerboard (CB) and checkerboard complement (CBn) patterns. Figure 6-42 shows their results. There is not adequate data for a single definitive fit, as they did not continue to high LET. We assume that the curves have the same limiting cross sections and shape. A linear fit to the set of curves produces a limiting cross section of 1.0 and very high values of L_c. A log fit produces a limiting cross section of 0.1, below the high cross-section points. The fits shown are for a log fit while forcing the curves through the high cross-section points. It is clear in this data display, independent of fit details, that the CB pattern has the critical charge reduced by a factor of 3, while the complement pattern has the critical charge doubled. $R^2 = 0.970$ for this set of curves.

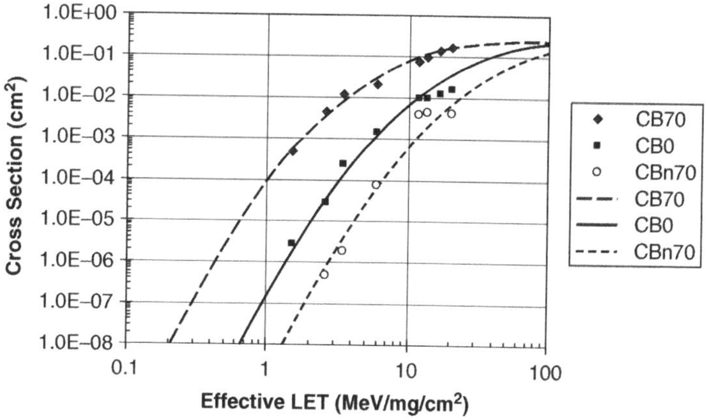

Figure 6-42 The effects of total dose exposures of 70 krad on 4 Mb memories with checkerboard (CB) and checkerboard complement (CBn) memory patterns. $cs = 0.25, s = 0.8$, and $L_c = 14.6$, 46.2, and 93. $R^2 = 0.97$ for this set of curves.

6.7.2.10 Single and Dual Port RAMs

Giot and co-workers measured the cross section on similar single and dual port SRAMs [Giot 2007]. This is a situation in which there is an actual change in area in adding the additional gates for the dual ports. Their results are shown in Figure 6-43. They observed that the relative area of the two devices varied significantly with LET. However, if we fit the data, assuming that there is no change in critical charge and only a change in area, we obtain SP $cs = 1.2, cs$ ratio $= 1.5, L_0 = 1.22, w = 2.08$, and $s = 0.61$. R^2 for this fit is 0.96 and the dosimetry standard deviation is 25%. There is no variation with LET, but there is a 50% change in the cross section. This large dosimetry error leads to the wrong conclusion unless carefully analyzed.

6.7.2.11 Sensitivity Patterns

Although we strongly recommend that data be displayed using either a log-log, or linear-log display, some may continue to use a log-linear display. In that situation, they should examine the data patterns. Figure 6-44 shows how the patterns are different for changing limiting cross section and for changing L_c. The previous examples have shown the families for changing L_c in the other displays. We obtain one pattern for a set of curves with the limiting cross section changing by factors of 5, but L_c stays constant. These are indicated by the series

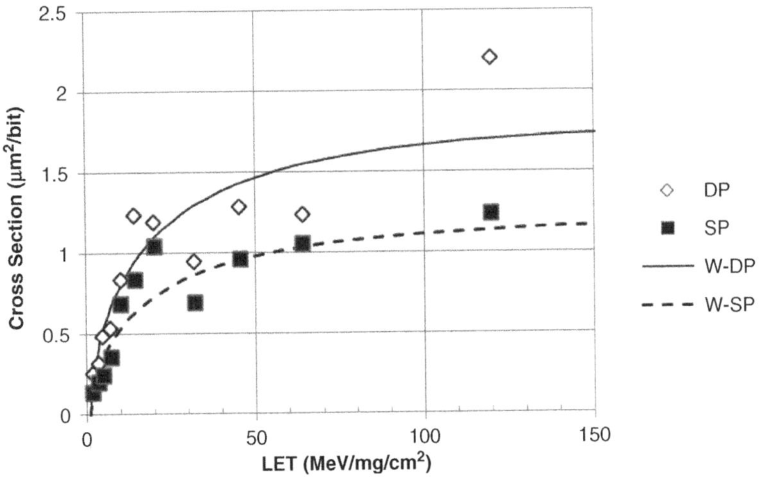

Figure 6-43 Single port (SP) and dual port (DP) SRAM single event cross section as a function of LET. This is an example of variation of cross section instead of critical charge.

of lines. We obtain another pattern for a set of curves with the median (L_c) changing, but constant limiting cross sections. These are indicated by the curves designated with symbols. For this set of curves, there is not a well-defined shoulder and the threshold appears to be shifting. Figure 6-45 shows the same data plotted on a log-log display. In this case the two families are distinct.

6.7.2.12 Variation of Upset Rate with Critical LET (Device Parameters)

The changes of device parameters have all been characterized by a change in the critical LET. We can use the figure of merit to quickly relate this to the corresponding changes of upset rate, independent of orbit.

$$FOM = \frac{\sigma_{HL}}{L_{0.25}^2}\left[\frac{(\text{MeV/mg/cm}^2)^2}{\text{cm}^2}\right] \tag{6.6}$$

$$R = C \times FOM \tag{6.7}$$

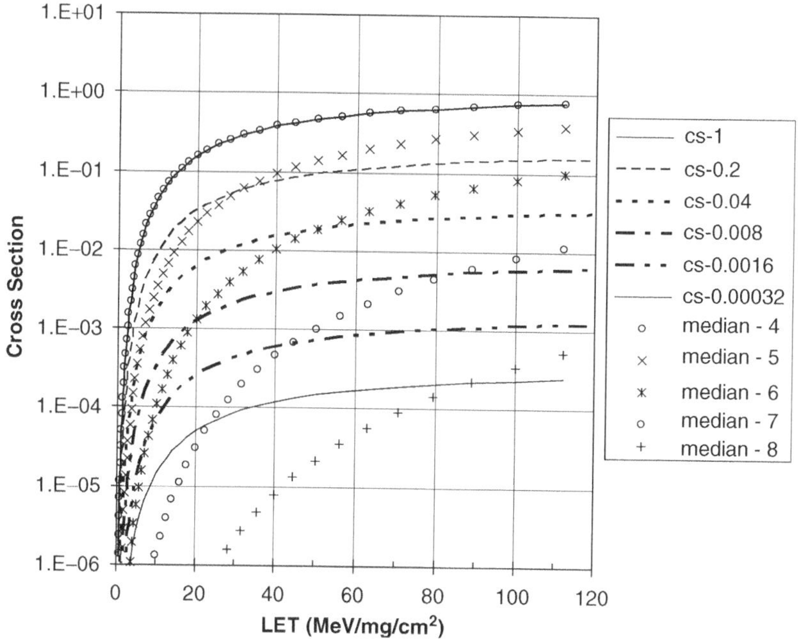

Figure 6-44 Limiting cross section and changing L_c patterns. The lines indicate the cross-section curves for devices with the same standard deviation ($s = 1$) and L_c, but with the cross section decreasing by factors of 5, starting at one. The symbols indicate the cross-section curves for devices with the same limiting cross section and standard deviation, but with $m (= \ln L_c)$ having successive values of 4, 5, 6, 7, 8. The patterns have different shapes, and the L_c variation shows increasing thresholds.

Consider that L_c is found to be proportional to a parameter F. If $L_c = aF$ then $L_{0.25} \approx F$. From this we can conclude the rate varies from an initial rate $R(0)$ as

$$R = \frac{R(0)}{a^2} \tag{6.8}$$

6.7.2.13 Parametric Studies Conclusions

It is not instructive to plot SEE cross-section data on a log cross section versus linear LET plot as that distorts the data and misleads the analyst. Although it is not strictly incorrect to fit the data with the

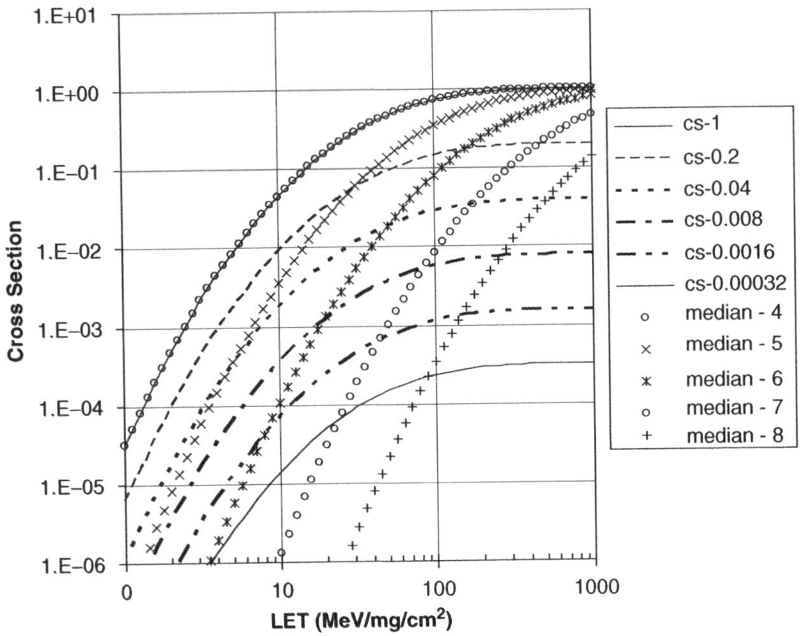

Figure 6-45 The curves of Figure 6-46 plotted on a log-log scale. In this case the two families of curves are clearly delineated and distinct.

Weibull function, much more can be learned about the data and the test device by using the lognormal distribution. When studying families of cross-section curves resulting from parameter changes, one should use the lognormal distribution, as it is most likely that the only parameter changing is the critical LET. It is also quickly obvious if the study is changing the shape of the curve or the limiting cross section.

These variations are obscured if the data is plotted on a log-linear scale, so the data should be examine using logarithmic horizontal scales. We have examined twelve different papers that studied various aspects of SEE sensitivity. The papers were important basic studies of physical mechanisms of SEEs, and examined the impact of operating conditions, pattern dependence, and so on. They plotted their data on log-linear plots and did not examine the lognormal nature of the data, so that they did not get the full benefit of their studies. They did not examine why their experiments should lead to changing cross sections rather than to changing critical charge. None of the papers examined the relationship between critical charge and their parameter changes.

We have reexamined results from these papers in terms of variation of critical charge. The critical charge for upset increases linearly with hardening resistance and voltage, and decreases linearly with temperature and frequency. Several other parameters have similar simple interactions with the critical charge. Use of the lognormal distribution for describing cross-section data enables a basic study of SEE mechanisms.

We examine the onset threshold region of the cross-section curve and show that this region has little impact on space upset rates. The apparent high sensitivity observed when changing the Weibull threshold is due to the fact that these variations change the shape and location of the entire cross-section curve.

6.8 INFLUENCE OF ION SPECIES AND ENERGY

There has been a continuing controversy since the work of Criswell and Koga about the influence of ion energy and ion type on SEU cross sections [Criswell 1984, Criswell 1987, Koga 1990]. There are a number of authors who say that they see no difference and that LET is the appropriate way of characterizing the beam. There have also been several papers claiming that LET is not adequate and that energy and ion species are important. (We are not discussing the papers that point out that effective LET is not appropriate for some geometries.)

The first studies of upsets at low and high energy accelerators that was very careful to use the same parts, test apparatus, and procedures was performed by Dodd [Dodd 1998]. Figure 6-46 shows one of their examples. It has been fit with Weibull parameters: $cs = 7.0\ \mu\text{m}^2, L_0 = 2.7$ MeV-cm^2/mg, $w = 16.9$ MeV-cm^2/mg, and $s = 1.7$. Note that in this display, two points that differ by a factor of 2 are still relatively close together. The standard deviations are: BNL 48%, GANIL 78%, and MSU 86%. They observed that the ratio of MSU data to BNL data at a LET of 24 varied depending on the MSU beam tune. These very large dosimetry errors are adequate to make the point that there is no appreciable difference between high energy and low energy data.

There is a recent paper using the high energy beams at Michigan State National Superconducting Cyclotron Laboratory (NSCL). They compared SEU cross-section results from BNL, TAASC, and NSCL (Figure 6-47). The results agreed very well [Ladbury 2004]. The data

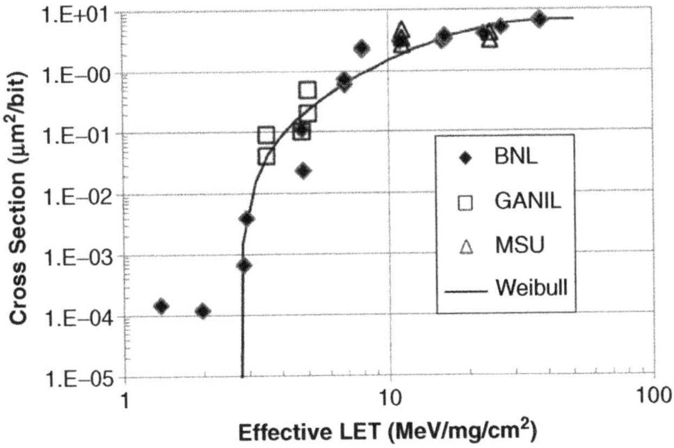

Figure 6-46 Low and high energy upset cross section data for Matra 6560E radiation-tolerant 1 Mbit SRAMs [Dodd 1998].

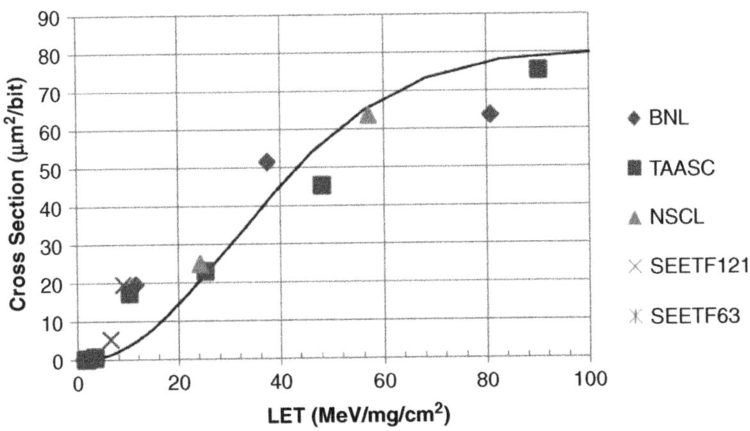

Figure 6-47 The same Matra HM65656 irradiated at TASC, NSCL, Michigan State University, and Brookhaven yields consistent cross-section versus LET curves over beam energies spanning a factor of 40 [Ladbury 2004].

is fit with $cs = 88.2\,\mu m^2/\text{bit}$, $L_0 = 1$ MeV cm²/mg, $w = 42.8$ MeV-cm²/mg, and $s = 1.1$. There is some spread at very low LETs, but the standard deviation is well behaved for the points above a LET of 10 MeV-cm²/mg. Table 6-1 shows the beam energies and standard deviations at the various facilities.

It appears that if one is restricted to ionization only, there is no difference in upset cross sections for high and low energy ions with

Table 6-1
Beam Energy and Dosimetry Standard Deviation Using Several Beams to Test the Same Device

Facility	Energy	Standard Deviation (%)
BNL	4 MeV/amu	24%
TAASC	10 MeV/amu	26%
NSCL	60 MeV/amu	17%
SEETF	63 MeV/amu	24%

the same LET. These results also let us compare the dosimetry for the different beams.

Dodd has another careful experiment examining the cross section at high and low energies for the same parts [Dodd 2007a]. We show some of the results in Figure 6-48. There is a good overlay of the two sets of results. They are fit with cs = 45.8 μm^2/bit, L_0 = 11.2 MeV-cm^2/mg, W = 35.5 MeV-cm^2/mg, and s = 0.93. For points above a LET of 15 MeV-cm^2/mg, the BNL data has a standard deviation of 7.3% and the TAMU data has a standard deviation of 21%. They took a great deal of low cross-section data below threshold and showed

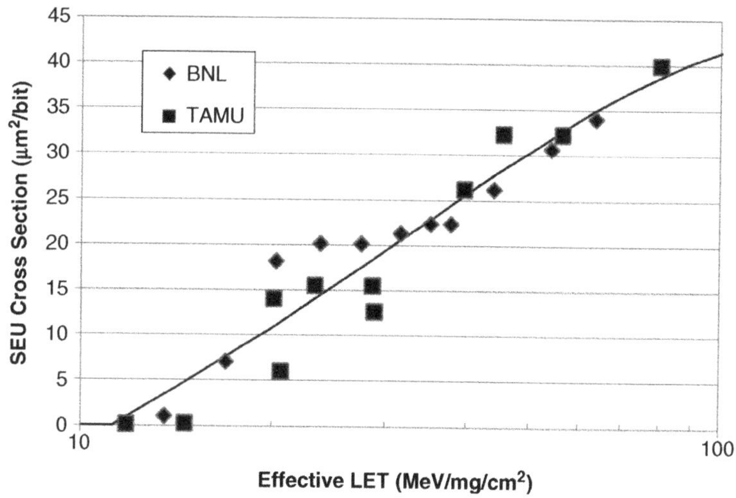

Figure 6-48 SEU cross sections measured for Block 3 of the Sandia 256 kbit sRAM taken with low energy (BNL) and high energy (TAMU) heavy ions [Dodd 2007a]. There is no difference between the two beams.

upsets from nuclear reactions, silicon elastic recoils, or beam impurities. Those results are not relevant for the present discussion. It is clear that the upsets from heavy ion ionization do not depend on ion or energy [Petersen 2010].

Reed shows that some of the results, not consistent with ionization, can be explained by introducing the events as due to secondaries from nuclear reactions [Reed 2007]. When one is using this data to calculate upset rates, one must do separate calculations for the two processes. This is discussed in Chapter 10.The physical modeling of the processes, such as that in the Monte Carlo Code MRED, can be important in separating these sources [Warren 2005, Reed 2007]. The apparent discrepancies appear to be either due to nuclear reactions or due to devices in which two bits must be hit to produce an upset.

6.9 DEVICE GEOMETRY AND THE LIMITING CROSS SECTION

6.9.1 Bulk CMOS

Dodd's work, displayed in Figures 6-21 and 6-22, showed that in his cases the saturation cross section was basically the drain area, although there could be a slight continual increase at higher LETs due to diffusion [Dodd 2001].

6.9.2 CMOS/SOI

Worley examined the limiting cross section for 2 μm SOI technology and found that it corresponded to the gate region [Worley 1990].

Dodd has carefully examined the limiting cross section for 0.6 μm CMOS/SOI RAMs [Dodd 2001]. The saturation cross sections are about 7 μm^2/bit for the 1M SRAM and almost 8 μm^2/bit for the 64K SRAM. The total sensitive gate area (assuming one sensitive n-channel and one sensitive p-channel gate per bit) in both SRAMs is the same: 0.64 μm^2/bit. On the other hand, the gate plus drain area per bit matches much more closely the measured saturation cross section 5.2 μm^2/bit for the 1M SRAM and 6.3 μm^2/bit for the 64K SRAM. The data in Figure 6-49 suggest that not only the gate regions of these SRAM are sensitive, but also the reverse-biased drain regions. This is in contrast with conventional wisdom (see references).

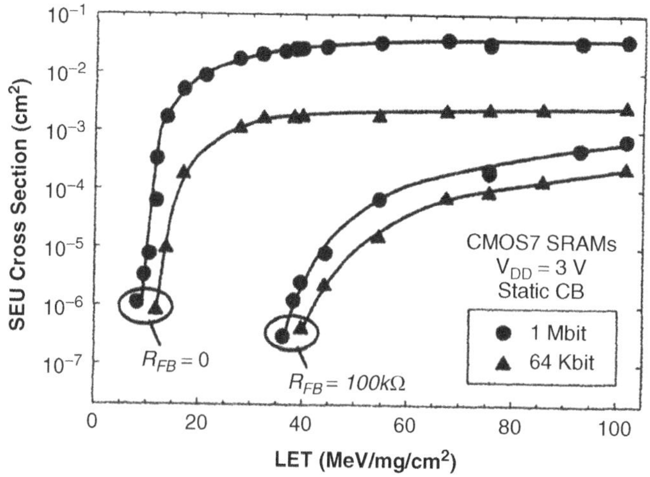

Figure 6-49 SEU cross section in 1M and 64K CMOS SOI SRAMs with and without feedback resistors [Dodd 2001]. Copyright © 2001 IEEE.

6.9.3 SRAMs

Koga carefully measured the cross sections of IDT 71256 SRAMs [Koga 1993b]. The cell cross section was 60 μm^2. This corresponded to the area of 64 μm^2 for the two drain areas.

It was originally assumed that the sensitive area was the gate (body) with dimensions determined by the gate width (W) and length (L) and silicon film thickness t_S. It has now become clear that the area extends over the entire transistor. Musseau and co-workers performed a charge collection experiment in which the cross section for the peak was much larger than the gate area [Musseau 1997]. This and the wide spread in charge collected are consistent with the later studies described in Sections 6.3.3 and 6.3.4. If only the gate was involved, the size of the track would be important in submicron devices.

SEU cross-section data have been reported for 0.25 μm commercial SOI and bulk SRAMs fabricated at IBM with the same basic layouts [Brothers 1997]. The measured saturation cross section in the SOI ICs was about 1/10 that of the equivalent bulk ICs, a reasonable ratio of gate to drain area [Dodd 2001]. Unfortunately, no transistor level design information was given and the results are not completely understandable with the limited information available [Dodd 2001, Petersen 2007].

6.10 TRACK SIZE EFFECTS

In the interaction of energetic ions with matter the energy is primarily transferred to target electrons. The high energy primary electrons (also called delta rays) decay in an electronic cascade, and finally produce dense plasma of electron–hole pairs. This ion track is located along the ion trajectory, and its radial extent, determined by the range of the most energetic delta rays, may reach some hundreds of nanometers [Xapsos 1992, Musseau 1996b].

Martin and co-workers used their code TRIPOS-E to study the electron transport around the ion track [Martin 1987]. They found that the radial part of the electron track was exponential. The track radius can exceed 0.5 micron. If the device separation is only 1 or 2 microns, resulting double-bit upsets can be several percent of the number of single bit upsets. We will see in the next section that diffusion can lead to multiple-bit upset at much larger distances than this.

In very small silicon volumes there may be a difference between the energy lost by the ion and the energy actually deposited [Xapsos 1992, Akkerman 2002b, Reed 2008]. Part of the energy deposited is carried outside the volume by the escaping delta rays. There is a corrective factor depending on the silicon thickness and the ion energy. However, a paper by Dodd concluded that the track structure is not important [Dodd 1998]. They calculate that the central cores of the track structures are nearly identical for low and high energy ions. Three-dimensional simulations confirm that charge collection is similar in the two cases. This is partially because the sensitive volumes do not have an abrupt boundary, and the initial track (effectively a line source) expands very rapidly by radial diffusion. Figure 6-50 shows the rapid decrease of the deposited dose as a function of radius for 100 MeV protons [Kobayashi 2004]. Heavy ion tracks have larger sizes (Figure 2.1) but the same rapid falloff.

Another argument that track sizes are not important for SEU is based on the observations that the total SEU cross sections for very small feature size devices (0.18 μm) correspond to the entire size of the memory cell [Hafer 2006b]. The parasitic bipolar interactions inside the cell mask the effects of the track. Sections 6.6 and 6.8 give other examples indicating that the sensitive region is large compared to the feature size.

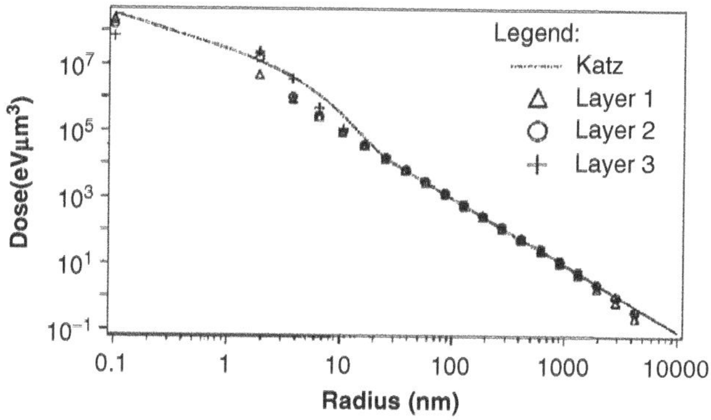

Figure 6-50 Deposited dose as a function of radius for 100 MeV incident protons as calculated by MRED and by the analytic theory of Katz [Kobayashi 2004, Reed 2008]. Copyright © 2008 IEEE.

6.11 CROSS-SECTION CURVES AND THE CHARGE COLLECTION PROCESSES

A large number of authors have investigated the interplay of the direct (funnel) and the diffusion charge-collection components of the upset process. Many of these are discussed in Section 6.6. The conclusion seems to be simple. There is a prompt fast charge collection immediately after the ion impact. The process is essentially multiplicative so that low LET ions that appear to deposit less than the critical charge are causing upsets. This is the high efficacy component at low LETs. There is then a slower diffusion component. For high LET ions this leads to the majority of the charge collection. The overall cross-section limit is then determined by the limits of device geometry.

Another way to look at the competition between direct and diffusion charge collection is to think of the device geometry. In devices with a very constrained vertical charge-collection region, there is little opportunity for diffusion, and direct collection dominates. At the same time, the vertical limits also constrain the direct component, so the efficacy never reaches large values. In devices with little vertical constraint the efficacy can be much higher, but the large relative volume leads to the diffusion component producing the major amount of charge collection. We can see these effects by examining the cross-section curves in terms of efficacy and diffusion.

Note that the devices with limited depth appear to have a narrower lognormal distribution than the devices with unlimited depth.

Dodd has shown that in very fast devices, such as unhardened CMOS, upset is primarily due to the fast transients. Charge collection by diffusion plays an insignificant part. However, in SRAMs with hardening with feedback resistors, upset takes longer to develop and diffusive charge collection sustained over time plays a much bigger role [Dodd 1997, Edmonds 1997].

6.11.1 Efficacy Curves and the Charge-Collection Process

The regions of high efficacy are regions of large amounts of charge collection. These correspond to the regions in which there is a large funnel contribution. The funnel effect is very strong, relatively short lived, and comes from only a small portion of the total area and therefore contributes a relatively small fraction of the total cross section.

Figure 6-51 shows the gain curves for SOS and EPI devices. The curves are for devices with a large variety of sizes and depths, but show nearly identical shapes. Figures 6-52 through 6-55 show the transition in the low cross section–low efficacy region from CMOS/SOI

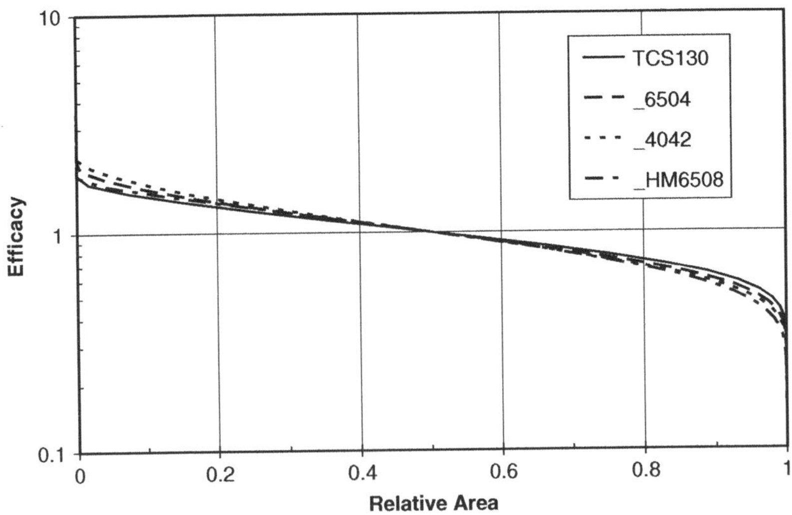

Figure 6-51 Charge-collection gain curves for SOS and EPI devices.

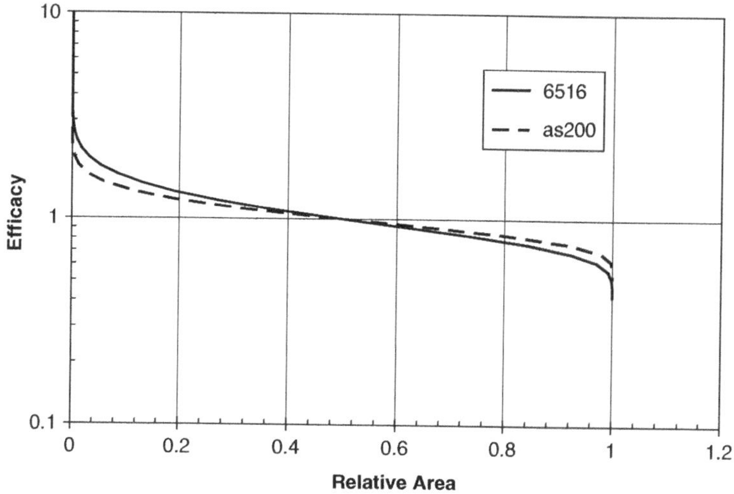

Figure 6-52 Efficacy for CMOS devices.

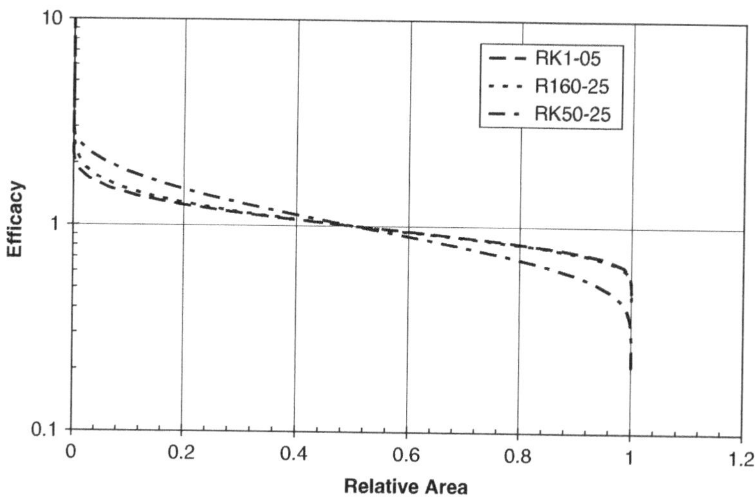

Figure 6-53 Hardened CMOS using resistive hardening.

devices where there is a very strong effect to bulk CMOS devices with moderate funneling contribution to MOS devices with minor funnel contributions. In general, the curves for a given family of devices resemble each other very closely. Other examples are given in [Petersen 1996c].

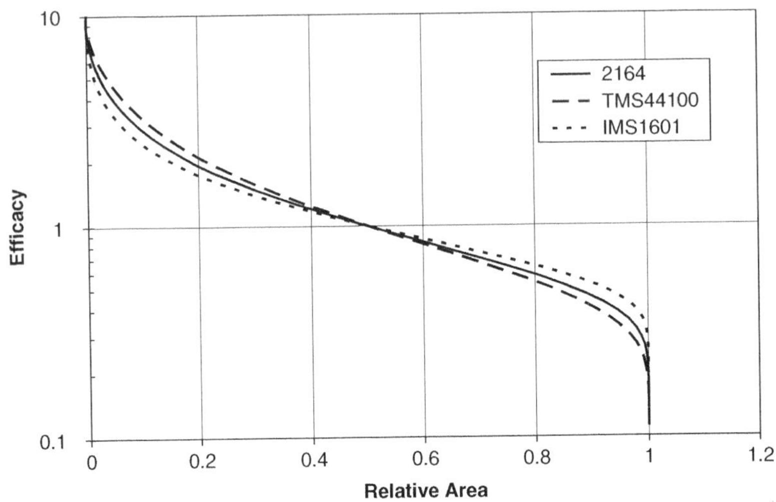

Figure 6-54 MOS devices.

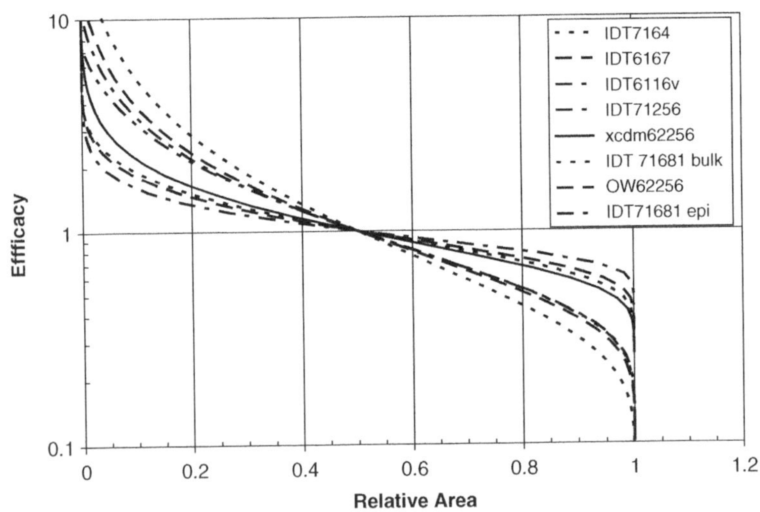

Figure 6-55 Resistive hardened MOS devices.

Section 6.11.4 shows an example of the charge-collection processes for a DRAM in which the funnel is very important, but not long lasting [Zoutendyk 1990].

The RMOS and DRAM devices are expected to have large amounts of charge collection by diffusion [Smith 1995]. This is an example of a process that leads to a low onset threshold at low cross sections, but does not increase or decrease the charge deposition volume. We observe two aspects in the characteristic curves. One is the amount of variation in the high and low gain regions. This corresponds to the tail in the Weibull distributions described by the shape parameter. The other is the slope of the center region. This basically corresponds to the width of the corresponding differential distribution. The concept of width becomes more pertinent when we introduce the lognormal distribution. The gain versus area plots used here compare and contrast the gain in various technologies.

The data normalizations that we use here have some resemblance to the ones discussed by Ecoffet [Ecoffet 1992]. He suggested that his data indicated a common shape. The approach here allows a more detailed examination of the shapes of the cross-section curves. This examination indicates that there are differences depending on technology.

6.11.2 Inverse LET Plots and Diffusion

Smith and Edmonds have suggested that the diffusion component of the cross-section curve is described by the exponential function [Kirkpatrick 1979, Smith 1992, Edmonds 1996]. Smith included a threshold term, while Edmonds did not. The diffusion terms, without the threshold, can be identified as a straight line on a plot of cross section versus inverse LET. Figure 6-56 shows the cross section curve for the IDT 71256 RMOS device. This has a very long internal time constant so the charge collection is dominated by diffusion. This is indicated by the long straight line with zero threshold. On the other hand, Figure 6-57 shows the results for several SOS and EPI devices. For convenience, the device results have been shown by lines indicating the lognormal fit to the data. In this case the exponential function (lines with symbols) has a threshold in the range of 25–50 MeV/mg/cm^2, the exponential limiting cross section is 25% to 50% high, and the exponential behavior does not extend below 25% to 30% of the limiting cross section. The exponential threshold is significantly above the Weibull threshold. In this case the device limiting cross sections are not described by the exponential function. For these devices there are clearly three different cross-section regions.

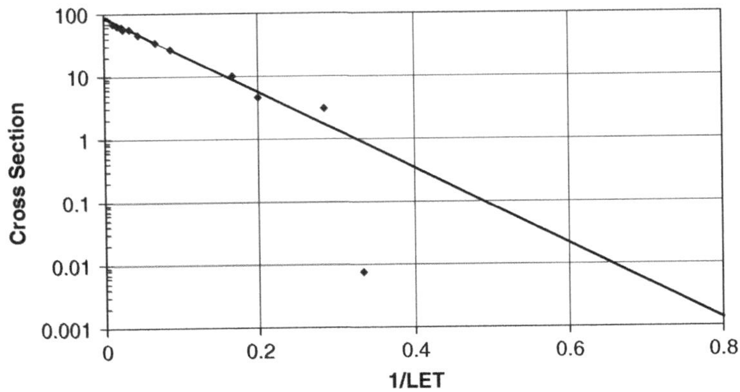

Figure 6-56 The cross section for the IDT 71256 plotted on an inverse LET scale. The straight line indicates that diffusion dominates the cross section down to approximately 2% of the limiting cross section.

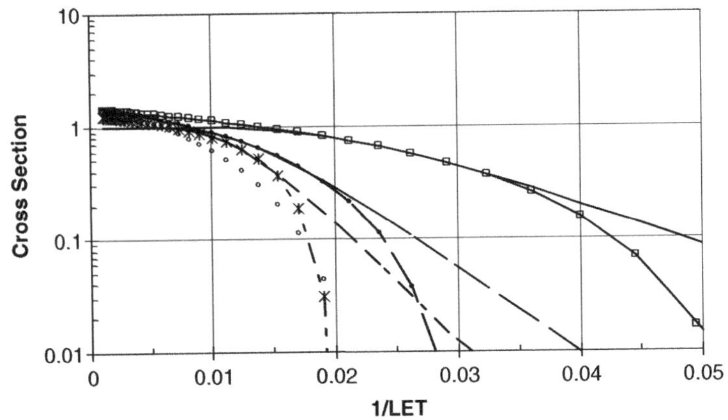

Figure 6-57 Cross-section curves for the TCS130, 6504, 4042, and HM-6508 SOS and EPI devices. The data, summarized by the solid lines, agrees with the exponential fit (lines with symbols) only for the top 70% of the limiting cross section. The cross sections are normalized so that the experimental limiting cross section is one.

6.12 SINGLE EVENT MULTIPLE-BIT UPSET

There are several types of single event multiple-bit upsets (MBUs) [Dodd 1999]. One type occurs when an incoming particle passes through adjacent cells. This was initially modeled by Bradford [Bradford 1980a, Bradford 1980b]. A complete analysis was given by Pickel in 1984 [Pickel 1984] and has been rediscovered several times since. This can occur for grazing particles and can produce a number of complicated patterns by intersecting various combinations of sensitive regions producing multiple upsets [Musseau 1996b] We will call these geometrical MBUs. The second type occurs when diffusion of charge to closely spaced junctions can upset more than one bit (diffusion MBUs) [Zoutendyk 1989, Dodd 1994]. This can occur in both SRAM and DRAM cells. A third type occurs in very small feature size devices where the ionization cloud overlaps two or more sensitive regions (overlapping MBUs). A fourth type is possible even if all of the previous modes have been eliminated. That is the statistical mode in which two random hits occur in different bits during a given time period (statistical MBUs). This is of concern when they occur in a single word with error detection and correction (EDAC).

The effects of MBU are typically alleviated by a combination of error-correcting codes that work on a word-by-word basis and layout rules that prevent physically adjacent bits from belonging to the same word of memory [Dodd 1999]. Still, single-word multiple-bit upsets can occur and pose a substantial threat to system integrity [Koga 1993b]. Multiple-bit upsets have been observed in on-orbit spacecraft data [Blake 1986, Campbell 1991, Underwood 1992].

There can also be multiple upsets due to upsets in peripheral circuitry. Occasionally long strings of bits upset due to "row fails" [Bisgrove 1986]. This type of error may be difficult to handle with ordinary error detection and correction approaches [McDonald 1989].

6.12.1 Strictly Geometrical MBUs

Figure 6-58 shows the geometry for a dual upset. Refer to the original publication for the complete geometrical analysis with and without the funnel [Pickel 1984]. The paper also lists a computer code for doing the calculations. A later treatment is available from Edmonds [Edmonds 1989]. A different treatment is given by Smith [Smith 1992].

The calculation of single hit geometrical multibit errors for bits that are near each other is complicated, as one has to consider the path lengths, energy deposition, and upset probability for each bit. The calculation is much simpler when the bits are widely separated. Consider the situation when we have the calculated error rate for a single bit,

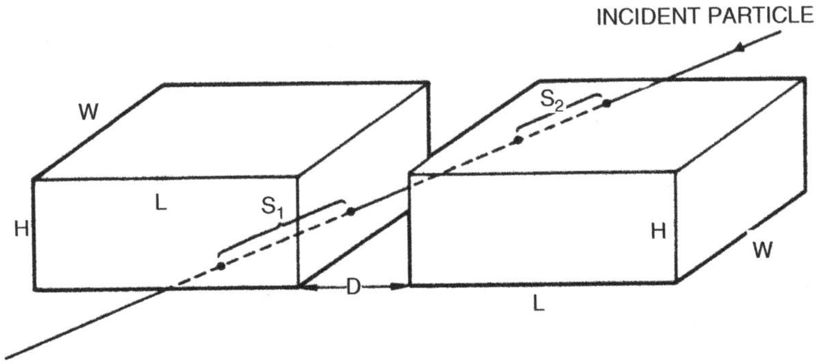

Figure 6-58 A high energy particle intersects two identical depletion regions of dimensions W, L, and A and separation D. Points of entry into and exit from the depletion regions are indicated by the dots on the path. The parts of the path that intersect the depletion regions have lengths S1 and S2, respectively [Pickel 1984].

and want to calculate the possible dual multiple-bit upset rate. This is of interest when we have single error correction and double error detection (SECDED). Consider that we know the calculated error rate of bit B and want to know the rate for combined bits A and B. We assume that the bits are widely separated to reduce the MBE rate, so the geometry is simpler than that of Figure 6-58 (see Figure 6-59). D is large enough so that the effect of corner hits is negligible.

The rate for bit B (R_B) is determined from the isotropic fluence over 4π steradians. However, the error rate for combined A and B is determined by the passage of ions through both bits. Therefore, only a fraction of 4π is available. The ions may enter from either the right or the left so we need to count both cases.

$$R_{A+B} = \frac{2 \times W \times H}{4\pi \times D^2} \times R_B \tag{6.9}$$

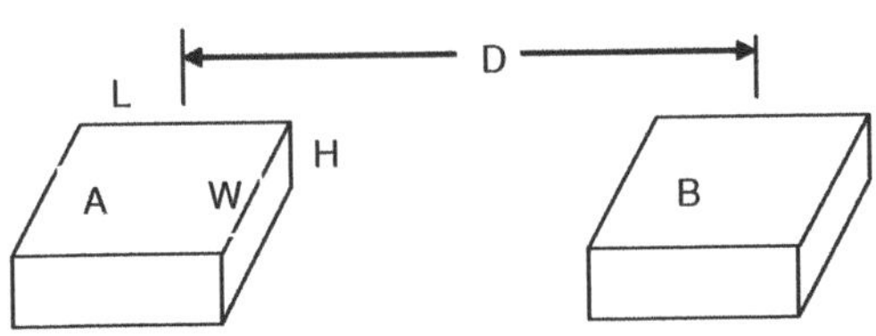

Figure 6-59 Geometry for dual multiple-bit errors (MBEs). We know the error rate for bit B, but the ion must pass through both sensitive volumes.

The height H is the same height that was used to calculate the upset rate R_B in device B. There are sometimes questions about the actual height. We will assume that the height has been determined from an examination of the discontinuities in cross-section data. For the present we will assume that $L = W =$ square root of the imiting cross section. We assume the RPP (rectangular parallelepiped) approximation so the shape does not depend on the angle of the incident particle. The same volume is always involved in the charge collection. The fact that volume B is small and thin is already included in the upset rate R_B. This approach appears to be same as that used by Koga in 1993 [Koga 1993a, Koga 1993b].

Figure 6-60 shows a calculation of the rates for a situation in which the bits have been widely separated in order to avoid geometrical multibit errors (MBEs). We reduce, but do not eliminate, the MBU errors. The results apply to any application for which MBE is important.

Multiple upset can occur due to the passage of an ion through a number of memory cells. Figure 6-61 shows a group of upsets observed in a space experiment outside the proton radiation belts [Buchner 2000]. Dreute has shown very long SEU tracks with high energy ions tilted in and near the plain of the device [Dreute 1994]. The number of upsets depends on the track length in the epilayer, which

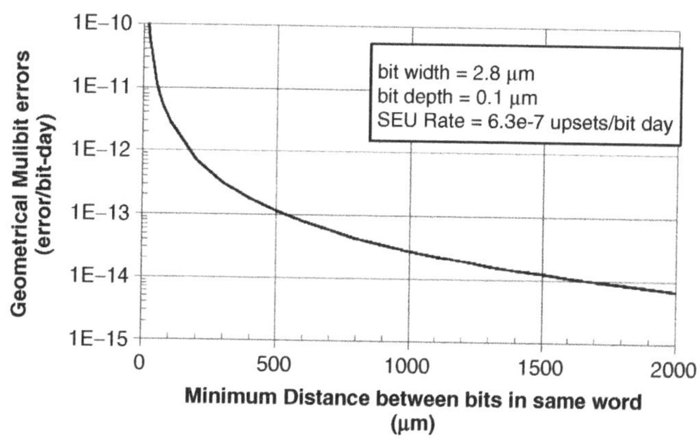

Figure 6-60 The geometrical multibit upset rate as a function of the minimum distance between bits in an EDAC word. If the distance is 1600 um, the rate is 1.1×10^{-14} MBE/bit-day, while if the distance is 16 μm, the rate is 1.1×10^{-10} MBE/bit-day (for our approximation).

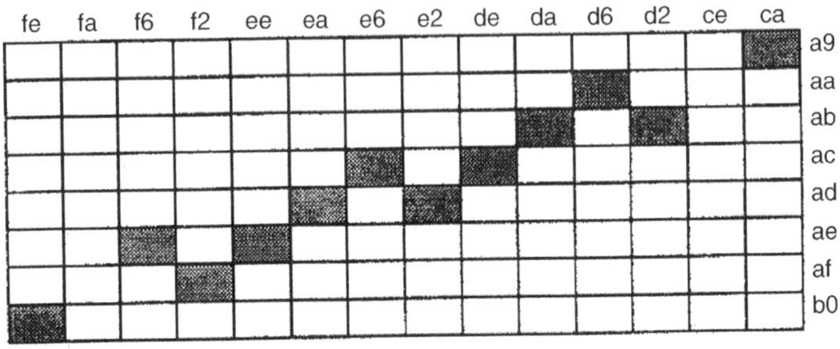

Figure 6-61 Line of hits produced in an initial checkerboard pattern outside the proton belts. The hits occur at an angle to the rows and columns [Buchner 2000]. Copyright © 2000 IEEE.

varies as tan φ, so there is a rapid increase as the angle increases from 85 degrees to 90 degrees. The number of upsets depends on the cell size and the depth of the epilayer. In the 4π environment in space, there are a significant number of particles that enter near and in the plane of the device, so these effects need to be considered. Approximately 9% of the particles incident on a sphere enter between 85 degrees and 95 degrees.

The long tracks can be either continuous or chopped, depending on the topology, and the memory pattern is continuous or checkerboard [Musseau 1996b]. These authors also found evidence that besides the normal drift and diffusion contributions in their CMOS devices, there may be some parasitic bipolar current between adjacent cells. They observed four adjacent bits upsetting when using very high energy laser pulses at normal incidence, but not in their heavy ion testing.

Of particular interest is the "single-word multiple-bit upset" SMU. This type of upset clearly depends on the physical arrangement and size of the memory cells. In particular, as cell sizes shrink and cell density increases, the prevalence of this type of upset can be expected to rise [Koga 1993b]. Koga considered several types of high density SRAMs with very similar SEU sensitivity. Their SMU sensitivity had large variation due to the variation of the spacing between logically adjacent bits.

6.12.2 Proton Induced Multibit Upsets

Geometrical upsets have also been observed with protons and are becoming increasingly common as devices shrink. They depend on proton energy, cell geometry, and the critical charge. They ordinarily have a strong angular dependence, as would be expected [Bisgrove 1986, Campbell 1991, Reed 1997, Buchner 2004, Ikeda 2005, Tipton 2006].

6.12.3 Dual Hits for Single-Bit Upset

It has now become common practice to design devices that are immune to single-node charge-collection mechanisms [Calin 1996]. However, with reduction in feature sizes, single-node hardening can be undermined by multiple-node charge-collection processes [Black 2005]. The space error rates for this type of device can be calculated by extensions of the geometrical two hit analysis discussed above [Edmonds, private communication]. Another very powerful tool for this type of problem is Monte Carlo analysis. The group at Vanderbilt has developed a software tool called MRED (Monte Carlo Radiative Energy Deposition) [Reed 2006]. This tool has been used by Warren and co-workers for the analysis of the upset rate in a hardened design latch [Warren 2007a]. The latch can be very sensitive for ions at some angles near the plane of the device, and the calculations match that behavior well.

Liu and Fulkerson have also modeled the situation where hits to two different areas are necessary for a single upset [Liu 2004, Fulkerson 2006, Liu 2006a].

6.12.4 MBU Due to Diffusion in DRAMs

Zoutendyk and co-workers have extensively studied multiple-bit upsets in DRAMs [Zoutendyk 1987a, Zoutendyk 1988, Zoutendyk 1989, Zoutendyk 1990]. Normal incidence ion tracks in a semiconductor chip produce an electron–hole pair charge distribution along the track. If the ion track intersects a charge-collection junction, prompt charge collection occurs by electric-field funnel-aided transport along the track to the junction. However, for a long track the electronic field is restored to its initial configuration (i.e., the funnel collapses) before all of the charge is collected. The remaining charge is transported laterally by diffusion as depicted in Figure 6-62. If an ion track does not intersect a junction, nearly all of the charge is transported laterally by diffusion. That portion of the charge that diffuses toward the surface of the chip may be captured by other

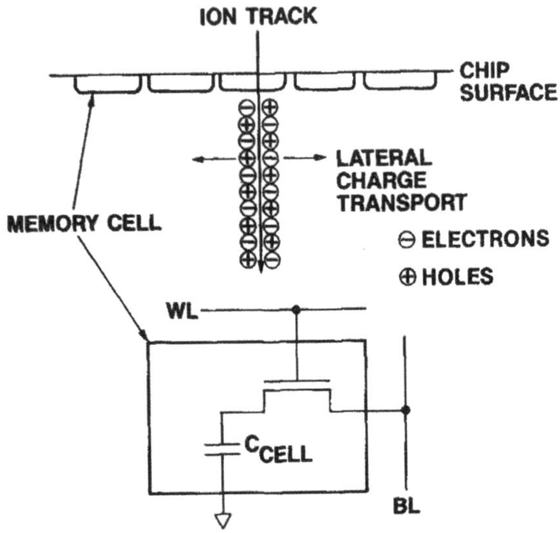

Figure 6-62 Lateral charge transport from an ion track in a DRAM cell. C_{CELL} = data-storage capacitor, WL = word line, BL = bit line [Zoutendyk 1988]. Copyright ©1988 IEEE.

charge-collecting junctions. Lateral charge transport leads to the proliferation of SEUs in the integrated circuit (IC) chips [Zoutendyk 1988]. Figure 6-63 shows a simulation of transport with a change in the drift field and then diffusion to the surrounding area [Zoutendyk 1990].

Figure 6-64 is a map of a bit-error cluster caused by a normal incidence Kr ion (E-295 MeV, LEYT = 40 MeV/mg/cm^2) during a single data scan cycle. This data is for a 256K DRAM. The circles are drawn to emphasize the approximate circular symmetry seen in the clusters. Similar patterns have been observed by a number of experimenters. [Makihara 2000].

Figure 6-65 shows the event cross section of a NEC 16 Mbit DRAM as a function of LET. In this case you want to plot the event cross section, not the SEU cross section. In the following sections where we discuss upset rate predictions, we are really talking about event rates. Table 6-2 shows the corresponding upset multiplicity, normalized to 1000 upsets at each LET. At the lowest LETs the upsets correspond entirely of SEUs, whereas at the highest LETs the upsets were exclusively MBUs. Similar tables of multiplicity are given by other authors [Koga 1991].

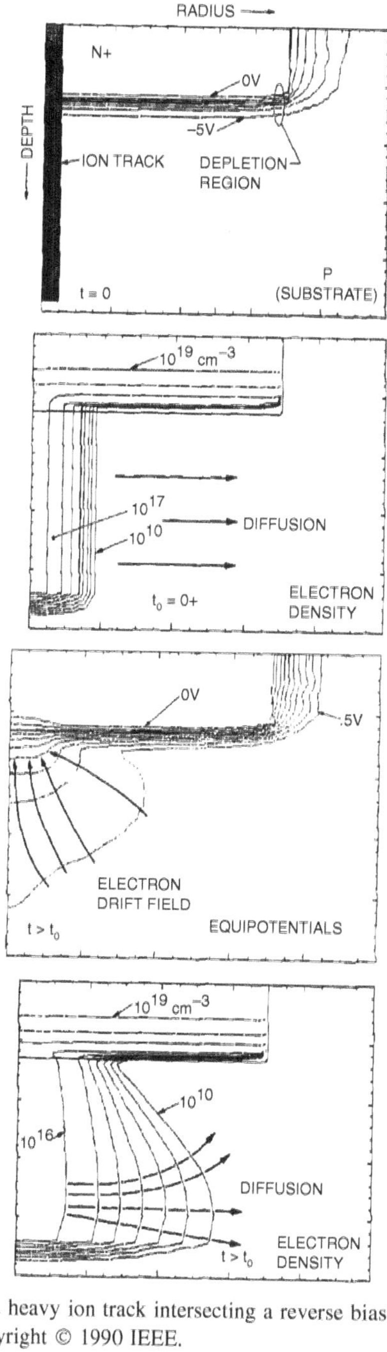

Figure 6-63 Simulated heavy ion track intersecting a reverse biased n^+/p junction [Zoutendyk 1990]. Copyright © 1990 IEEE.

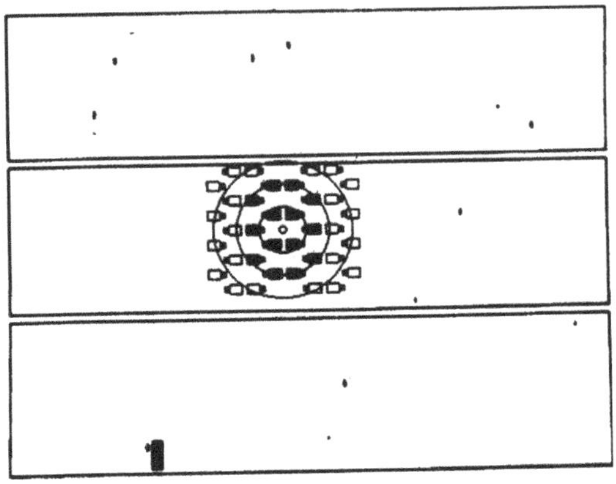

Figure 6-64 Multiple-bit error cluster showing the correct orientation and relative scale of the memory cell data-storage capacitors. Bit-error locations are represented by the solid capacitor areas, surrounded by the outlines of adjacent capacitors with no errors. The number of errors depends on the voltage, in this case 4.0 volts [Zoutendyk 1989]. Copyright © 1989 IEEE.

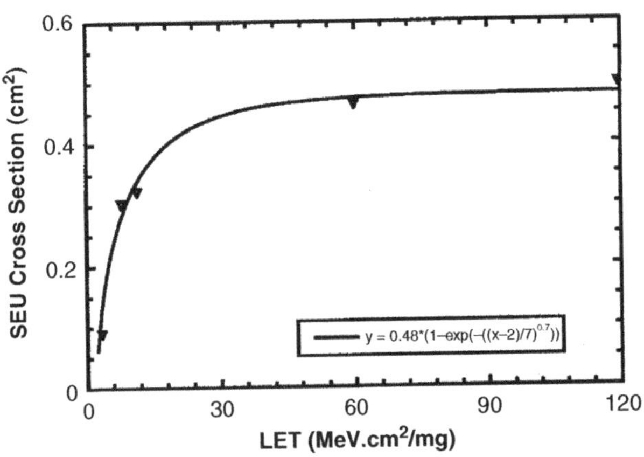

Figure 6-65 Event cross sections for a NEC 16 Mbit DRAM. The events produce multiple upsets and the multiplicities are listed in Table 6-2 [Buchner 2000]. Copyright © 2000 IEEE.

It is possible to use mixed-mode analysis to examine MBU multiplicity [Song 1988]. They examined high speed 64K resistor loaded NMOS SRAMs (RMOS). They observed the conventional error cross section to exceed the physical cross section. The extreme SEU sensitivity of this memory cell, shown in Figure 6-66, stems from the

Table 6-2
Multiplicity of the MNUs for Normally Incident Ions as a Function of Ion LET[a]

	Mutiplicity										
LET	1	2	3	4	5	6	7	8	9	10	11
2.9	1000	0	0	0	0	0	0	0	0	0	0
5.5	940	60	0	0	0	0	0	0	0	0	0
14.2	903	55	37	5	0	0	0	0	0	0	0
28.6	32	297	405	254	11	0	0	0	0	0	0
61.8	0	17	17	6	23	174	267	256	145	81	12

[a]The data are normalized to 1000 at each LET.
SOURCE: [Buchner 2000]. Copyright ©2000 IEEE.

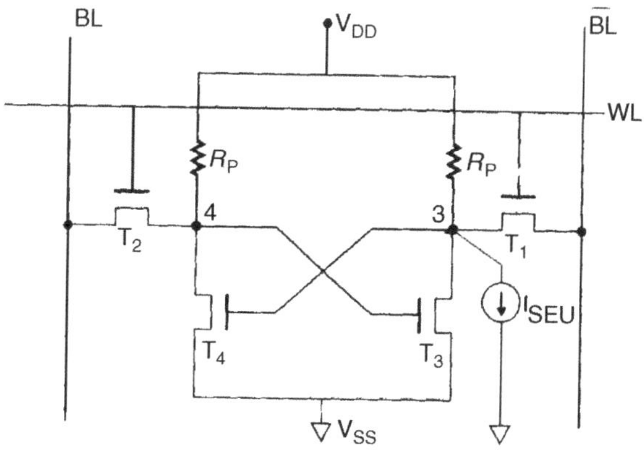

Figure 6-66 Circuit diagram of RMOS memory cell [Song 1988]. Copyright © 1988 IEEE. The large R_p leads to long time constants, so that diffusion current from high LET particles leads to upsets in adjacent cells as well as the struck cell.

use of very high pull-up resistance R_p, on the order of giga-ohms, to maintain the stand-by current at the equivalent device density CMOS level. The high R_p leads to a very long circuit time constant, on the order of microseconds. In this regime, slow diffusion currents contribute to upsets. As the LET is increased, more than one cell can be upset by one ion hit.

Song and co-workers used the two-dimensional PISCES code and the Monte Carlo electron slowing down card TRIPOS-E to calculate SEU pulse shapes. These were performed for low LET hits directly on a drain and high LET hits midway between two drains. These pulse shapes were used as input to SPICE circuit simulations. These calculations were used to obtain the cross section for four bit errors consistent with the observed values.

6.12.5 Hits to Adjacent Sensitive Regions

Dodd and co-workers extended mixed-mode MBU calculations with three-dimensional analyses [Dodd 1994]. They emphasized the need to include external circuitry and loading effects in addition to charge collection. Their work carefully examined the transition from funnel assisted drift to diffusion. They used DAVINCI to perform their mixed-mode calculations. Figure 6-67 shows a schematic of the approach. Figure 6-68 shows their results. The authors give a careful analysis of the model limitations.

6.12.6 Multibit Upset in FPGAs

SRAM based field programmable gate arrays (FPGAs) are very appealing for space based processing tasks. However, single event upsets on orbit are a challenge. They can modify both the data

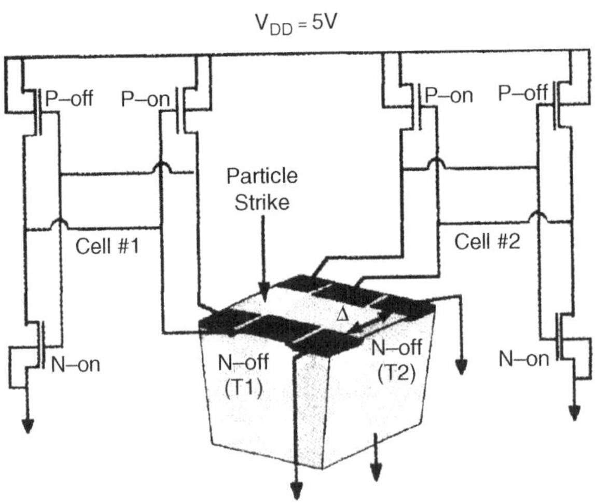

Figure 6-67 Mixed-mode device/circuit used in multiple-bit simulations of adjacent CMOS SRAM cells [Dodd 1994]. Copyright © 1994 IEEE.

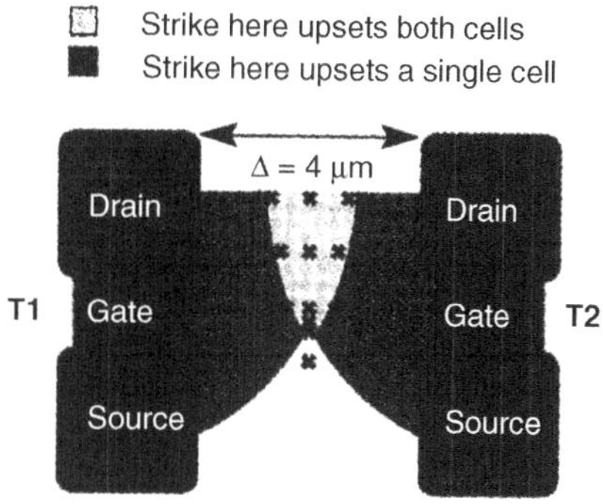

Figure 6-68 Upset map of double SRAM cell structure for normally incident 100 MeV Fe. The simulation strike locations are indicated by X's. The regions are approximate, between the two devices, as inferred from the simulation locations [Dodd 1994]. Copyright © 1994 IEEE.

being processed and the functionality of the circuit. Triple modular redundancy (TMR) is often used to mitigate the SEU effects. MBUs complicate the issue as they violate the assumption that only one error exists in the system at a time, since a MBU can affect redundant circuit copies. In recent FPGAs, clusters of 5 or 6 upsets are observed. These effects can trigger domain crossing errors (DCEs), where an error in two or more redundant copies (domains) of the TME circuit causes the vote to select the wrong value [Quinn 2005, Quinn 2007]. Quinn calculated that there would be a small, but appreciable, number of worst case TMR defeats. The reader is guided to the extensive literature on SEUs in FPGAs for further information [Kastensmidt 2007].

6.12.7 Calculation of Upset Rate for Diffusion MBUs

For the calculation of upset rates for diffusion based MBUs, we need to distinguish between the upset rate and the event rate. The event rate is the number of occurrences of one or more upsets. The upset rate is the total number of upsets. The standard IRPP calculation is appropriate for the event rate, as it calculates the rate at which the particle environment manages to hit a sensitive volume and cause one or more upsets. The appropriate geometry is the volume of one sensitive volume. The total upset rate is obtained by combining this with a multiplicity factor.

6.12.8 Geometrical MBE Rates in EDAC Words

It is now common to design memory chips that include on-chip single error correction and double error detection (SECDED). Although single bit errors are handled gracefully, the system may strain to handle double-bit errors, leading to unwanted down time. The double- or multibit error (MBE) case may arise from a single hit affecting adjacent bits in the same word, a single hit passing through two bits in the same word, or from the statistical case of two separate hits to bits in the same word during a single scrub cycle. Bits in the same error detection and correction (EDAC) word are often physically separated as much as possible. This eliminates the first case and minimizes, but does not eliminate, the second case. We want to compare the rates from the geometrical dual hits and the statistical dual hits.

If there are several bits at similar distances, the rates for each coincidence should be added for the total coincidence rate. It should be possible for a chip designer to use his/her geometrical layout to calculate the geometrical MBE rate for a representative word. Consider the layout in Figure 6-69.

Assume that the bits are laid out on a grid with a basic x axis spacing D_0 and y axis spacing defined by the aspect ratio ($a = dy/dx$). We can then add all possible contributions in an EDAC word. We rewrite the basic equation in terms of the spacing D_0 and the integer

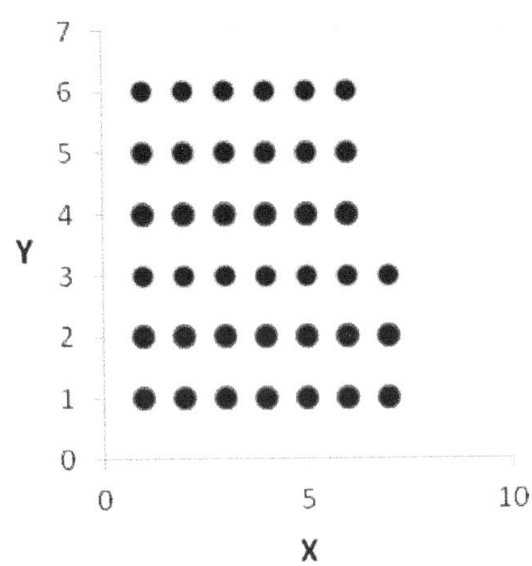

Figure 6-69 A possible layout of EDAC bits so that there is maximum separation between bits.

steps reflecting the varying distances between the bits.

$$R_{\text{Word}} = \frac{A_{\text{word}}}{2} \times \frac{W \times H \times R_{\text{B}}}{2\pi \times D_0{}^2} \qquad (6.10)$$

$$A_{\text{word}} = \sum_{x_0} \sum_{y_0} \sum_{x_1} \sum_{y_1} \left(\frac{1}{(x_1 - x_0)^2 + (a(y_1 - y_0))^2} \right) \qquad (6.11)$$

A_{word} is the sum of the contributions from all bits in the word. The sum over x_0 and y_0 and over x_1 and y_1 both step through all of the bits in the array. When scanning a single row or column, we only count adjacent pairs of bits. If we have bits "a," "b," and "c," and examine the coincidence of "a" and "c," they have already been counted in the "a"–"b" count as the bits are very well aligned. This equation double counts the contributions, which has already been considered in the basic equation, so the result must be divided by 2. The bracketed term is the square of the distance from x_0, y_0 to x_1, y_1 in terms of D_0.

The average bit MBE rate can now be determined by dividing A_{word} by 39 (the number of bits in a word). Note that A will be independent of D_0. We have now averaged the MBE rate over all of the bits in the EDAC word. The basic bit rate of 1.1e−14 for two bits at 1600 microns base spacing has been increased by a factor of 2.15 to a rate of 2.37 e−14 MBE/bit-day for an aspect ratio of one. If we had counted all of the pairs in the rows and columns, the rate would be 3.9 e−14. The difference corresponds to more than two bits in a word being upset.

Figure 6-70 shows the slow dependence of $A_{\text{bit}}(= A_{\text{word}}/39)$ on the row/column aspect ratio. We assume that the x (column) distance is constant and vary the y (row) distance.

In SRAMs where the bits in a single word are spaced by many columns or rows in a regular pattern, a single MBE event is likely to cause MBEs in a very large number of words. Take, for example, a SRAM where the bits in a single word are spaced 128 columns apart: Word A will have bits in columns 0, 128, 256, and so on; while Word A+1 will have bits in columns 1, 129, 257, and so on. A grazing angle strike that causes a MBE in word A by flipping the bits in columns 128 and 256 would also have traversed all the bits in the same row between these two columns. Furthermore, this strike is also likely to continue to remain in the same row for some distance beyond these

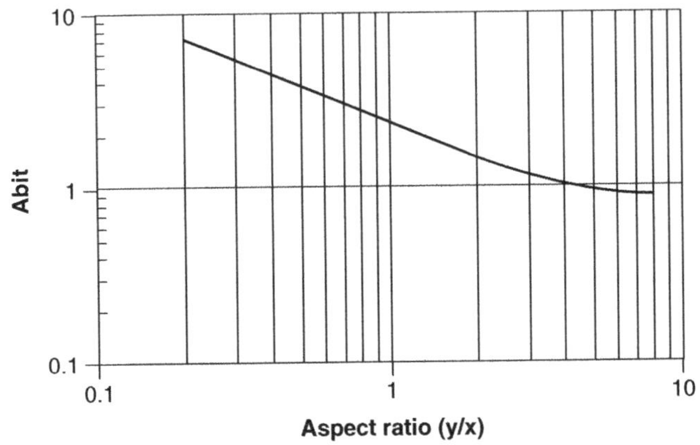

Figure 6-70 Dependence of the MBE rate on the aspect ratio for the layout of Figure 6-69.

bounds. If the strike goes beyond the known range of 128 columns by one additional column, then Word $A+1$ also has a chance of flipping. On average, it is expected that when at least one MBE occurs, then there will be 10 to 100 MBEs. Cross sections for upsets for hits along a word row are very, very low, but, if they occur, they may completely corrupt the information on the chip. The designers may want to assure themselves that this cannot be a fatal error.

Let us consider another layout in which the EDAC word is located with 20 bits in a row in the top half of the chip and 19 bits in a row in the bottom half, as shown in Figure 6-71.

The MBE rate per pair of bits is 5.5 e-14 MBE/day and the MBE rate is 5.2 e−14 MBE/bit−day. We are counting only adjacent bits as a hit to bits "a" and "c" would have produced a MBE for bits "a" and "b." The row to row distance is ten times the column to column distance, so the row−row coincidences are a second order effect and are not considered.

6.12.9 Statistical MBE Rates in the Space Environment

The EDAC is ordinarily implemented with a scrub routine that walks the error detection circuitry through the EDAC words so that they are periodically read and single-bit errors corrected if necessary. Each scrub reports if any dual errors are detected. The number of multibit

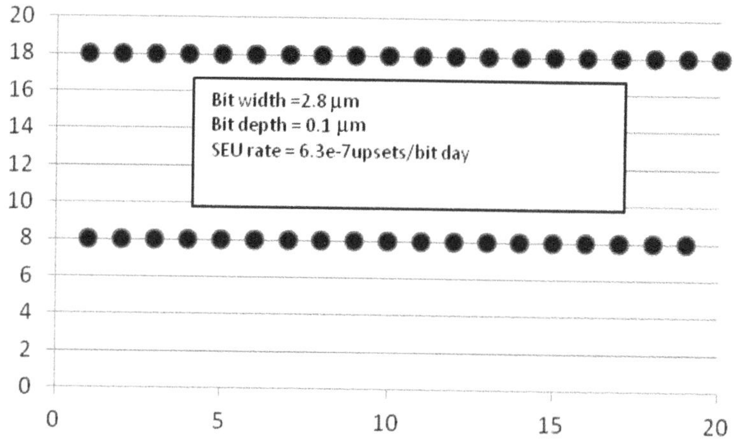

Figure 6-71 A possible layout, assuming a bit separation of 720 μm, with a single word located in the four quadrants of a memory chip.

errors due "simultaneous hits" detected in any environment depends on the single-bit error rate and the scrub rate. We are interested in predicting the rate of multibit errors as a function of the environmental rate and the scrub frequency.

Hafer has examined the statistical probability of statistical multibit errors [Hafer 2006a]. A similar analysis has been presented by Tausch [Tausch 2009]. We will present Hafer's results. Let σ be the total number of single events during the test period and N be the total number of EDAC words in the memory. The probability of a MBE occurring in a word during a specific time interval is calculated as

$$P_{\mathrm{MBE}}(W_e > 1, \sigma \geq 2) = \sum_{i=2}^{\sigma} \binom{\sigma}{i} \left(\frac{1}{N}\right)^{i} \left(1 - \frac{1}{N}\right)^{\sigma - i} \tag{6.12}$$

Where W_e is the number of MBWs, σ is the probability of I bit upsets in a word and is the probability of the word not being upset in $\sigma - i$ events. This can be rewritten as 1 minus the probability of a correctable error:

$$P_{\mathrm{MBE}} = 1 - \sum_{i=0}^{1} \binom{\sigma}{i} \left(\frac{1}{N}\right)^{i} \left(1 - \frac{1}{N}\right)^{\sigma - i} \tag{6.13}$$

For the space environment

$$N \gg 1 \quad \text{and} \quad N \gg \sigma \tag{6.14}$$

so that

$$P_{\mathrm{MBE}}(W_e > 1, \sigma \geq 2) \cong \left(\frac{\sigma}{N}\right)^2 \tag{6.15}$$

From this the average number of MBs during a specified test interval without data scrubbing is determined as

$$E_{\mathrm{ave-ns}} = P_{\mathrm{MBE}} \times N = \frac{\sigma^2}{N} = \frac{\#SEU^2}{\#EDAC\,words} \tag{6.16}$$

With data scrubbing, the average number of MBEs during a specific test interval becomes

$$E_{\mathrm{ave-s}} = E_{\mathrm{ave-ns}} \times \frac{t_{\mathrm{scrub-dev}}}{t_{\mathrm{interval}}} \tag{6.17}$$

$t_{\mathrm{scrub-dev}}$ is calculated from the scrub frequency and the number of EDAC words. The scrubbing steps from 39 bit EDAC word to 39 bit EDAC word at the scrub frequency. Therefore, the scrub time per device I ($t_{\mathrm{scrub-dev}}$) is scrub time per word (= 1/frequency) times the number of EDAC words, which is the total memory size on the chip.

The result can be written as

$$MBE_{\mathrm{err\ rate}} = \left(\frac{SEU_{\mathrm{errs}}}{\mathrm{dev-second}}\right)^2 \times (t_{\mathrm{scrub-dev}}) \times 1/N \tag{6.18}$$

We prefer to write the equation in term of errors/bit-day.

$$\frac{MBE}{\mathrm{bit-day}} = \left(\frac{SEU\ \mathrm{errors}}{\mathrm{bit-day}} \times \frac{\#\,\mathrm{bits}}{\mathrm{dev}}\right)^2 \times \frac{\mathrm{dev}}{N_{\mathrm{EDAC\ words}}} \times \frac{t_{\mathrm{scrub-dev}}\ \mathrm{seconds}}{\mathrm{second/day}} \times \frac{\mathrm{dev}}{\#\,\mathrm{bits}} \tag{6.19}$$

We will consider “16 Mbit” memories of 5.24E+05 EDAC words using 20 Mbit, corresponding to a 16 Mbit chip without EDAC. Table 6-3 shows the expected space MBE rate for a representative sample of the two types of “16 Mbit” chip.

Table 6-3
Statistical MBE Rate Calculation "16 Mbit" Memory Chips

Errors/bit-day		6.30E−07
# bits		20447232
errors/dev-day		1.29E+01
Squared		1.66E+02
# words	39 bit	5.24E+05
MBE/device	No scrubbing	3.17E−04
seconds/day		86400
Frequency	Hz	1.11E+05
scrub time	seconds	9.01E−06
scrub time	seconds/dev	4.72E+00
MBE/device day	Scrubbing	1.73E−08
MBE/bit-day	Scrubbing	8.46E−16

Note that the resulting MBE rates are significantly lower than the geometrical rates shown in Figure 6-60 or the rates associated with Figures 6-69 and 6-71. There is a factor of 60 between these rates and the corresponding device in Figure 6-71. The geometrical MBE rates in general will be between 25 and 250 times the statistical MBE rates for devices in which the bits have maximum separation. If the bits in a single MBE word are physically closer, the geometrical rates will be higher, as shown in Figure 6-60.

6.12.10 Impact of Geometrical Errors on System Performance

System soft error performance analysis requires a convergence of failure rate metrics and outage impact. We will follow the discussion of Silburt [Silburt 2008]. Failure rates are measured in FIT (failures in time) defined by

$$1FIT = \frac{1 \text{ Failure}}{10^9 \text{ hours}} \tag{6.20}$$

System outages are measured in DPM (defects per million) where for a given time period:

$$DPM = 10^6 \times \left(\frac{\sum \text{outage durations}}{\text{duration period}} \right) \tag{6.21}$$

The popular term "5 9's" availability refers to 99.999% uptime and translates to 10 DPM. For a given component, c, with a soft failure rate of f_c FIT, and each failure causing an outage of t_c seconds, its DPM contribution would be

$$DPM_c = \frac{t_c \times f_c}{3.6 \times 10^6} \tag{6.22}$$

DPM is a convenient measure to work with when analyzing system reliability since independent component contributions to outage can simply be added together,

The double error detection capability of the SECDED ECC generates a trigger for a reset as a common generic response. This is often supported with localized reset islands, in which a subset of hardware can be isolated and reset under software control without impacting the surrounding circuitry. A target of the fast reset may be less than 10 ms [Silburt 2008]. Table 6-4 shows the results of reliability calculations for some of the cases that we have considered.

Silburt quotes possible specifications, for ground-based systems, as 30,000 FIT when the recovery time is 0.01 s and 1000 FIT and 0.05 DPM when the recovery time is 180 s. If the total memory size approaches one gigabyte, we are exceeding these limits with the memory alone.

An alternate approach to system performance is to follow the approach of Hafer and discuss mission risk assessment [Hafer 2008]. Consider 5.2×10^{-14} MBE/bit-day. This would be 3.88×10^{-4} device-years/MBE for a 16 Mbit EDAC memory. Modern small computers often have 1 gigabyte of memory. That means that there would be 5.2 mean years between MBE. A 5.2-year mission would likely have 1 MBE. It may be worthwhile to be certain that this cannot be a fatal error.

We observe that for situations in which the bits are primarily in well-separated rows, the rate will be nearly the primary bit rate as calculated using Equation (6.9). We can use this relationship to plot the various measurements of reliability as a function of the bit spacing for 16 Mbit EDAC chips. This is shown in Figure 6-72.

We have assumed a very regular row–column layout. This may not be true with actual devices that also have EDAC logic interspersed on the layout. Some row or column spacings may be reduced to several hundred microns. If several bits are closer than 300 μm, the MBE

Table 6-4
Reliability Calculations for 2 bit Spacing and for Geometrical MBE in Single 16 Mbit EDAC Devices (containing 20 Mbit) and for 1 gigabyte Memories

		16 μm	1600 μm	Figure 6-69	Figure 6-71	Statistical
MBE rates for 20 Mbit	errors/bit-day	1.10E−10	1.10E−14	2.37E−14	5.20E−14	8.46E−16
	errors/dev-day	2.25E−03	2.25E−07	4.85E−07	1.06E−06	1.73E−08
	errors/dev-hour	9.37E−05	9.37E−09	2.02E−08	4.43E−08	7.21E−10
	errors/dev-year	8.21E−01	8.21E−05	1.77E−04	3.88E−04	6.32E−06
	Years/MBE	1.22E+00	1.22E+04	5.65E+03	2.58E+03	1.58E+05
FIT	Failures/10^9 h	9.37E+04	9.37E+00	2.02E+01	4.43E+01	7.21E−01
DPM for t_c	0.01	2.60E−04	2.60E−08	5.61E−08	1.23E−07	2.00E−09
DPM for t_c	180	4.69E+00	4.69E−04	1.01E−03	2.22E−03	3.60E−05
1 Gbyte	errors/day	1.12E+00	1.12E−04	2.42E−04	5.32E−04	8.65E−06
	errors/mem-hour	4.69E−02	4.69E−06	1.01E−05	2.22E−05	3.60E−07
	errors/mem-year	4.10E+02	4.10E−02	8.84E−02	1.94E−01	3.16E−03
	years/MBE	2.44E−03	2.44E+01	1.13E+01	5.15E+00	3.17E+02
FIT	failures/10^9 h	4.69E+07	4.69E+03	1.01E+04	2.22E+04	3.60E+02
DPM for t_c	0.01	1.30E−01	1.30E−05	2.80E−05	6.15E−05	1.00E−06
DPM for t_c	180	2.34E+03	2.34E−01	5.05E−01	1.11E+00	1.80E−02

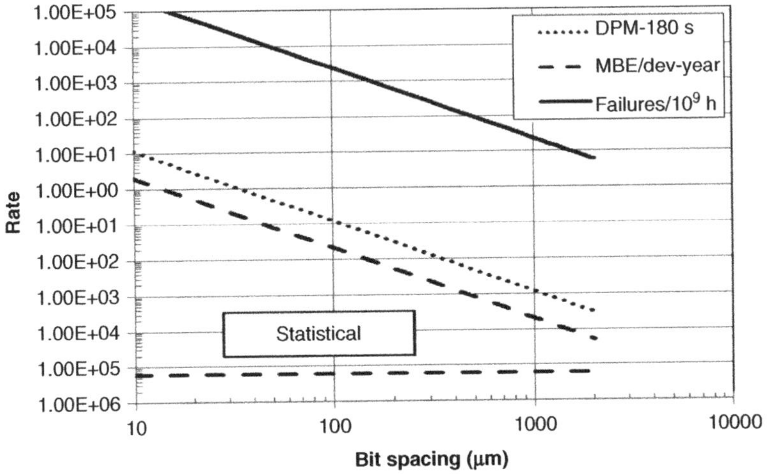

Figure 6-72 Reliability values for geometrical MBE as a function of bit spacing in 16 Mbit EDAC words for the device in Figures 6-60 and 6-61. The MBE/dev-year values can be compared with the value of 6.32 e−6 MBE/dev-year for statistical MBE.

rate will be greater than 1×10^{-12} MBE/bit-day. It appears that we should make a careful analysis of geometrical MBE in EDAC memory systems.

6.12.11 Statistical MBUs in a Test Environment

The above equations for statistical upsets were developed for the space environment. The high upset rates in an ion beam test may render the approximations in Equation (6.14) invalid. In this case you have to use Equation (6.13). This is what Hafer did to obtain the results in Figure 6-73.

6.13 SEU IN LOGIC SYSTEMS

Another type of problem arises in dynamic logic systems. This is the situation when a single-bit error propagates through a system of logic modules and interferes with the normal passage of signals or leads to many apparent errors at the outputs of the system [May 1984, Newberry 1992]. The majority of the discussion in this book relates

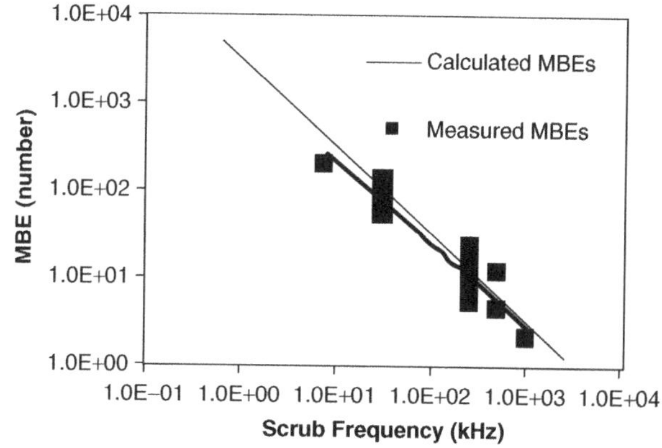

Figure 6-73 Measured MBE data compared to calculated MBEs with a flux of 400 ions/cm^2 for 100 seconds [Hafer 2006a]. Copyright © 2006 IEEE.

primarily to latching circuits, especially memories. However, logic circuits, such as those of microprocessors, have a large number of gates that may be susceptible to upset, with the upset signal then being processed through a chain of gates until it reaches a latch or an output pad [Petersen 1988, Kastensmidt 2007]. The historical development of logic has been such that individual logic cells have more capacitive loading than memory cells and so are less susceptible to upset. However, as logic circuitry has progressed to small feature size, it becomes more susceptible. In the limit of the same susceptibility as memory cells, logic and memory chips of the same chip size will have roughly equivalent chip upset rates [Diehl-Nagle 1984]. When processors do present problems, the problems can be serious.

There are three different types of errors that may show up in logic chips. These are bit upset, lost data word, and lost data path. Newberry made an extensive study of these errors and of the hardening techniques that are applicable [Newberry 1986, Newberry 1988, Newberry 1992]. The lost data path errors can be very serious. Sometimes they can be corrected by a chip reset, but in other situations the chip must be powered down and then restarted from initial conditions before it will work properly again [Guenzer 1981].

Figure 6-74 shows one portion of the Newberry analysis [Newberry 1988]. The alphanumeric designations in each box merely represent the logic-cell type. The figure may be thought of as a time-lapse diagram of an error propagating through interconnecting

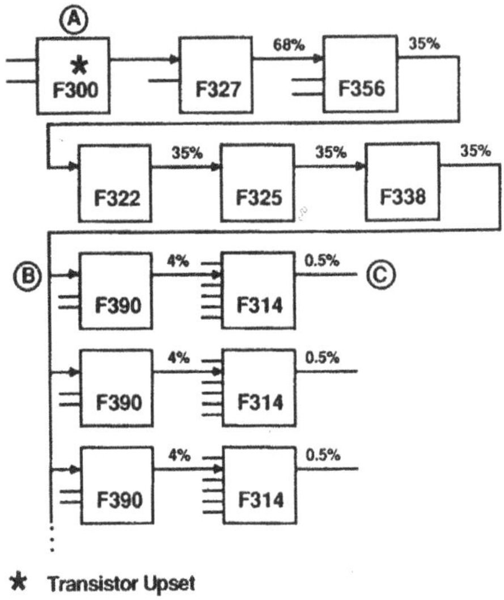

Figure 6-74 Error propagation through logic devices [Newberry 1988].

cells. That is, with each clock cycle the error moves through one or more cells. In this manner, it may take the error originating at point A six clock cycles to reach point B and an additional two clock cycles to reach point C. The percentage values shown in Figure 6-74 represent the probability of an error prorogating between those cells and therefore can serve as a measure of the path sensitivity to error propagation.

As logic circuits on a single chip become more complex, the detection and handling of errors becomes much more difficult. Figure 6-75 shows error propagation in an Intel 80186 microprocessor [May 1984]. It takes a number of machine cycles for an error to reach a bond pad. By the time it does so, it has arrived at several bond pads simultaneously so there is no way to trace its path. Some errors (moles) may remain dormant for thousands of cycles.

Even though events such as this can show up as many errors at the chip outputs, the upset rate calculation is done using the rate for a single bit. It is possible, with some effort, to build SEE hard microprocessors, such as the RAD750 computer built by BAE systems.

The logic blocks in modern spacecraft are as likely to be implemented in FPGAs as in logic blocks (Figure 6-74) or microprocessors

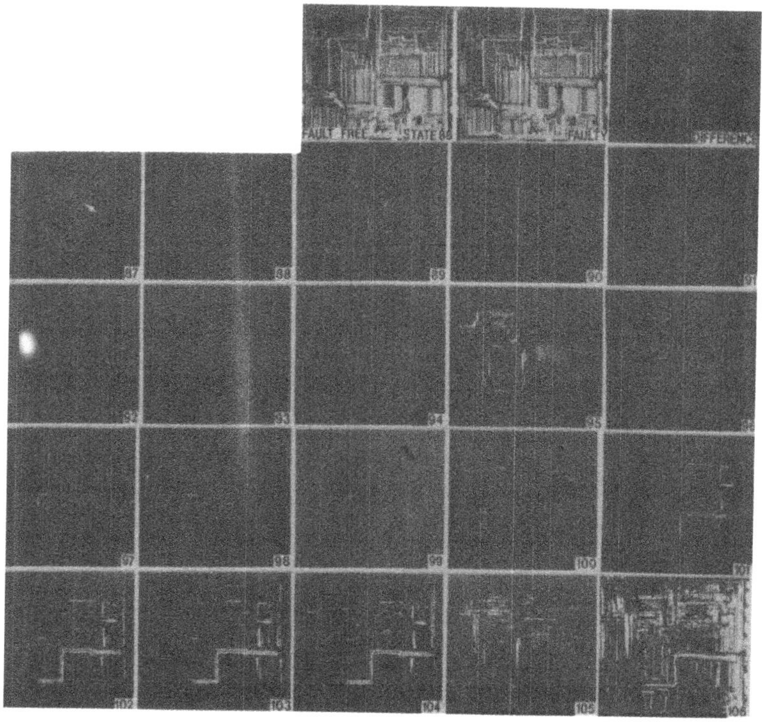

Figure 6-75 Error propagation in the Intel 80186 microprocessor. These photographs are obtained by using two electron microscopes to compare the voltages present in a chip without errors to those present in a chip that has had an error introduced. The differences show up as a pattern of highlighted areas. A number of internal machine cycles pass before the error has propagated to an exterior pad (normally a large number of pads nearly simultaneously) [May 1984]. Copyright © 1984 IEEE.

(Figure 6-75) [Quinn 2005, Quinn 2007, Kastensmidt 2007, Berg 2007, Swift 2008, Rezgui 2008].

6.14 TRANSIENT PULSES

In Figure 6-1 we showed an early measurement of a typical single particle transient pulse. It was soon demonstrated that the pulses in multilayer structures were more complicated [Knudson 1984]. In recent years, transient pulses have become very important in logic circuits, such as those discussed in the last section, and there has been a resurgence of studies of pulse characteristics. Buchner has presented two short courses on the subject [Buchner 2001, Buchner 2005]. In Sections 6.7.2.3 and 6.7.2.7 we presented analyses of sample data involving transients [Johnson 2000, Benedetto 2005]. It has now become possible to study the transient pulse structure and this has been reported in a number of papers, such as those by Gadlage and by Ferlet-Cavrois [Gadlage 2004, Ferlet-Cavrois 2006].

CHAPTER 7

Cosmic Ray Single Event Rate Calculations

Chapter 1 introduced the basic concepts that are used to describe single event effects. A basic problem is that there is an inconsistency in calculating upset rates. The upset cross section for both proton and heavy ions is defined as

$$\sigma = \#\text{upsets}/\text{fluence} \tag{7.1}$$

If you invert this, you have the upset rate calculation for protons:

$$\text{Rate} = \Sigma\, \sigma(E)F(E) \tag{7.2}$$

However, this is not valid for heavy ions. The heavy ion upset rate depends on the energy deposition, not the number of hits. This invalid approach is called the integral flux method and is not to be confused with the effective flux approach of Binder.

The heavy ion rate calculation involves the probability of depositing an adequate amount of energy, which depends on the device dimensions and the environment. There has been an evolution of approaches for considering these factors. We will present a historical treatment of the development and use of the concepts.

Single Event Effects in Aerospace, First Edition. By Edward Petersen.
© 2011 the Institute of Electrical and Electronics Engineers, Inc.
Published 2011 by John Wiley & Sons, Inc.

7.1 INTRODUCTION TO RATE PREDICTION METHODS

Binder, Smith, and Holman presented the first paper to report on single event effects observed in space [Binder 1975]. It was not readily accepted by the radiation effects community. All of the traditional radiation effects were bulk effects, depending on total radiation fields. This paper was describing an effect caused by individual particles in space. They described the primary radiation environment as that due to cosmic ray iron nuclei, allowing for the effects of the spacecraft shielding. They published rates describing charge collection in a thin disk, although the authors did parallel calculations using a thick disk and actual ion energy loss. The calculated critical charge was determined using circuit modeling and the authors performed laboratory verification using scanning electron microscope measurement techniques. They calculated the upset rates by integrating the flux of ions that are able to cause upset over the area of the disk. This paper introduced all of the basic concepts that are involved in upset rate calculations.

May and Woods brought single event effects to the attention of the general electronics community in 1978 [May 1979]. The authors were attempting to explain why the latest generation of Intel memory chips was upsetting—something not previously observed. They read the paper by Binder and co-workers and realized that radioactive contaminants in device packaging could be emitting alpha particles, which in turn could cause upset. They then verified the effects by testing, showing conclusively that alpha particles could cause the upsets. This paper in turn alerted the space radiation effects community to the fact that single event effects are real.

7.2 THE RPP APPROACH TO HEAVY ION UPSET RATES

Pickel and Blandford introduced the Heinrich form of the LET spectrum, which summed the contribution of all of the ions in space [Heinrich 1977, Pickel 1978]. They also considered an integral over the average path length in a rectangular parallelepiped (RPP) device geometry.

Bradford made several important early contributions to the field. His first paper introduced an approximate form of the integral path length distribution of a RPP [Bradford 1979]. The author proposed an exact formula for rate calculations using the integral chord length distribution combined with the differential LET spectrum. The paper had several misprints and did not receive a great amount of attention. In 1980, Bradford performed a rate calculation using an approximate differential Heinrich spectrum and the integral chord length distribution [Bradford 1980a]. He also performed an analysis of multiple-bit upset due to multiple hits.

The basic approach in either the Pickel or Bradford rate calculations is based on the following concepts.

1. The upset is caused by charge deposition in a circuit (Q). We will temporarily assume that the charge deposited equals the charge collected.
2. There is a unique charge that must be exceeded for the circuit to upset: the critical charge Q_c.
3. We need to calculate the number of occasions $Q > Q_c$. Q depends on the energy loss of the particle

$$Q = dE/dX \times dS$$

   So it depends on the path length dS.
4. The number of occasions depends on the number of ionizing particles. This is determined by the LET (dE/dX) distribution of the particles in the space environment.
5. The number of occasions depends on the size of the device as in Figure 7-1.
6. Combine target size, LET distribution, path length distribution, and critical charge.

The upset rate integral over the RPP has the basic form in the Bradford formulation:

$$N = \frac{S}{4} \int_{L_0}^{L_{max}} I(L) C \left(\frac{\Delta E}{\rho L} \right) dL \qquad (7.3)$$

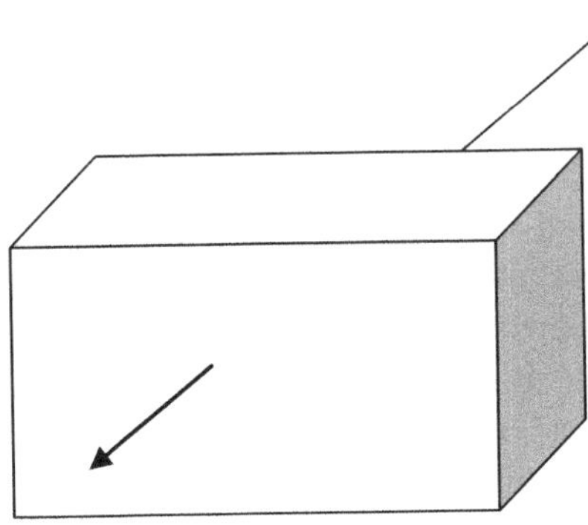

Figure 7-1 Device geometry.

where

$$N = \text{upset rate for a RPP cell}$$
$$S = \text{surface area of the RPP}$$
$$\Delta E = \text{critical energy for upset}$$
$$\rho = \text{the material density}$$
$$L = \text{linear energy transfer (LET)}$$
$$L_0 = \text{minimum } L \text{ that will produce upset}$$
$$L_{\text{max}} = \text{the maximum LET in the environment}$$
$$I(L) = \text{differential LET spectrum}$$
$$C(s_{\text{min}}) = \text{integral path length distribution}$$
$$s_{\text{min}} = \Delta E/\rho L$$

The model assumes that the device has a unique critical charge Q_c at which it upsets. The device also has a constant depth, so that critical charge corresponds to a unique value of effective LET. The basic curve of cross section versus effective LET is then a step function as shown in Chapter 2, Figure 2-66. The relationship of critical charge to critical energy for silicon is $Q_c = \Delta E/22.5$ (pC/MeV). The term $\Delta E/\rho L$ determines the minimum path length necessary for upset for each LET value.

The two important inputs for this calculation are the LET spectra and the path length distribution. In Chapter 2, Figures 2-18 through 2-25 presented various LET spectra. Figures 2-22 and 2-23 showed the differential LET spectrum that is used in this calculation. The other important distribution is the path length distribution. Figures 7-2

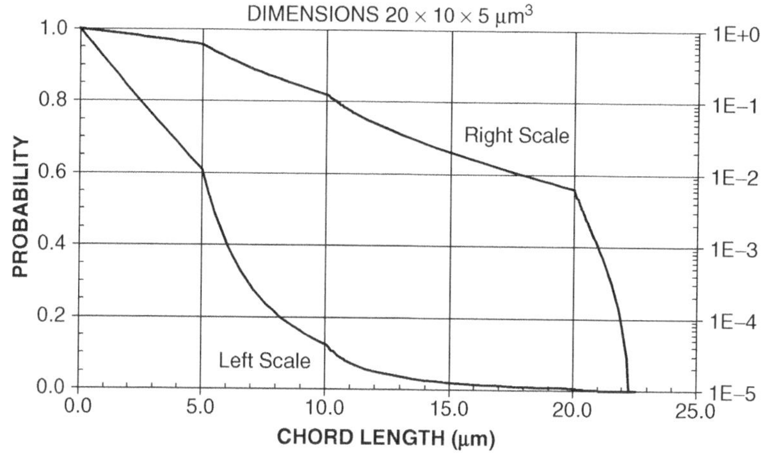

Figure 7-2 Integral chord length distribution for a device 20 × 10 × 5 microns.

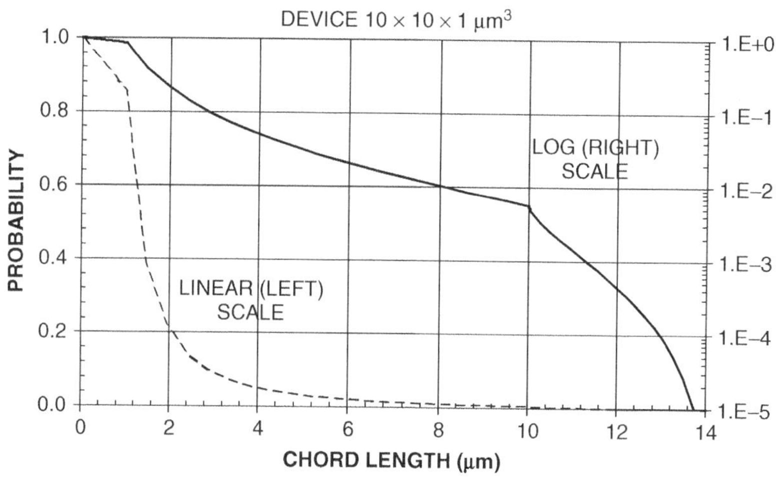

Figure 7-3 Integral chord length distribution for a volume 10 × 10 × 1 micrometers.

and 7-3 show representative integral path length distributions of the form used in Equation (7.3). The distributions scale with distance, so the distribution for a volume 4 × 2 × 1 will have the same appearance as the distribution for the volume 20 × 10 × 5. Figures 7-4 and 7-5 show the corresponding differential forms of the path length distribution used in other versions of the RPP integral, to be discussed later.

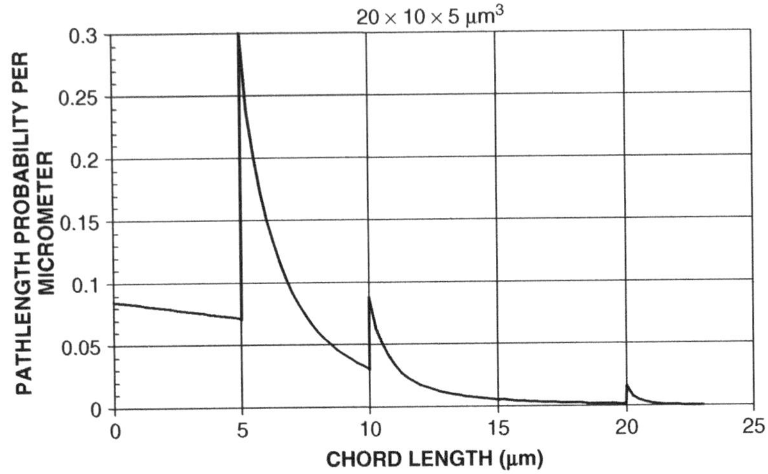

Figure 7-4 Differential path length distribution for a volume 20 × 10 × 5 micrometers.

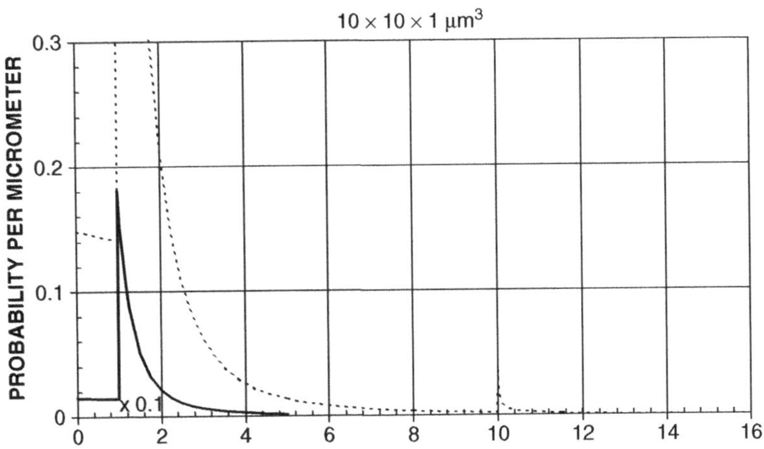

Figure 7-5 Differential path length distribution for a device 10 × 10 × 1 micrometers.

Figures 7-6 and 7-7 show the upset rate as a function of critical charge for a device 10 microns × 10 microns × 1 micron. It is easy to generate a similar curve in terms of effective LET. We can use the relationship 1 pC/micron = 98 MeV/mg/cm^2. Figure 7-8 shows the upset rate as a function of effective LET for the 10 × 10 × 1 device. Note that there can be upsets at effective LETs past 200 even

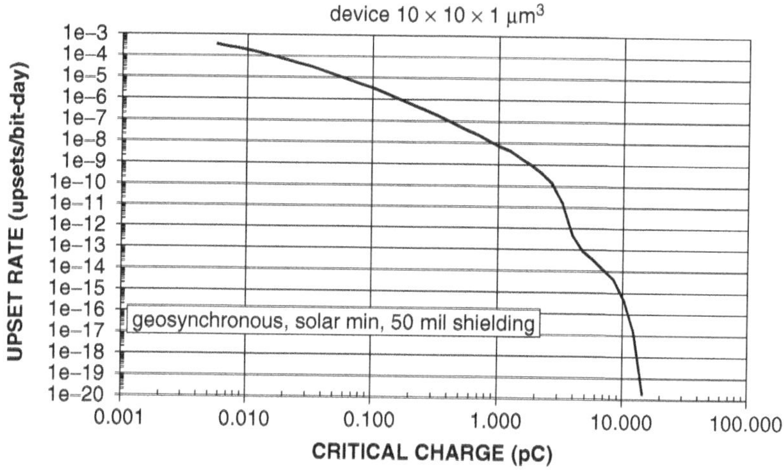

Figure 7-6 Upset rate as a function of critical charge for a device with dimensions $10 \times 10 \times 1\ \mu m^3$.

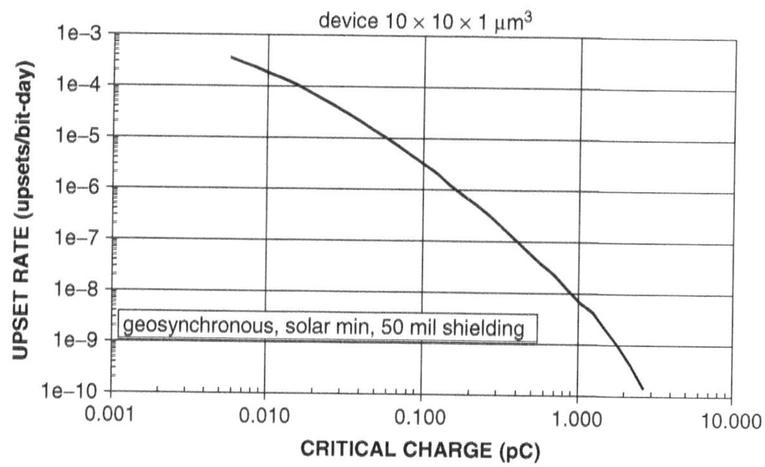

Figure 7-7 Expanded plot of upset rate as a function of critical charge.

though the LET spectrum, as shown in Chapter 2, Figures 2.23 and 2.24, effectively does not extend past 30 MeV/mg/cm^2. This is due to the ions that pass nearly in the plane of the device. The drop at 30 MeV/mg/cm^2 in the LET spectrum does not have a corresponding drop in the upset rate versus the effective LET due to the large number of paths longer than 1 micron, as shown in Figures 7-5 and 7-3. If

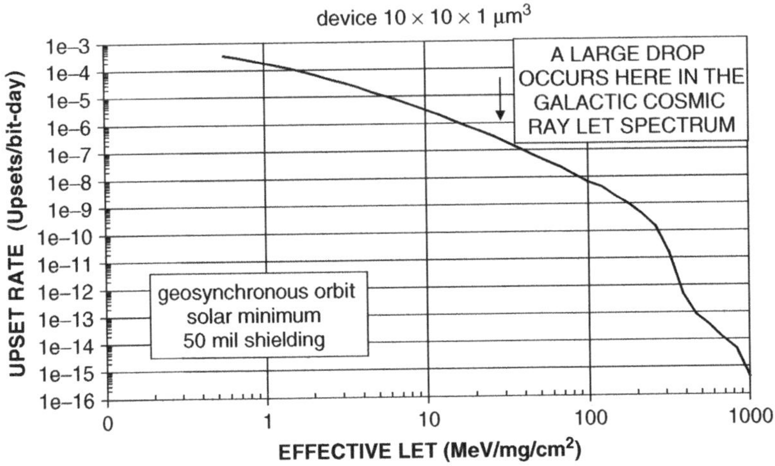

Figure 7-8 The upset rate calculated as a function of critical LET assuming that the cross section is a step function of LET.

you attempt to use the integral flux calculation, you will not have any contribution past the drop at 30 MeV/mg/cm^2. This tells you immediately that it is wrong. Figure 7-8 also shows upsets past effective LETs of 200 MeV/mg/cm^2. Therefore, it is wrong to say that there is no concern if the threshold is above 200.

Pickel and Blandford introduced a comprehensive code for upset rate calculations [Pickel 1980]. The approach was to integrate over the integral LET spectrum, the differential path length distribution, with the integration variable being the path length. It used the Heinrich integral LET spectrum and also a worst case spectrum. They used the RPP volume with an exact differential path length distribution derived by Petroff. The calculations were then used to calculate the upset rate as a function of critical charge. They assigned the critical charge as the minimum charge necessary to produce upset, and based the results on both experiment and circuit modeling. This is the basic RPP approach that has been used by the community. The computer code that they wrote to evaluate the upset rate was called the CRIER code. Pickel and Blandford later provided a report summarizing their CRIER calculation, added a funnel length in it, and also presented a detailed analysis of multiple-bit upsets [Pickel 1984].

In 1981 Jim Adams wrote a report presenting an updated Heinrich LET spectrum for the galactic cosmic rays [Adams 1981]. It presented solar minimum, solar maximum, and the 90% LET spectra.

Petersen and co-workers introduced approaches based on the Bradford formalism, using an integral chord length distribution and a differential LET distribution [Petersen 1982, Shapiro 1982]. This paper publicized the Adams 90% differential LET spectrum, and it became the standard environment for device comparisons. They introduced two calculations: (1) an approximate approach using two simple analytic terms to describe the LET environment, based on Burke's Burke 1981 work, and the two chord length terms from Bradford [Burke 1981, Bradford 1979]; (2) a numerical integration over Petroff chords to get the integral chord length distribution and then integrate this with the Adams differential LET spectrum to obtain upset rates. This is an alternative basic RPP approach. The two approaches for evaluating the upset rate for a RPP are completely equivalent. Both approaches need to consider the complexities of integrating over discontinuities—either in the differential LET spectrum or in the differential path length distribution. Figures 7-2 and 7-3 show representative integral chord length distributions.

Bendel simplified the Petroff differential distribution and then obtained an analytic expression for the integral chord length distribution [Bendel 1984a]. This is given in Appendix B, Section B.8 and is the form used in the calculations since that time.

Adams developed a code based on the integral LET and differential path length approach but changing the integral to be over LET [Adams 1983, Adams 1986]. This was named the CREME code. The major contributions of the 1986 report were the updated environments. This report includes solar activity, interplanetary weather conditions, orbital parameters, geomagnetic transmission, and material shielding. The environments included in this code became the standard for all upset rate calculations, even if other upset rate codes were used.

Shapiro published an upset rate calculation code using the Bradford approach with the analytic chord length distribution, the differential Adams LET spectra, and the integral over LET [Shapiro 1983]. Then he and Rabe furnished a modification that included additional chord length to allow for the funnel effect [unpublished]. The codes were called CRUP and HCRUP.

7.3 THE INTEGRAL RPP APPROACH

Petersen pointed out the inadequacy of the single low LET threshold for upset rate calculations [Petersen 1983a]. It is necessary to integrate the LET spectrum with the cross-section curves to properly allow for changes of sensitivity across the device (sensitive region). This allows for the difference between the "ideal" curve shown in Chapter 2, Figure 2-66, and the actual curve as shown in Figure 2-67. This is called the integral RPP (IRPP) approach. The paper also presented the figure of merit equation for quick evaluation of device upset rates. This was obtained by using a simple approximation to the differential 10% spectrum and the Bradford two-term path approximation. The paper discussed the inclusion of charge funneling in the rate calculation and some of the effects of the finite device depth on the device measurements.

The integral RPP calculation allows for the variations of sensitivity by integrating over a distribution of upset rates corresponding to the variation of cross section versus LET. The process is to convert the effective LET to critical charge assuming a single sensitive volume thickness, so that the critical charge at any part of the curve is the LET times the thickness. You then calculate a differential upset rate at a number of Q_C (LET) bins. Assume the RPP chord length distribution, the area equal to the maximum cross section and the areal dimensions obtained from the square root of the area, and the RPP thickness used to calculate the critical charge. Each individual rate is then calculated by integrating the RPP chord distribution with the LET distribution of the environment. Figure 7-9 shows a possible set of steps for the integration. Figure 7-10 shows two ways of writing the integration procedure. The individual differential rates are then summed to obtain the total rate. Table 7-1 shows the spreadsheet calculation of the upset rate. Figure 7-11 shows the contributions to the total upset rate from LET bins. The device upset rate depends on contributions over the entire curve. Figure 7-12 shows the cumulative upset rate rising slightly faster than the corresponding Weibull cross-section curve. Both curves show very little contribution from

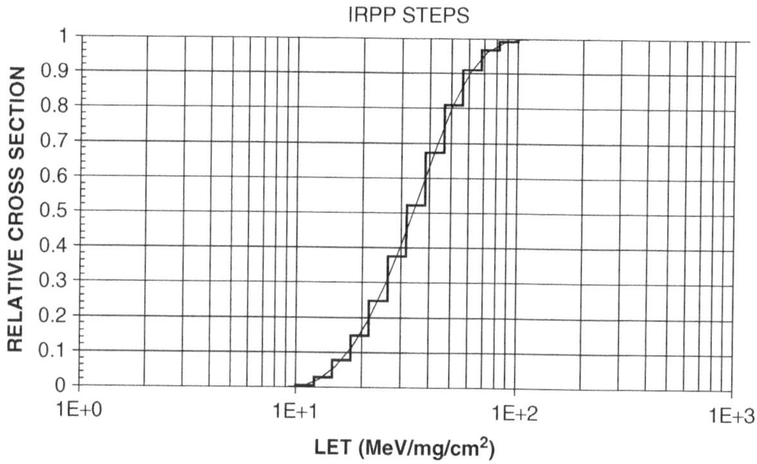

Figure 7-9 The introduction of steps on the cross-section curve for integral upset rate calculations.

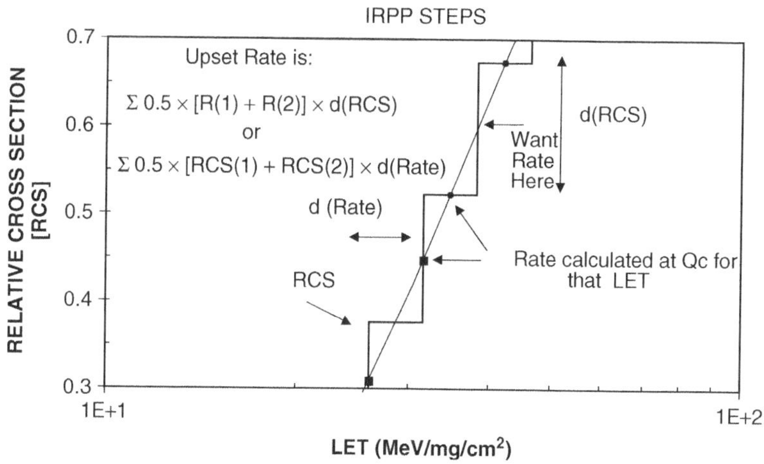

Figure 7-10 Calculation of upset rate using IRPP method.

the low cross-section region just above the threshold. Another way to view this calculation is to say that you are folding together the cross-section curve shown in Chapter 2, Figures 2-67 and 2-68, with the upset rate curve shown in Figure 7-8.

Harboe-Sorensen and Shoga were the first to apply the integral RPP approach to upset calculations [Harboe-Sorensen 1986, Shoga

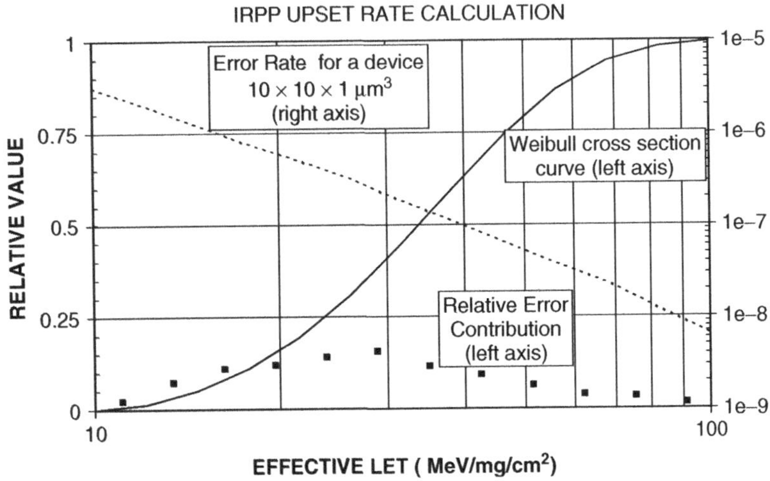

Figure 7-11 Contributions to the upset rate from the bins in relative cross section.

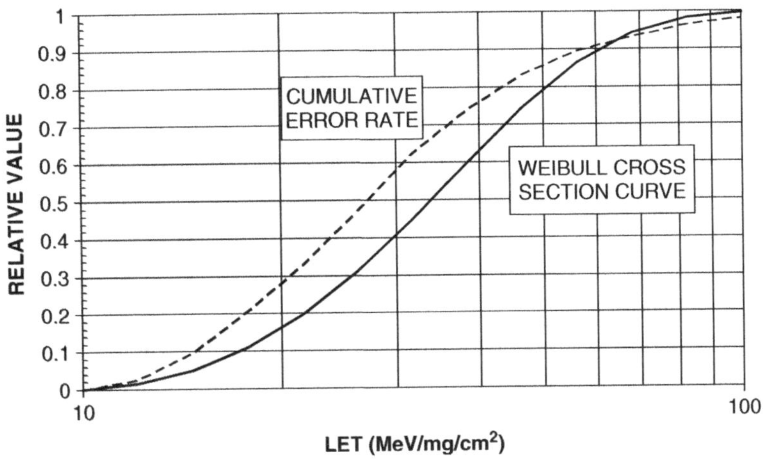

Figure 7-12 Cumulative upset rate contributions compared with the Weibull distributions as a function of LET.

1987]. They obtained good agreement with space data. Luke also recommended the integral RPP approach [Luke 1988].

Figures 7-11 and 7-12 show the contribution to the upset rate as a function of LET. In Chapter 3, Figure 3-5, we showed Figure 7-11 with a logarithmic scale.

Table 7-1
Spreadsheet Showing the Two IRPP Modes of Calculation of Upset Rate Indicated in Figure 7-10[a]

Upset Rate Calculation Using the IRPP Method					
LET steps	pC/1 micron	rate(Qc)	Weibull = RCS	RCS*d(rate)	rate*d(RCS)
8.256	0.084	4.68E−06	0.000	0.00E+00	0.00E+00
10.003	0.102	3.10E−06	3.06E−07	2.42E−13	1.19E−12
12.12	0.124	1.99E−06	0.014	7.93E−09	3.64E−08
14.68	0.150	1.23E−06	0.050	2.44E−08	5.74E−08
17.79	0.182	7.69E−07	0.109	3.67E−08	5.92E−08
21.55	0.220	5.07E−07	0.195	3.99E−08	5.49E−08
26.11	0.266	3.20E−07	0.309	4.72E−08	4.71E−08
31.63	0.323	1.83E−07	0.447	5.18E−08	3.47E−08
38.32	0.391	1.09E−07	0.598	3.87E−08	2.21E−08
46.43	0.474	6.35E−08	0.744	3.05E−08	1.26E−08
56.25	0.574	3.71E−08	0.865	2.12E−08	6.04E−09
68.15	0.695	2.31E−08	0.944	1.26E−08	2.40E−09
82.57	0.843	1.20E−08	0.984	1.07E−08	6.94E−10
100.04	1.021	6.54E−09	0.997	5.41E−09	1.24E−10
121.20	1.237	4.22E−09	1.000	2.32E−09	1.47E−11
146.84	1.498	1.99E−09	1.000	2.23E−09	8.73E−13
177.90	1.815	9.84E−10	1.000	1.01E−09	1.75E−14
215.53	2.199	4.11E−10	1.000	5.73E−10	1.03E−16
261.13	2.665	1.45E−10	1.000	2.66E−10	1.01E−19
316.37	3.228		1.000	1.45E−10	7.11E−24
			Sum =	3.336E−07	3.336E−07

[a]This calculation combines the RPP upset rate calculation indicated in Figures 7-6 and 7-9 with the device characteristic shown in Figure 7-9 and previously in Chapter 2, Figures 2.67 and 2.68.

Petersen and co-workers assembled a review paper that performs a critical discussion of the foundations of the RPP approach [Petersen 1992b]. It suggests the use of the 50% cross section to represent the critical charge and presents results from the literature that supports this view. It proposes the use of the integral Weibull distribution to fit the experimental cross-section data. The paper emphasizes the importance of integration over the experimental cross-section curve, the integral RPP approach. The paper proposes standards for basic experimental data and proposes approaches for interpreting data for SEU hard parts.

A 1993 review discussed some of the geometrical effects important for rate calculations and makes a comparison of various approaches to upset rate calculations [Petersen 1993]. The geometrical effects show up in the laboratory. There can be distortion of cross-section data due to finite depth effects, and it may be necessary to correct the effective LET scale due to funneling effects. These corrections can be derived from the data and applied to produce consistent data that fall on the Weibull curve. They do not change the integral RPP rate predictions, which are based on the Weibull curve through the points taken at normal incidence. The paper pointed out the importance of the truncation of the funnel and of the inclusion of the funnel in rate predictions. These effects were discussed in Chapter 5, Section 5.3.

The paper introduced the modified figure of merit equation for approximate rate calculations. It then discussed a variety of rate prediction techniques and presented rate results for a group of representative devices.

7.4 SHAPE OF THE CROSS-SECTION CURVE

7.4.1 The Weibull Distribution

Marshall suggested that device characteristic upset curves might be described using the Weibull integral function [Petersen 1992b]. Petersen verified this for a number of devices. The Weibull distribution is

$$F(L) = 1 - \exp\{-[(L - L_0)/W]s\} \qquad L > L_0 \qquad (7.4)$$

$$= 0 \qquad L < L_0$$

where L is the effective LET, L_0 is the device threshold, W is the width parameter, and s is the shape parameter. This function is then normalized to the limiting cross sections of the device: CS. Figures 2.67 and 2.68 in Chapter 2 introduced a sample Weibull function.

Appendix D lists the characteristics of a number of representative devices taken from the literature. Figures 7-13 and 7-14 show the cross-section curves for some of these devices. The curves are shown both for the traditional logarithmic cross-section scale and for the linear scale that appears more appropriate for the Weibull shapes.

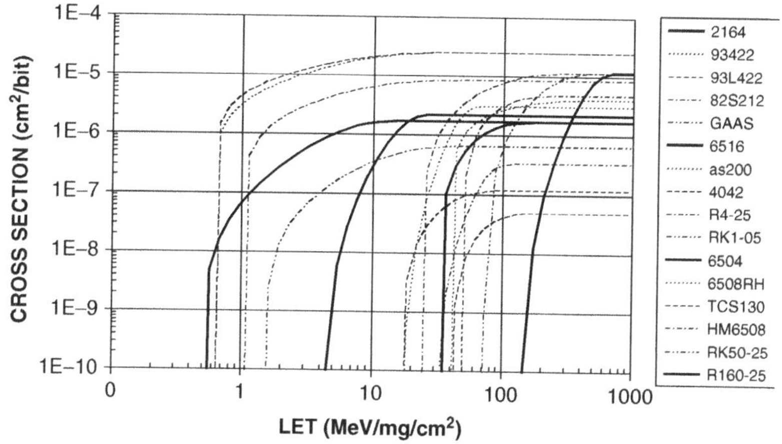

Figure 7-13 Upset cross section for representative devices.

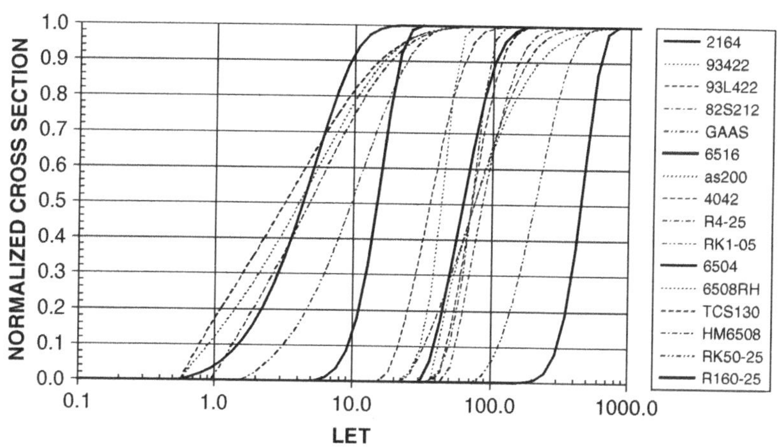

Figure 7-14 The cross sections of Figure 7-13 plotted to show the Weibull shapes.

The plot of cross section on a linear scale gives a much better indication of the shape of the cross-section curve. The plot of the cross section on a linear LET scale is misleading in two ways. (1) It tends to emphasize the low cross-section points. These are difficult and time consuming to obtain. As we saw in Figures 7-11 and 7-12, the low cross-section points have little impact on the error rate calculated. (2) The logarithmic cross section–linear LET plot tends to give a false impression of saturation. The experimenter makes much better use of her time if she investigates the saturation cross section and does not attempt to make a careful measurement of the "threshold."

Some physicists immediately think of energy straggling when they first encounter the cross-section curves. In Section 3.9 we showed energy straggling curves. They are much narrower than these curves. Furthermore, energy straggling depends primarily on the incident ion. We observe that the shape of these curves depends primarily on the device.

7.4.2 Lognormal Distributions

Now that we believe (from Section 2.8) that the cross-section curve results from a charge-collection gain distribution across a transistor, we are led to another possible mathematical form to describe the distribution: that is the cumulative lognormal distribution [Petersen 1996c]. See Equation (6.1) and Appendix B, Section B.2. This distribution was previously introduced by Massengill to describe the cross-section curve for SOI devices [Massengill 1993a]. Figure 7-15 shows a fit of the cumulative lognormal distribution with Koga's data for the IDT 71256 [Koga 1988]. There is a very good fit over the data shown. The lognormal distribution has a long upper tail,

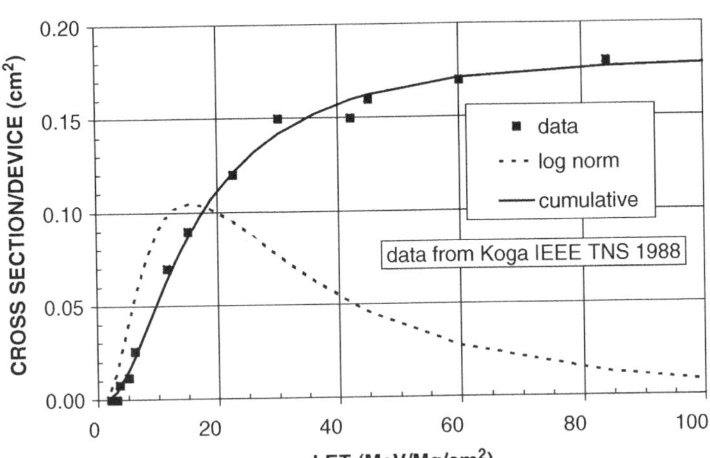

Figure 7-15 A fit of the cumulative lognormal function with SEU cross section data taken by Koga [Koga 1988]. This data has little spread, with 9% fractional standard deviation.

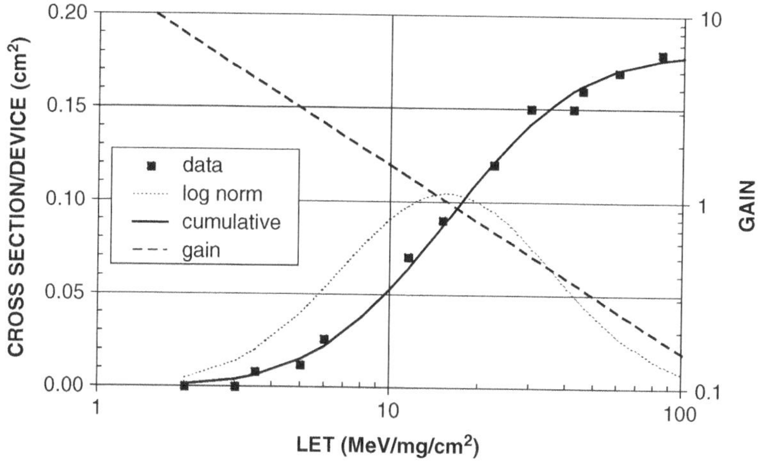

Figure 7-16 The lognormal distribution applied to the IDT 71256. This is plotted so as to emphasize the normal shape of the distribution on log coordinates.

so that may explain why some parts have not appeared to saturate. Figure 7-16 shows the same data in such a way as to emphasize the normal distribution of the gain. The internal gain is the ratio of the LET at the peak of the distribution to the LET at a point on the curve. There must be high gain at low LETs for the deposited charge to generate the critical charge. In this type of plot the gain can be plotted as a straight line diagonally across the plot.

7.4.3 Exponential Distributions

An exponential distribution has been used to fit SEU data [Smith 1992]:

$$\sigma = \sigma_0 e^{-K/(L-L_0)} \tag{7.5}$$

Edmonds [Edmonds 1996] suggested an exponential without the threshold term:

$$\sigma = \sigma_0 e^{-k/L} \tag{7.6}$$

The exponential form fits some data, but does not present uniformly good fits. Figures 7-17–7-19 show comparisons of Weibull and exponential functions. Figure 7-20 shows three different curves

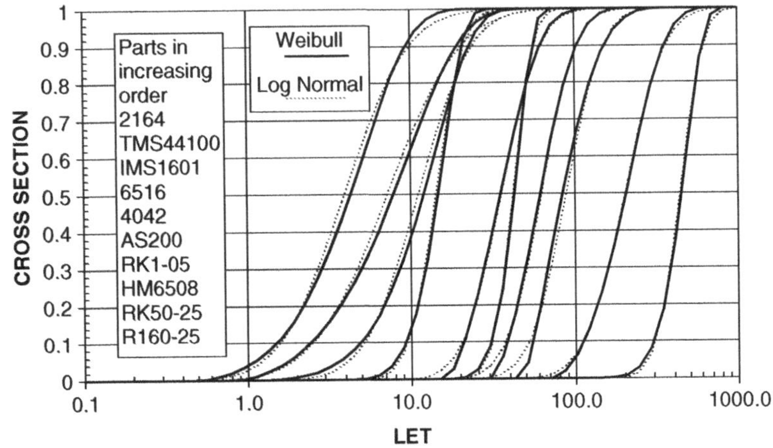

Figure 7-17 Comparison of Weibull and lognormal distributions for MOS, CMOS, and CMOS EPI devices.

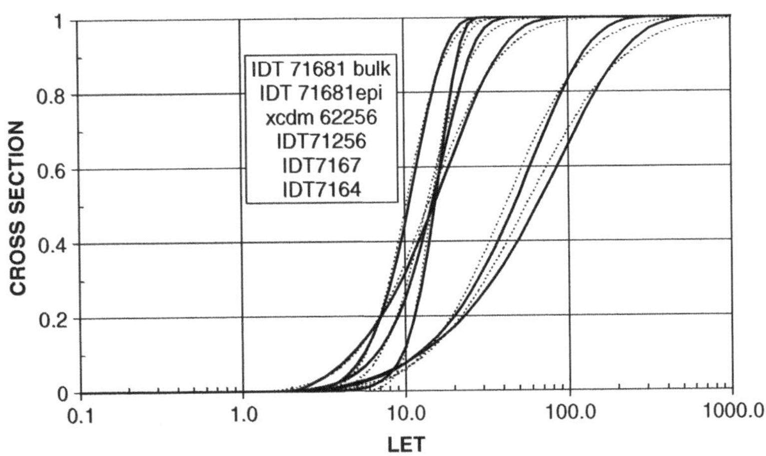

Figure 7-18 Comparison of Weibull and lognormal distributions for RMOS devices. The IDT 71681 data is from [Campbell 1985].

fit to the same set of data. In this case the fits are reasonably good over the data shown. If you compare fits matching the curves at the threshold and 50% points, the comparisons are often poor. This is due to the fact that the thresholds correspond to field and funnel effect charge collection, whereas the high LET corresponds to diffusion and is better described by the exponential function.

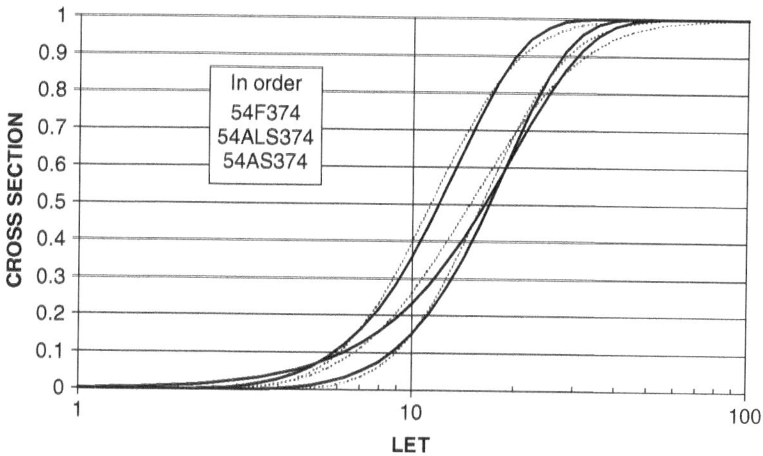

Figure 7-19 Comparison of Weibull and lognormal distributions for advanced bipolar devices.

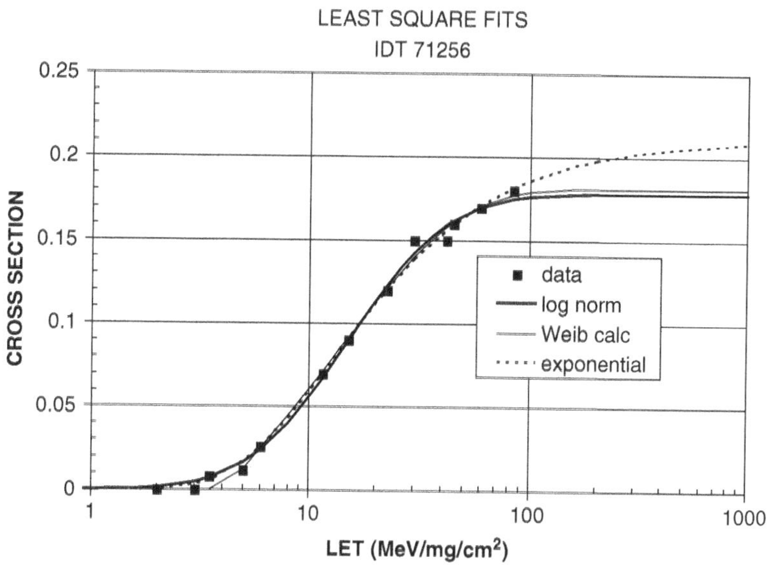

Figure 7-20 Three fits to data. The exponential and lognormal forms do not have threshold terms in this plot. Other devices may require a threshold with the exponential fit.

7.5 ASSUMPTIONS BEHIND THE RPP AND IRPP METHODS

7.5.1 Device Interaction Models

Device interaction models are needed to relate the measured test results to the predicted ion environment and make a prediction of SEU rate. The fundamental assumption in all SEU rate models is that there is a sensitive region within a circuit element, the sensitive volume (SV), that can collect charge generated by the passage of a heavy ion. A SEU occurs when the electrical disturbance in the circuit passes some critical threshold so that the circuit responds in an uncommanded way. Possible circuit effects are switching to another state for bistable memory elements (memory upset) or propagation of a transient to following circuits which lead eventually to toggling of a bistable memory element (logic upset).

The interaction of ions with the SV of active circuits is a very complicated process that is not amenable to analysis without many simplifying assumptions. Several approaches to the problem have been developed, but in the following discussions we show that all the approaches evolve from the same concepts and mathematical basis. This discussion is based on an analysis by E. C. Smith that was included in [Petersen 1992b]. A parallel approach from a general framework has been presented by Weller [Weller 2009]. His framework encompasses both the RPP and IRPP models and the physical models that are encompassed in the Monte Carlo codes. He specifies the assumptions that differentiate the IRPP and Monte Carlo approaches.

7.5.2 Critical Charge

The subject of charge collection from ion tracks is well documented in the literature (see *IEEE Transactions of Nuclear Science* over the last 30 years). For a SEU to occur, a circuit is disturbed by the charge generation from a passing ion to the point of changing logic state. For the circuit there is a set of current pulses that can cause the logic to change. The minimum amount of charge in a current pulse at the circuit node is the critical charge Q_C. If the current pulse produced by the ion is short compared to the integration time constant of the circuit (including parasitic capacitances), the critical charge is independent of the pulse shape and is a constant of the circuit. In other instances Q_c will depend on the circuit and the shape of the current pulse. In any event, critical charge is a circuit parameter. If there are amplification processes in the device, it is not a simple circuit parameter.

The charge generated by an ion in the SV may be different from the charge collected at the circuit node. The principal charge-collection processes are drift in the depletion region, enhanced collection by drift through the funneling process, and diffusion. From the circuit standpoint, it does not matter what process is responsible for the charge that is collected within the circuit response time; that is, drift, funneling, and fast diffusion all can contribute to the collected charge causing upset. The processes may enter into circuit modeling of critical charge due to their different temporal characteristics.

7.5.3 Mathematical Basis of Rate Equations

When a device is exposed to the heavy ion environment in space, the rate of upset, or other ion-induced effects, can be expressed in spherical coordinates as

$$Rate = \sum \iiint Kt(Z,A,E,\vartheta,\varphi)D(Z,A,E,\vartheta,\varphi)\, dE \,\sin\vartheta \,d\varphi \,d\vartheta \tag{7.7}$$

where D is the flux of ions with atomic number Z and mass number A, and differential in energy E and solid angle. We choose a spherical coordinate system with a polar axis along the normal to the plane of the device, so we can define the direction with the azimuth angle θ, and polar angle, φ.

The term Kt has the units of area and we designate it the transport kernel. It will combine the energy deposited in the sensitive volume with the critical charge for upset. It is zero if the deposited charge cannot cause upset. It is nonzero if the deposited charge can cause upset.

The sum in Equation (7.7) has hundreds of terms, one for each isotope in the periodic table. However, since stopping power or linear energy transfer (LET) is primarily dependent on Z, practical consideration requires only 92 elements. A model for the differential flux environment for each of the 92 elements is available in the CREME [Adams 1986] set of computer programs (in which 7 have an abundance of 0). However, the transport kernel cannot be readily determined. Simplifying assumptions are required to make the problem amenable to analysis. In the following, we consider the various assumptions.

Assumption 0: The energy deposited in a sensitive volume is equal to the energy loss of an energetic ion passing through that volume as calculated using its LET. The charge generated by the ion then corresponds to the energy deposited. This assumption may not be necessarily true for very short path lengths [Xapsos 1992]. This is due to the finite range of the secondary electrons, which can cause a portion of the energy lost by the ion to escape the sensitive volume. Dodd concluded that this is not a major effect [Dodd 1998]. See the discussion in Section 6.9.

Assumption 1: Ions with the same linear energy transfer (LET) have the same effect. This allows a great simplification of the analysis since ions can be grouped and treated by LET. An approach for developing a LET spectrum was first provided by Heinrich [Heinrich 1977]; models exist in the CREME code for calculating the LET spectrum (Heinrich spectrum) for any given orbit, interplanetary environment, and shielding case. For some time it was thought that exceptions would arise for situations in which the same ion at two different velocities has the same LET but different track structure. Early studies were conducted at several laboratories [Dussault 1995, Duzellier 1995, Connell 1996]. However, we saw in Section 6.8 that this appears to be a valid assumption.

Assumption 2: The change in LET along an ion track in the region of interest is negligible. This assumption is needed to preclude problems associated with the fact that ions with the same LET may have different ranges, and therefore deposit different amounts of energy in the same volume. Range effects are more of a potential problem with accelerator simulations, as discussed elsewhere in this book, than with the high energy environment encountered in space. This is sometimes referred to as the cosmic ray approximation.

Under Assumptions 1 and 2 the SEU rate can be calculated with a single triple integral as

$$Rate = \iiint Kt(L, \vartheta, \varphi)D(L, \vartheta, \varphi)\, dL \sin\vartheta\, d\varphi\, d\vartheta \qquad (7.8)$$

where L is the LET of the ion, K is an effective cross section that we designate variously as beam cross section or upset cross section or sigma (σ), and the other terms are as defined for Equation (7.7).

Here it will be more convenient to use the cosine to the polar angle, $\mu = \cos\theta$, than the angle itself. Furthermore, we define the cross section by

$$S(L, \varphi,\ \mu) = K(L, \theta, \varphi)/\mu \qquad (7.9)$$

Then the rate, Equation (7.7), becomes

$$Rate = \iiint S(L, \varphi, \mu) f(L, \varphi, \mu)\mu\, d\varphi\, d\mu\, dL \qquad (7.10)$$

The degree of simplification to this point is great. For predicting SEU rates, the required task consists of experimentally determining the cross section as a function of the direction of incidence and of the LET, and integrating it with the flux appropriate for the environment of concern. The basis of most current models for prediction of SEU rates is Equation (7.10). However, further simplifying assumptions are required for any practical calculation.

The direct approach for determining the rate of SEUs in a given SV for a known flux of ions distributed in LET is to determine the number of ions, for each value of LET, which intersect the SV in such a way that the threshold charge, Q_c, is generated. The issue is complicated by the fact that the SV is a three-dimensional shape and its dimensions can be a function of ion LET and angle of incidence on the SV in some instances. Since generated charge is directly proportional to the path through the SV, ion path length must be taken into account. In addition, there are complicated processes associated with charge collection that need to be considered.

To make practical use of Equation (7.10), the following assumptions are usually made.

Assumption 3: The charge generation is equal to the product of the LET of the ion and a chord of the region, perhaps augmented by a funnel region or a diffusion region. We designate the augmented chord as the "effective chord."

With this assumption, rate prediction models have been developed based on chord lengths through the sensitive volume. The most common models assume a rectangular parallelepiped (RPP) shape for the SV.

7.5.4 Chord Length Models

Consider the sensitive volume. For each direction in space, θ, φ, there is then an area projected in the plane normal to θ, φ—$S(x, \theta, \varphi)$—such that the charge-collection path exceeds a value X. If X exceeds the largest path length in the SV for that direction, then $S = 0$. If the critical charge is Q_c, and the ion has LET L, then an upset will occur if $x \geq Q_c/L$. The number of upsets from direction θ, φ from ions with LET in dL about L is then given by

$$dR = D(L, \theta, \varphi) S(Q_c/L, \theta, \varphi) \sin\theta \, dL \, d\theta \, d\varphi \tag{7.11}$$

where rate is the event rate and the other terms are defined above. This approach is assuming a single critical charge corresponding to a single charge deposition. The traditional RPP calculations then allow for varying the amount of charge deposition leading to a single critical charge by integrating over the cross-section curve in the IRPP method. Connell arrived at the same results by introducing a Weibull probability of upset at this point in the development [Connell 1995a, Connell 1995b].

Assumption 4: The charge collection path is independent of the LET. When this assumption is not valid, the projected area, S, is a function of L, as well as of θ, φ, and Q_c.

With this assumption, the event rate is

$$Rate = \iiint f(L, \vartheta, \varphi) S(Qc/L, \vartheta, \varphi) \sin\vartheta \, dL \, d\vartheta \, d\varphi \tag{7.12}$$

The model can be made more tractable to analysis with additional assumptions.

Assumption 5: The sensitive volume is a convex body. The charge collected from an ion track is that generated along the chord defined by the path through the sensitive volume. Many sensitive volumes are not simple convex shapes but are "L" shaped or even more complicated. However, there is little error made in the rate calculation for assuming a rectangular parallelepiped volume, as will be discussed in a following section.

Assumption 6: The particle flux is isotropic at the device and therefore the LET spectrum is the same for all directions. This is not strictly true because massive parts of the spacecraft may shadow certain directions. Also, for satellites in low earth orbit, the geomagnetic cutoff is not the same in all directions and indeed some directions are shadowed by the solid earth. Letaw and coworkers have investigated the effects of this assumption on SEU rates in the General Purpose Computers of the Space Shuttle and find that this assumption had negligible effect on the final SEU rates [Letaw 1985]. In principle, the effective flux approach of Binder can be used for nonisotropic flux, but this has not been applied [Binder 1988]. The Connell HICCUP model can also be used in this case [Connell 1995a, Connell 1995b].

There are then several simplifications to the rate integral. The integral over directions involves only the projected area for upset. The projected area as a function of path can be converted to a distribution of chords in the convex sensitive volume. If the distribution of chords in a sensitive volume is known, the event rate in space can be calculated.

The approach to calculation of single event effect rates by heavy ions depends on the concepts of a sensitive volume for charge collection, with a distribution of chords, and a critical charge deposited in that volume that can cause the effect. The first paper that discussed rate calculations used a thin disk [Binder 1975], but the standard volume for upset rate calculations soon became the rectangular parallelepiped—the RPP [Pickel 1978, Bradford 1979, Pickel 1980, Adams 1983]. This approach has been incorporated into the standard codes for calculating heavy ion upset rates in space. The new code CREME-MC may allow other geometries.

The various RPP codes are based on one of two approaches: an integral chord length distribution convolved with a differential flux distribution or a differential chord length distribution convolved with an integral flux distribution. The integral chord/differential flux and differential chord/integral flux approaches produce the same results. The differential chord/integral flux approach may be preferred because the differential LET spectrum contains two Dirac delta functions for each element included in the LET spectrum. These spikes are awkward to plot since they (formally) go to infinity. They occur at the points where $d/dE(dE/dx)$ has zeros. The differential chord approach also includes delta functions when the chord length matches one of the RPP dimensions.

Note that the charge-collection effects attributed to the "funnel" can be included in the above model in either of two ways. One can assume a RPP shape for the sensitive volume, whose dimensions include the augmented chord length. The depth of the sensitive volume then includes a funnel contribution. The preferred approach, to be discussed later, is to add the funnel length to the charge-collection path, but not change the basic RPP volume. Most of the codes allow a separation of an intrinsic sensitive volume and an additional path length attributed to the funnel.

It is possible to use geometries other than the RPP geometry. Langworthy explained the shape of the cross-section curve using a depletion region geometry analysis and calculated the path length distribution for a hemisphere [Langworthy 1989a, Langworthy 1989b].

7.5.5 Bradford Formulation

The sensitive volume is represented by a parallelepiped with dimensions a, b, c. The minimum dimension is a. The Bradford formulation was used in the CRUP program written by Phil Shapiro as discussed in [Petersen 1982]. In order to obtain the upset rate, one calculates the probability that a cosmic ray particle passing through the sensitive volume will create ionization along it path greater than the critical charge Q_c that will change the logic state of a device. The energy deposited traversing a chord of length s through the sensitive volume,

E_{dep}, is related to the linear energy transfer (LET), L, by

$$E_{dep} = L\rho s \tag{7.13}$$

where ρ is the density of the material. When E_{dep} is greater than ΔE, a critical energy, a sufficient number of electron–hole pairs are created to cause an upset. The critical charge is related to ΔE, the critical energy by $Q_c = \Delta E/22.5$ in Si, where Q_c is in pC and ΔE is in MeV. Thus, an event can occur if the chord length associated with the particular passage of a particle of LET L is greater than s_{min}, where

$$s_{min} = \Delta E/\rho L \tag{7.14}$$

The number of events is calculated using a formulation due to Bradford:

$$N = \frac{S}{4}\int_{L_0}^{L_{max}} \Phi(L)C\left(\frac{\Delta E}{\rho L}\right) dL \tag{7.15}$$

$C(s_{min})$ is the integral chord length distribution, the probability of a particle traversing the sensitive volume with chord length greater than s_{min}. S is the total surface area of the volume

$$S = 2(ab + ac + bc)$$

The differential LET distribution $\Phi(L)$ is obtained from the differential distribution in energy $\Phi(E)$ using the transformation

$$\Phi(L) = \Phi(E)\, dE/dL$$

The particle flux Φ is integrated over 4π steradians.

The minimum value of L that will produce an event L_0 is

$$L_0 = \Delta E/\rho s_{max}$$

where s_{max} is the diagonal of the parallelepiped

$$s_{max} = \sqrt{a^2 + b^2 + c^2}$$

L_{max} is the maximum of the LET spectrum.

It is useful to restate the integration in terms of a critical LET L_c, rather than the critical energy ΔE. The critical LET L_c is defined so that no L less than that can cause an upset when passing through the cell at normal incidence, that is, passing through a path length a. The incident flux can cause upset for a given L, for path lengths greater than L_c/L times a. The larger L is, the shorter the path length can be. The minimum L that can cause an upset is that corresponding to the maximum possible path length. This value is given by L_0. For cases near the minimum L that can cause upset, the corresponding path lengths are very long, corresponding to a large value of $L_c a/L$.

1. Restate basic equation in terms of L_c:

$$\Delta E = L_c \rho a$$

again $s_{\min}$, the minimum path length that will produce upset, is

$$\Delta E/\rho L = L_c \rho a/\rho L = L_c a/L$$

$$L_0 = \Delta E/\rho s_{\max} = L_c \rho a/\rho s_{\max} = L_c a/s_{\max}$$

$$N = \frac{S}{4} \int_{L_0}^{L_{\max}} \Phi(L) C\left(\frac{L_c}{L} a\right) dL \tag{7.16}$$

$$L_0 = \frac{a}{s_{\max}} L_c \tag{7.17}$$

2. Now we want to examine the introduction of the funnel effect. We assume a funnel length f. ΔE is now $L_c(a+f)$ and $s_{\min}$ has changed:

$$(s_{\min} + f)L = \Delta E = L_c(a+f)$$

$$s_{\min} = \frac{L_c}{L}(a+f) - f \tag{7.18}$$

Note that $s_{\min}$ can be less than zero; therefore, you need to check the limits and cutoff at zero. This says that there are upset contributions from some paths that just nick a corner or edge of the RPP!! Note that this could explain test results on 6504 that seemed to say that the device upset wherever it was hit.

L_0 also changes:

$$\Delta E = L_0(s_{max} + f) = L_c(a + f)$$

$$L_0 = \frac{a + f}{s_{max} + f} L_c \tag{7.19}$$

These two changes appear consistent with description of the changes made in the change from CRUP to HCRUP. That is, f is subtracted from the minimum chord length, for a shift in the chord length table, and f is added to s_{max} in calculating L_0.

7.5.6 Pickel Formulation

This formulation was used in the CRIER code written by Jim Pickel [Pickel 1980]. The discussion is based on the special issue paper [Pickel 1996].

$$N(E_c) = A_p \int_{s_{min}}^{s_{max}} \Phi(L_{min}) f(s) ds \tag{7.20}$$

$$s_{min} = E_c / L_{max} \tag{7.21}$$

$$s_{max} = \sqrt{a^2 + b^2 + c^2} \tag{7.22}$$

$$L_{min} = E_c / s \tag{7.23}$$

A_p is the average projected area $= \frac{1}{2}(ab + ac + bc)$.

Φ is the integral omnidirectional flux distribution.

$f(s)$ is the differential path length distribution.

E_c is the threshold energy for generating Q_c, the critical charge for upset. If we assume a step function at L_c, then $E_c = L_c/a$, where a is the thickness of the RPP. Experimentally, Q_c is calculated from E_c, which is calculated from L_c.

s_{min} is the minimum path length to deposit critical energy E_c.

L_{max} is the maximum LET in the environment.

s_{max} is the maximum path length in the RPP.

L_{min} is the minimum LET that can deposit critical energy in path length s.

We can again introduce the funnel in this formulation. L_{min} and s_{min} will change:

$$L_{min} = E_c/(s+f)$$

$$L_{min} = \frac{a+f}{s+f} L_c \tag{7.24}$$

s_{min} is the minimum path length in the RPP that can lead to upset. It needs to be calculated realizing that the funnel also contributes, but is outside the RPP.

$$L_{max}(s_{min}+f) = E_c = L_c(a+f)$$

$$s_{min} = \frac{L_c}{L_{max}}(a+f) - f \tag{7.25}$$

Note that this value can be less than zero; we can get contributions from the corners.

7.5.7 Adams Formulation

Adams reformulated the integral, keeping the integral flux, differential path length distributions, but changing the integral to be over LET. This is the approach used in the widely used CREME program [Adams 1983, Adams 1986, Tylka 1996, Tylka 1997]. We again introduce the critical charge Q_c, the critical energy E_c, and the critical LET, L_c; where a is the thickness of the RPP. If Q_c is experimentally determined, then it is found from the critical LET and the thickness. Later when we introduce the integration over the measured cross section, that cross section is presented as a function of effective LET. A given effective LET corresponds to the critical LET for upset at a device thickness a.

$$E_c = 22.5Q_c \text{ MeV/pC in silicon}$$

$$= L_c \times a$$

The minimum LET that can produce an upset is

$$\begin{aligned} L_{min} &= E_c/s_{max} \\ &= L_c \times a/s_{max} \end{aligned}$$

The integral will be from L_{min} to the largest LET in the spectrum, L_{max}:

$$L_{max} = 1.05 \times 105 \text{ MeV cm}^2/\text{g}$$

For each value of L, there is a corresponding path length at which upset occurs. We want to evaluate the probability distribution at that point:

$$s = E_c/L(I)$$

Adams used the relationship

$$\int f(y)dx = \int \frac{f(y)dy}{dy/dx} \tag{7.26}$$

$$L = E_c/s$$

$$dL/ds = -E_c/s^2 = -L^2/E_c$$

The minus sign disappears in the change of the order of the integration limits. Then the Pickel formulation in Equation (7.20) becomes

$$N(E_c) = A_p E_c \int_{L_{min}}^{L_{max}} \frac{1}{L^2} f(s(L))\Phi(L)dL \tag{7.27}$$

where $s(l) = E_c/L$ and $L_{min} = E_c/s_{max}$. Here S is the surface area of the sensitive volume $2(ab+ac+bc)$ and the flux is expressed in units of particles/m^2 ster-s.

We can introduce the funnel in this formulation. If we take the measured L_c as the basic measurement, then the longer path length furnished by the funnel means that the critical energy and the critical charge are larger.

$$E_c = L_c(a+f) \qquad Q_c = E_c/22.5 = L_c(a+f)/22.5 \qquad \text{(pC/MeV)}$$

L_{min} will be lower:

$L_{min} = E_c/s_{max}$ becomes $L_{min} = L_c \times (a + f)/(s_{max} + f)$

s will change:

$s = E_c/L(I)$ becomes $s + f = E_c/L(I) = (a + f) \times L_c/L(I)$

The resulting Adams formulation with funneling then uses the expressions

$$E_c = L_c(a + f) \tag{7.28}$$

$$s(L) = (a + f)\frac{L_c}{L} - f \tag{7.29}$$

$$L_{min} = \frac{a + f}{s_{max} + f} L_c \tag{7.30}$$

7.5.8 Formulation of Integral RPP Approach

First we will present the approach for integrating over the experimental cross-section curve assuming that it represents a range of critical charges. The basic upset rate equation for a single critical charge was given in Equation (7.16). In that equation, we expressed the rate in terms of a critical LET L_c. This basically assumes that the cross-section curve is that of Chapter 2, Figure 2.66. We now assume that the cross-section curve represents a variation with LET as shown in Chapter 2, Figures 2.67 and 2.68. That is, there is a range of critical LETs. We need to sum over these contributions. We introduce a second integration variable, L_w. The cross-section curve is expressed as $w(L_w)$. Then L_c in the above equation becomes L_w

$$N = \frac{S}{4} \int w(L_w) \int_{L_0}^{L_m} \Phi(L) C\left(\frac{L_w}{L} a\right) dL\, dL_w \tag{7.31}$$

$$L_0 = \frac{a}{s_{max}} L_w \tag{7.32}$$

We now want to see how this equation appears when we assume that there is a single critical energy that causes upset. We introduced the evidence for this in Section 2.8. The cross-section curve is caused by changes of gain, or charge collection, or depth across the cell. The depths are relatively well defined. The relationship can be expressed as changes of depth [Petersen 1996a]. That is probably a good analogy, but a poor explanation to use with cross section and rate calculations.

We need to introduce a factor so that, at low cross section values, there is a multiplier in the energy deposition (other than length) that corresponds to L_0/L. We need another factor that corresponds to the reduced cross section. At high energies the ratio will still be L_0/L so that less energy is deposited. We again use the corresponding cross section.

There are two energies of interest:

1. The critical energy for upset, E_c. This is a single energy. We can determine it from L_c, which is the LET at which the cross section is 0.5. This is the energy that must be available at the node for the circuit to upset. There is a corresponding Q_c, the deposited charge for upset.

2. The energy that must be deposited for upset at some particular part of the cross-section versus LET curve. This is a function of the location on the cross-section curve. On the low cross-section portion of the curve this is less than E_c due to the gain of the circuit (which we have called efficacy). We assume that the cell always has the same depth. Then

$$E_{dep} \times L_c/L = E_c$$

The L in this case corresponds to the point on the cross-section curve. We need a different symbol for it, as it is not the same L that is integrated over in the rate calculation. We will call it L_w

$$E_{dep} \times L_c/L_w = E_c$$

$$E_{dep} = L_w E_c/L_c$$

At low L_w, E_{dep} can be less than E_c due to the extra gain that is present.

The minimum path length for upset, which determines path length distribution,

$$s_{min} = E_c/\rho L$$

is now a function of L_{W}. s_{min} will be less for low L_{W}:

$$\begin{aligned}
s_{\mathrm{min}} &= E_{\mathrm{dep}}/\rho L \\
&= E_{\mathrm{c}}L_{\mathrm{W}}/\rho L L_{\mathrm{c}} \\
E_{\mathrm{c}} &= L_{\mathrm{c}}\rho a \\
s_{\mathrm{min}} &= E_{\mathrm{c}}L_{\mathrm{W}}/\rho L L_{\mathrm{c}} \\
&= L_{\mathrm{c}}\rho a L_{\mathrm{W}}/\rho L L_{\mathrm{c}} \\
&= L_{\mathrm{W}}a/L
\end{aligned}$$

Note that, surprisingly, L_{c} has disappeared. We have the same relationship that we had assuming that L_{c} varies as L_{W}.

L_0, the minimum L at which upset can occur, will also change. At low L_w, it will be lower due to the gain mechanism:

$$\begin{aligned}
L_0 &= E_{\mathrm{c}}L_{\mathrm{W}}/\rho s_{\mathrm{max}}L_{\mathrm{c}} \\
&= L_{\mathrm{c}}\sigma a L_w/\rho s_{\mathrm{max}}L_{\mathrm{c}} \\
&= L_{\mathrm{W}}a/s_{\mathrm{max}}
\end{aligned}$$

Again, we have the same relationship that we had for varying L_{c} !!?

$$N = \frac{S}{4}\int w(L_{\mathrm{W}}) \int_{L_0}^{L_{\mathrm{m}}} \Phi(L) C\left(\frac{L_W}{L}a\right) dL\, dL_{\mathrm{W}} \tag{7.33}$$

where

$w(L_{\mathrm{W}})$ = differential Weibull function

$L_0 = L_{\mathrm{W}}a/s_{\mathrm{max}}$

L_{m} = maximum L of Φ (L) spectrum

7.5.9 HICCUP Model

Connell presented a different view of the IRPP calculation [Connell 1995b, Connell 1996]. His method allows for the fact that the measured cross section may depend on the orientation of the chip. He also uses an elegant approach to the geometrical effects in laboratory measurements that we discussed in Section 5.3 [Connell 1995a]. The geometrical effects were previously introduced by Petersen [Petersen 1992b, Petersen 1993]. His upset rate calculation uses an integral over the differential space flux and an angle integrated experimental cross section. The results of his calculation correspond with previous IRPP calculations [Petersen 1993, Connell 1995b].

Connell also introduces possible effects of track structure in the HICCUP model [Connell 1996]. He concluded that more analysis is needed before definitive conclusions can be reached about these effects. We presented such an analysis in Section 6.8.

7.5.10 Requirements for Use of IRPP

We can now summarize the information that we need for an upset rate calculation. If we describe the cross-section curve with the Weibull function, we need the four Weibull parameters. If we describe the cross-section curve with a lognormal function, we need a minimum of three parameters. This description may be slightly improved by the addition of a low LET cutoff. For the upset rate calculation we need the dimensions of the RPP volume. We can ordinarily assume that the surface is a square, so that the defining dimension is the square root of the limiting cross section. If diffusion plays a large role, then we may want to separate the area and the limiting cross section [Smith 1995]. The depth of the RPP needs to be determined in some way. If we do not know it, we can use the conservative assumption that it is one-tenth of the dimension of a side. We also need to know the funnel length. The conservative estimate is to assume that there is no funneling. We have seen (Chapter 5) that it is possible to determine these two parameters, as well as the Weibull parameters, from the experimental cross-section measurements.

7.6 EFFECTIVE FLUX APPROACH

Binder presented an alternative to the RPP calculations [Binder 1988]. He recognized that the upset rate could be calculated independent of path length distributions if the incident spectrum was converted to a form such that it is a function of effective LET. In this form it can be integrated directly with the measured cross section versus effective LET to obtain the upset rate. He used a two-term approximation to the differential LET spectra, based on [Petersen 1982]. He cut the integration off at a 10 micron path length and considered both a single threshold and a simple fit to the cross-section curve. He obtained analytic results similar to the figure of merit approach. He compared his calculations with space results and obtained good agreement for the 93L422 and two other relatively soft devices.

This approach was initially suggested in Binder, Smith, and Holman in 1975 [Binder 1975]. However, it was not developed in detail until Binder's Binder 1988 paper. The concept of effective flux was introduced originally to give a consistent view to experimental results. Cross sections are reported as a function of effective LET to allow easy normalization to results obtained at normal incidence. Binder recognized that the approach can also be applied to the description of the environment. The concept combines the increased number of particles that enter a sphere as the angle approaches a reference plane, but also the decreasing area that is intercepted. The resulting flux as a function of effective LET can be integrated directly with the cross section as a function of effective LET to determine the upset rate. This approach is basically independent of any assumptions about device geometry. However, some related assumption has to be made in the cutoff of the integration of the effective flux so that the calculation cuts off before reaching the plane of the device and assuming that very long path lengths are available.

This approach has been demonstrated to be functionally equivalent to the RPP approach by Watts and by Pickel [both unpublished]. The cross section can be expressed as cross section per bit or cross section per device. In principle, this approach can be used for nonisotropic flux. The approach does not appear to allow for side penetration, but that can depend on the implementation.

Scott developed a code for an approximate calculation of upset rates using the Binder effective flux approach [Scott 1989a, Scott 1989b]. The code calculates an approximate differential LET spectrum by taking coarse differences in the integral spectrum to determine the number of ions as a function of effective LET. The calculation cuts off at 80 degrees. The documentation did not specify the maximum effective LET to be calculated. This can vary depending both on the user and on the LET table used as input. The results depend on this parameter. The code then performs a numerical integration over the differential flux versus effective LET and the cross section versus effective LET to obtain the upset rate. This approach, using a LET cutoff at 110 MeV/mg/cm^2, was adopted in NASA documents SSP 30512 and SSP 30513 for use with the International Space Station Alpha (ISSA). Petersen explained the inconsistencies reported between the Scott effective flux calculations and integral rectangular parallelepiped calculations [Petersen 1993, Petersen 1995]. The two types of calculation yield consistent results when the effective flux calculation is performed allowing for the appropriate geometry. The approach using the Scott type of calculation is to not use a fixed angle cutoff as Scott did, but to use an angle cutoff that corresponds to the depth to width ratio.

7.7 UPPER BOUND APPROACHES

Edmonds was also concerned about the geometry assumptions that have to be made in the standard RPP calculations [Edmonds 1989]. He presented a superposition method for upper bound estimates. You calculate the maximum RPP rates at a series of steps on the cross-section curve, and then find the fraction of upsets at each step. He used this to calculate a term he called the effective flux, which represents the maximum upset rate contributions as a function of LET. He then adds up contributions to the total rate from each step. This is a variation of the integral RPP method. It is not equivalent to the Binder effective flux approach.

Chlouber presented another upper limit approach [Chlouber 1990]. It uses the integral LET, differential path length calculation. It considers the maximum upset rate possible if the device geometry is not known. The results are presented as a figure of merit type equation.

The development of upper bound approaches came about because of the problems with determining the depth of some devices. The depth is an important parameter in the RPP calculation of upset rate. In bulk CMOS devices especially, there is uncertainty about the depths of the collection volume. There is also uncertainty about the possibility of an additional apparent charge-collection depth due to funneling or diffusion processes. The upper bound approaches are one way of making a conservative rate calculation independent of depth. The problem with them is that the upper rates occur for very thin devices. Bulk devices are never very thin. If the devices are very thin, such as with SOI or SOS technology, the depths are reasonably well defined. Therefore, the upper bound approaches are much more conservative than is necessary.

7.8 FIGURE OF MERIT UPSET RATE EQUATIONS

The initial figure of merit approximation was introduced by Petersen in 1983 [Petersen 1993]. He recognized that there was a simple approximation for the space environment that could be introduced into the Bradford upset rate formulation. This is shown in Figure 7-21.

He then combined this with an approximate integral chord length distribution assuming a rectangular parallelepiped of dimensions *a, b, c* with $a, b > 3c$. The surface area S is introduced:

$$S = 2(ab + ac + bc) \quad \text{and} \quad r = (a^2 + b^2 + c^2)^{1/2}$$

Then the upset rate is

$$R \approx \frac{S}{L_c^2}\left[1 - \frac{9}{10}\frac{c}{r}\right]^{0.2}$$

This expression is simplified by writing S in terms of surface area and factors varying slowly as ratios of dimensions. If the ratios are

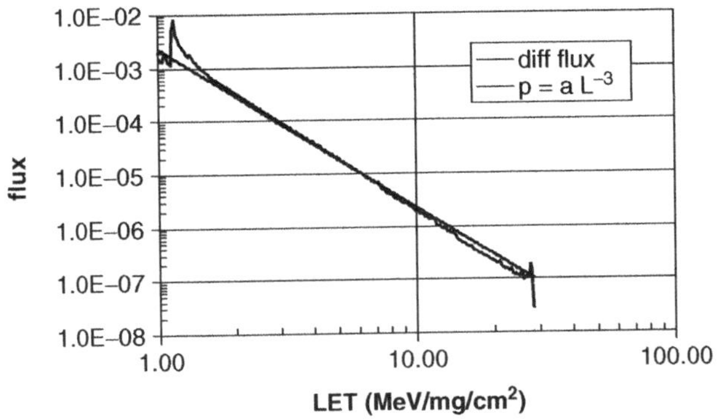

Figure 7-21 The differential flux in space can be approximated by a simple power expression.

approximated as constants,

$$R \approx \frac{a \times b}{L_c^2}$$

The upset rate can be expressed approximately in terms of device characteristics. If the device has been examined experimentally so that the limiting cross section σ_L and the critical LET L_c are known, we can write

$$R \approx \sigma_L / L_c^2 \tag{7.34}$$

If the device characteristics are known from device modeling, then the equation can be expressed in terms of the surface area dimensions a and b, the depth c, and the critical charge Q_c:

$$R \approx a \times b \times c^2 / Q_c^2 \tag{7.35}$$

The original figure of merit was based on the Adams 10% environment. The current figure of merit is based on the solar minimum environment and other slightly less conservative parameters [Petersen 1993]. (You may occasionally see it referred to as the modified figure of merit.) This is a good approximation for the upset rate in solar minimum geosynchronous orbit and useful for comparing relative performance of different devices. The paper pointed out that the figure of merit could be used for nongeosynchronous orbits, but did not give details about the application.

The figure of merit equation for geosynchronous orbit is

$$R = 200 \times \frac{\sigma_L}{L_{0.25}^2} \left[\frac{\text{upsets}}{\text{bit} - \text{day}} \frac{(\text{MeV/mg/cm}^2)^2}{\text{cm}^2} \right] \tag{7.36}$$

where σ_L is the limiting cross section and $L_{0.25}$ is the LET at 25% of the limiting cross section. The 25% factor arises as the cross-section curve is not a step function and approximates integration over the curve. As data for more devices became available, the coefficient is slightly modified as shown below.

If experimental data is not available, but device modeling data is, then the following expression can be used:

$$R = 2 \times 10^{-10} \frac{a \times b \times c^2}{Q_c^2} \left[\frac{\text{upsets}}{\text{bit} - \text{day}} \frac{\text{pC}^2}{\mu\text{m}^4} \right] \tag{7.37}$$

where a and b are the surface dimensions, c is the depth, and Q_c is the critical charge. These expressions were originated as terms for comparing device performance. With the present coefficients, they do a reasonable job of predicting upset rates in the geosynchronous solar minimum environment, as will be shown next. Sometimes the heavy ion cross-section curves are not available (and therefore $L_{0.25}$), but the threshold LET is. In this case you can use an approximation due to Rudy [unpublished]: $L_{0.25} = 2.77 L_{\text{th}}^{0.88}$.

7.9 GENERALIZED FIGURE OF MERIT

The figure of merit was developed for use with heavy ion induced upsets at geosynchronous orbit. The 1993 paper pointed out that the expression could be used for other orbits, but did not give any details. Petersen in 1995 discussed the general use of the figure of merit expression [Petersen 1995]. This paper points out that the coefficient in the figure of merit equation can be adjusted to fit an exact calculation of the rate for a large soft part, and then used to estimate the heavy ion upset rates for all other parts in a particular orbit.

The figure of merit is defined as

$$FOM = \frac{\sigma_{\text{HL}}}{L_{0.25}^2} \left[\frac{(\text{MeV/mg/cm}^2)^2}{\text{cm}^2} \right] \tag{7.38}$$

where σ_{HL} is the limiting heavy ion cross section per bit at large LET and $L_{0.25}$ is the LET at 25% of the limiting cross section. It is common to describe the heavy ion cross-section curve using the cumulative Weibull distribution, with threshold L_0, width w, and shape parameter s. In this case the $L_{0.25}$ term can easily be calculated:

$$L_{0.25} = L_0 + w \times 0.288^{(1/s)} \tag{7.39}$$

The upset rate R in a particular orbit is calculated by introducing an orbit-specific rate coefficient C with units of upsets/bit-day:

$$R = C \times FOM \tag{7.40}$$

7.9.1 Correlation of the FOM with Geosynchronous Upset Rates

The literature now contains a number of reports of heavy ion and proton cross-section curve parameters. The primary reference for our comparisons is the compilation of Calvel [Calvel 1996]. We will also use some Weibull parameters compiled by Petersen [Petersen 1996a, Petersen 1996b]. Additional devices are added from a study of 3.3 V 16 Mbit DRAMs [Harboe-Sorensen 1998]. Appendix D at the end of the book lists the device parameters for the devices used in this document.

There have been several critical reviews validating the FOM approach [Barak 1999b, Edmonds 2000, Hiemstra 2003, Normand 2004]. With the limiting cross section and $L_{0.25}$, the designer can calculate the upset rate in any orbit.

Figure 7-22 shows a plot of the calculated upset rates as a function of the FOM for a number of devices. We have calculated the rates for both the recommended solar minimum environment and the sometimes required 90% worst case environment. The upset rates are calculated using the IRPP (integral rectangular parallelepiped) standard codes and the Weibull curves for the parts [Petersen 1992b].

The CREME96 solar minimum results could be interpreted as falling on one curve with slight curvature. We will stay with the spirit of the figure of merit approach by interpreting it as having two straight line sections. The break is somewhat arbitrary but will be made as the parts transition from hardened parts to unhardened parts. The hardened parts have resistive hardening, or are SOI–SOS technologies, or are on EPI, or use proprietary DRAM hardening.

Fitting a straight line curve to the data points gives the rate coefficient C for that set of data. Both environments have a very good linear correlation of the upset rate with the FOM over several orders of magnitude. We minimize the sum of the squares of the differences of the logarithms of the calculated upset rate and of the line. The width factor of the fit is the ratio of the prediction to calculation

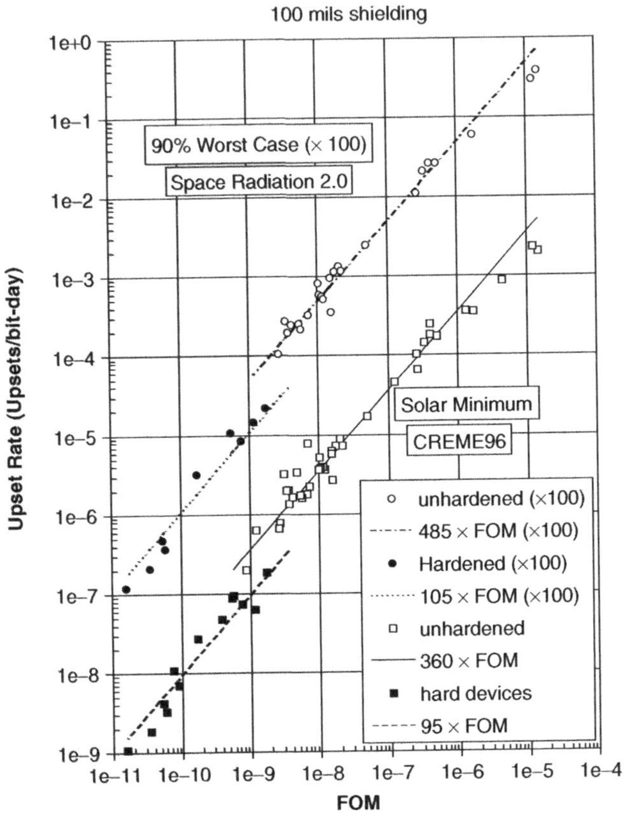

Figure 7-22 Geosynchronous upset rates as a function of the figure of merit. The scaling factors (485, 105, 360, and 95) are the rate coefficients relating the upset rate to the figure of merit (FOM). The results for the 90% worst case have been displaced by two decades to make the chart more readable.

corresponding to the logarithmic width (standard deviation) of the fit. For the solar minimum case the width factors for hard and soft parts are 1.56 and 1.52. There are slight shifts in the coefficients as we move from hardened parts to unhardened parts.

These two environments have slightly different LET spectra. This is reflected in the slightly different ratios of the rate coefficients between hard and soft parts in the two different environments. Sometimes one wants the 90% worst case upset rate predictions and has the CREME96 solar minimum predictions. We observe that there are upset rate ratios of 1.35 for unhardened parts and 1.10 for hardened parts between these two environments.

We have examined the LET fraction of 25% used in defining the figure of merit by using it as another parameter in the fit to the upset rates. The least squares fit is much worse if we use 50%. The fit is best at 12.5%, but the width factor only changes from 1.56 to 1.51. As this is much smaller than the overall uncertainty, the fraction has not been changed.

The rate predictions show good correlation with the FOM, but very poor correlation with the LET threshold. Upset rates may vary over four orders of magnitude for devices with nearly the same thresholds. Devices with the LET threshold at 0 may be among the hardest parts. This is discussed in Chapter 11. The lack of relevance of the LET threshold to upset rate has been pointed out previously, but occasionally needs to be repeated.

Figure 7-23 shows the figure of merit coefficient as a function of altitude and inclination for a number of circular orbits. In this case the relative upset rates for the 93L422 were used to determine the relative figure of merit coefficients. The coefficient varies from its normal value of 200 to a minimum value of 4.5 for low equatorial orbits. Figure 7-23 gives a good view of the general relative dependence of upset rate on altitude and inclination for any part.

This approach can be used for orbits not easily read from Figure 7-23. Choose the orbit, solar weather conditions, and shielding. Calculate the upset rate for a large soft device using the IRPP approach.

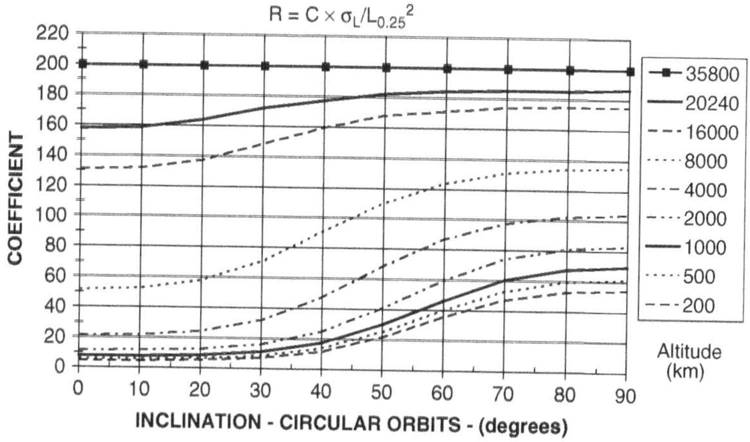

Figure 7-23 Coefficient for generalized figure of merit equation in various orbits.

Determine what coefficient should be substituted for 200 in the figure of merit expression in order to predict that rate. Then use the figure of merit expression with that coefficient to estimate the rates with other parts. A set of part parameters that might be useful for the initial calculation is $L_0 = 1.0, w = 6, s = 0.8, L_{0.25} = 2.26, CS = 8.7 \times 10^{-6}$, and depth $= 1$ micron.

7.9.2 Determination of Device Parameters

Calculation of the upset rate using either the standard IRPP codes or the figure of merit requires the parameters that describe the LET behavior of the cross section. These are obtained by fitting the Weibull curve to the measured cross sections. The Weibull parameters should be obtained using an exact fitting procedure, rather than a best case or worst case fit by eye. One simple approach is to use routines such as the Excel Solver (Microsoft Corporation) routine that varies one set of parameters to optimize another parameter. In this case one sets up a calculation of a Weibull curve with a trial set of coefficients. The fit can be performed using either a logarithmic scaling or a linear scaling of the data points and of the calculated curve, and calculating the differences of the observed and calculated points. The parameter that is minimized in the optimization routine is the sum of the squares of the differences. One needs to start with adequate data (data that reaches saturation) and realistic initial parameters; then this approach works very well.

The logarithmic scaling stresses the low LET points, while the linear scaling emphasizes the high LET points. If the error bars are not known for the low LET points, so that all points are equally weighted during the fit, the fits can be poor. On the other hand, the linear fit through the high LET points sometimes does a very poor job of fitting the threshold region. A good approach may be to find the threshold with the logarithmic scaling, and then using a linear fit, while forcing the threshold value.

A good cross-check on the data is to also fit it with an integral log-normal distribution [Petersen 1995]. It is not clear at present whether the cross-section curve should be described by the Weibull distribution or the lognormal distribution. If the two fits appear very different, there may be problems with the data. (Not enough high LET data?)

Sometimes the data do not appear to be described by the Weibull curve. There appears to be a Weibull curve fit through the points obtained at normal incidence, but the points obtained at other angles to vary the effective LET seem to be either above or below that curve. These effects are nearly always due to geometrical effects during the measurements. They have been discussed thoroughly in the literature [Petersen 1992b, Petersen 1993, Connell 1995a, Connell 1995b, Petersen 1997b]. The fitting procedure should be modified to include these effects for any device 200 μm^2 or smaller. See Chapter 5, Section 5.3.

It is actually an advantage to have geometrical effects as they contain information about the depth of the device. If one does not see these effects, the depth of the device does not have much effect in the IRPP calculations, and can be assumed to be 1 micron for large devices [Pickel 1996]. If these effects are present, the depth of the device is important, but can be calculated. In these cases the geometrical corrections to the Weibull calculation are also included in the spreadsheet and the optimization is expanded to also include the depth of the device.

The data should always include the highest LET beam available. It is a common fault to use a maximum LET in the range of 30–60 MeV/mg/cm^2, where the normal incidence LET is often still on the rising portion of the LET curve. The data at higher effective LETs with this beam appears to saturate on a log plot, but actually are not adequate to determine the saturation cross section. A curve fit that does not include geometrical effects overestimates the Weibull width parameter and underestimates the limiting cross section when the data actually includes geometrical effects. This leads to low estimates of the figure of merit and of the upset rates. One needs data with a higher LET ion in order to determine the saturation cross section for many parts.

On rare occasions it is also necessary to include the effects of diffusion [Smith 1995]. One does have to be careful in all cases that one is counting single event events, not multiple upsets due to one event.

7.9.3 Calculation of the Figure of Merit from Tabulated Parts Characteristics

The heavy ion upset rate calculations performed for this chapter used the Weibull parameters obtained by fitting the full LET cross section curve. Sometimes the only data readily available is that in the periodic

JPL–Aerospace compilations. These summaries list a threshold and a limiting cross section [Nichols 1988, Nichols 1995, Nichols 1997]. They do not use a consistent definition of limiting cross section or threshold. A common limiting cross section is the value at a LET of 37 MeV/mg/cm^2. They use at least two definitions of threshold. One is assigned at a LET where no upsets are observed with a fluence of 10^7 ions/cm^2. Although the cross-section value will depend on the number of bits in the device, this bears some resemblance to the Weibull threshold. The second is at a LET corresponding to 1% of the limiting cross section.

Rudie has suggested that these thresholds can be used with the figure of merit if we use the relationship in Equation (7.41) [Rudie 1996, unpublished]. This relationship is obtained from an empirical fit of the two parameters.

$$L_{0.25} = 2.77 L_0^{0.88} \tag{7.41}$$

We can examine the use of this approach by examining the upset rate calculated as before with the Weibull parameters, plotting this as a function of a FOM′ calculated using the threshold and limiting cross section in conjunction with the Rudie approximation. We then compare these results with the result obtained using the FOM rate coefficient. We will do this for the geosynchronous solar minimum case presented earlier, so that the rate coefficient is 360. Figure 7-24 shows the result for the lower definition of threshold. The results are conservative in that the average predicted value (360 × FOM′) will be about a factor of 3 higher than the actual rate (∼114 × FOM′). However, the very wide width of a factor of 3 indicates that about one in six devices will have the upset rate underestimated.

Figure 7-25 shows the results using the 1% threshold. The predicted rates using the rate coefficient of 360 will be a factor of 1.5 high relative to the average IRPP value (∼239 × FOM′). However, the width of a factor of 1.7 indicates that slightly more than one in six devices will be underpredicted.

These results show that it is possible to use the tabulated values in combination with Equation (7.41). The results in general will be conservative, although one should recognize that one in six cases might not be conservative. It is still preferable to obtain the full set of device parameters, but in many situations this approach will be adequate.

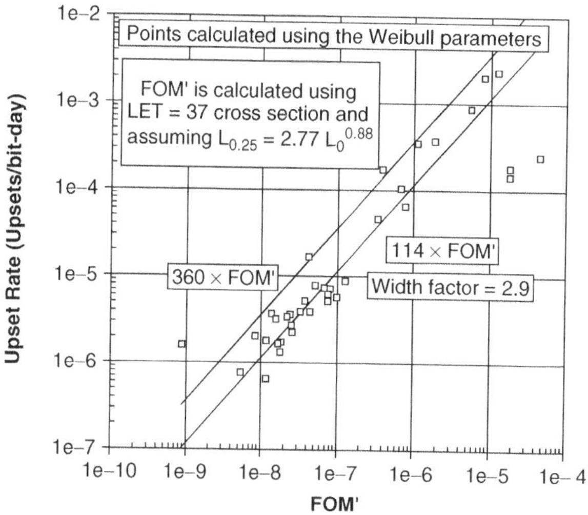

Figure 7-24 The upset rates from IRPP calculations compared with rates obtained using the standard FOM rate coefficient (360 for this orbit) and a FOM′ calculated from tabulated threshold near onset and limiting cross section.

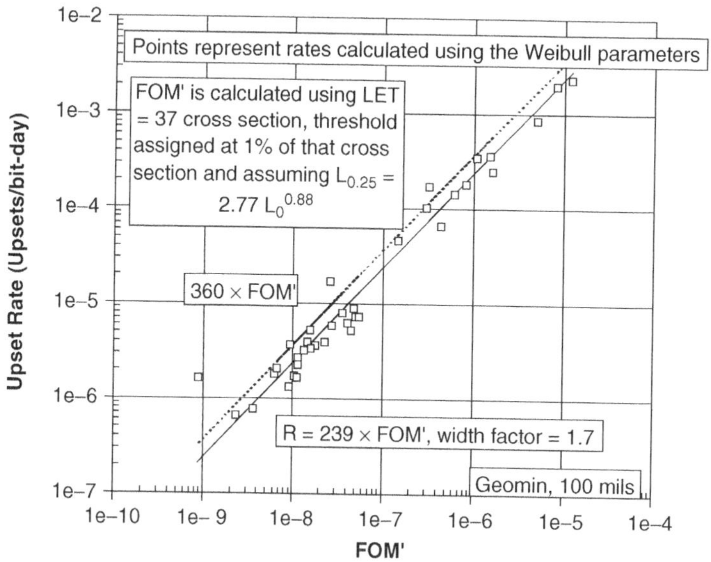

Figure 7-25 The upset rates from IRPP calculations compared with rates obtained using the standard FOM rate coefficient (360 for this orbit) and a FOM′ calculated from tabulated threshold, at 1% of the limiting cross section, and the limiting cross section.

Section 11.4 will discuss the general problem of estimating rates using thresholds.

7.9.4 Rate Coefficient Behind Shielding

All of our comparisons above have been made assuming 100 mils of shielding. This brings up a question about the rate coefficient when there is shielding involved. We will calculate the upset rates behind shielding for several cases and see if any general statements can be made. Figure 7-26 shows the rate coefficients for unhardened and hardened devices at geosynchronous orbit. There is a very slow decrease in upset rate with increased shielding.

Figure 7-27 shows similar curves for both protons and heavy ions at 500 km. In all cases the rate decreases by about a factor of 10 behind 6.5 inches of aluminum shielding. If we define C_0 as the rate coefficient with 100 mils of shielding, a very approximate equation for the change of the rate coefficient with depth is

$$C = 2C_0 - 0.5\, C_0 \log(t) \tag{7.42}$$

where t is the shielding thickness in mils.

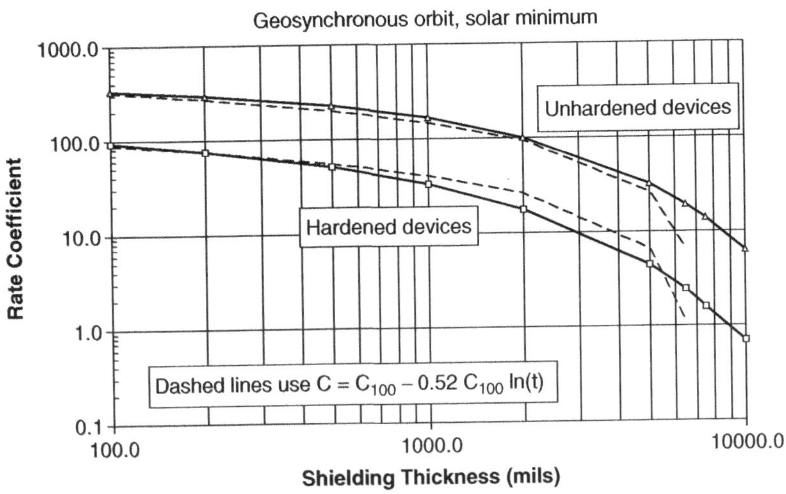

Figure 7-26 The variation of the rate coefficient with shielding at geosynchronous orbit. C_{100} = rate coefficient C at 100 mils.

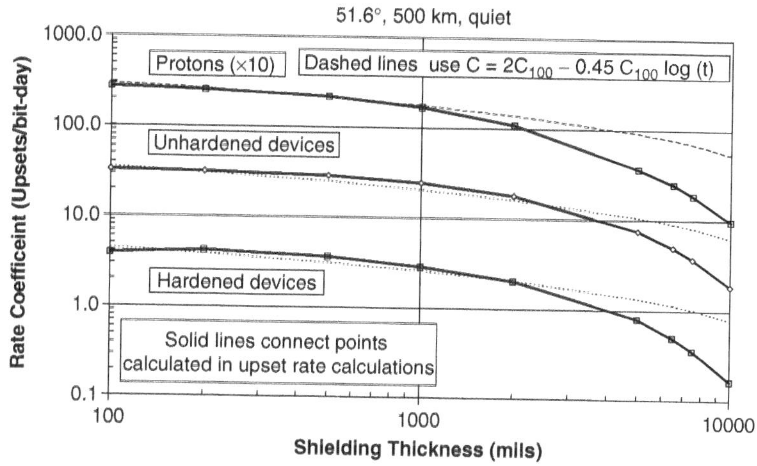

Figure 7-27 The variation of the rate coefficients with shielding for low Earth orbit. C_{100} = rate coefficient C at 100 mils.

According to Smith, typical satellite shielding corresponds to 0.3 inch to 1 inch of aluminum [Smith 1994]. The accompanying rate factors would be 75% to 50% of C_0. An improved value of C can be obtained by combining the depth distribution with Equation (7.42) or the type of information in Figures 7-26 and 7-27.

7.10 THE FOM AND THE LOG NORMAL DISTRIBUTION

Petersen suggested that the integral lognormal distribution could be used to describe the heavy ion cross-section curve [Petersen 1996c]. In this case the parameters are the limiting cross section and

$$m = \text{the mean of}\ln(\text{LET})$$

$$s = \text{the standard deviation of}\ln(\text{LET})$$

$$z = (\ln(\text{LET}) - m)/(\sqrt{2}s)$$

The cumulative lognormal distribution is available in Excel. It can also be expressed in terms of error functions:

$$F = IF(z < 0, 0.5\ \text{erfc}(-z), 0.5 + 0.5\ \text{erf}(z)) \qquad (7.43)$$

The $L_{0.25}$ values used in the figure of merit can be calculated from the lognormal parameters:

$$\ln(L_{0.25}) = m - 0.674\,s \tag{7.44}$$

We mentioned in Section 7.9.1 that the fits were slightly better if the fraction of 25% were changed toward 12.5%. If it were set at 16%, the term in Equation (7.44) would be

$$\ln(L_{0.16}) = m - s$$

The lognormal function may be appropriate for describing both the heavy ion and the proton cross-section curves.

7.11 MONTE CARLO APPROACHES

7.11.1 IBM Code

The most financially significant Monte Carlo code has been the IBM code SEMM [Murley 1996, Srinivasan 1994, Srinivasan 1996]. This was the basis of the IBM SEU hardening program for cosmic ray induced errors described in [Ziegler 1996a, Ziegler 2004]. The hardening for neutrons, of course, led to devices that were also very hard for cosmic rays as shown by heavy ion testing by Calvel [Calvel 1996].

The SEMM code has been updated into SEMM-2 [Tang 2004]. This allows it to apply to smaller feature size devices and to include the enhancements in the Nuclear Databases. The latest version includes modeling of material effects that are introduced by back-end-of-line (BEOL) processes [Tang 2007].

7.11.2 GEANT4

The Monte Carlo code GEANT4 was initially developed by the high energy physics community for modeling particle transport and nuclear interactions. It contains a large number of tools that are useful for single event analysis and has been tested and extended by several groups. A discussion of the various physics tools is given in the NSREC short course by Santin [Santin 2006]. One of the first papers was by Inguimbert and Duzellier and found good agreement with CREME86 calculations [Inguimbert 2004]. Truscott examined neutron and proton interactions using GEANT4 [Truscott 2004]. He discusses the importance of nuclear reactions in passivation and tungsten layers above the sensitive silicon region.

There is now a large program, and large group, at Vanderbilt University applying GEANT4 to single event effects [Warren 2007b, Schrimpf 2008, Weller 2009]. We will not attempt to list all of the resulting papers. The project leaders are Weller, Mendenhall, Reed, and Warren and an examination of the author index of recent issues of *IEEE Transactions on Nuclear Science* for these authors will be instructive. In Chapter 10 we will give an example of one of their calculations and show that the predicted upset rate agrees very well with the IRPP upset rate calculation. The Vanderbilt group is now combining GEANT4 with CREME in the code CREME-MC.

7.11.3 Neutron Induced

The Monte Carlo approach has been used extensively for the study of neutron induced upsets. See references in [Palau 2002]. Palau studied the importance of the reaction location. In this situation the upset depends on the time to complete flipping. They use simulation to develop the relationship between magnitude and duration of the pulse capable of flipping the cell. Figure 7-28 shows a sample of their results. This is the "tmax" approach.

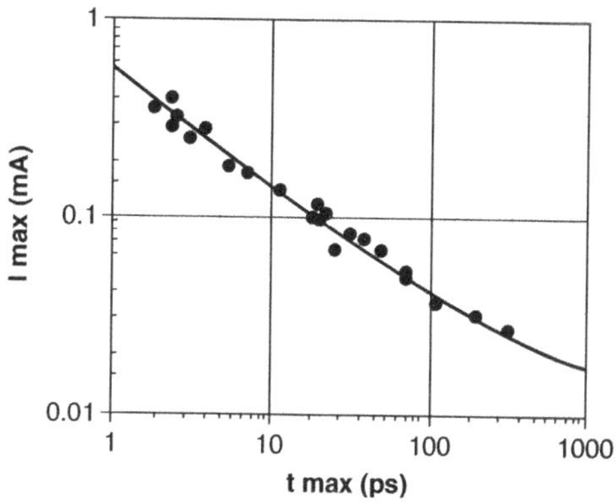

Figure 7-28 Relationship between peak current and time at the peak [Palau 2002]. Copyright © 2002 IEEE.

7.12 PRIVIT

The PRIVIT method was introduced as an alternative to the IRPP method used in CREME [Zebrev 2005]. However, the authors compared results using this method and using CREME96 and demonstrated that they were not equivalent [Zebrev 2005]. Therefore, this approach is not recommended. Previous comparisons have shown that the very different CREME96 and effective flux approaches gave comparable results [Petersen 1995].

7.13 INTEGRAL FLUX METHOD

We earlier introduced the concept of upset cross section using the equation

$$\sigma = \#\text{upsets}/\text{fluence} \tag{7.45}$$

It appears that this can be inverted so that

$$\text{Rate} = \Sigma\sigma(E)F(E) \tag{7.46}$$

This is a valid approach for proton upsets, as the upset rate depends only on the number of events. However, this is not valid for heavy ion upsets. It is called the integral flux method and calculates the upset rate by integrating the measured cross section as a function of ***effective LET*** with the LET spectrum as a function of ***LET***.

$$\text{(Invalid)}\quad \text{Rate} = \Sigma\sigma(L_{\text{eff}})F(L) \tag{7.47}$$

The approach is basically treating the single event problem as if it is completely parallel to the nuclear cross-section problem. As we showed in Section 2.10, this is not true for heavy ion upsets. Heavy ion upsets depend on the amount of ionization in the device, and therefore on the path length in the device. This method does not recognize that. ***The method is conceptually wrong and should not be used***. It does give reasonable numbers for large soft devices, but breaks down for harder and smaller devices. The LET spectrum essentially cuts off at 30 MeV/mg/cm^2. Therefore, any contributions from long path lengths and high *effective LETs* are ignored [Petersen 1995]. The pioneers in the field recognized that this approach was wrong. Still, it has somehow slipped into the literature even though there are no articles advocating it.

CHAPTER 8

Proton Single Event Rate Calculations

The basic physics of the upset interaction is the same for heavy ions and protons. Both types of upsets are caused by the corruption of a device after it collects charge generated by the ionization of a passing heavy ion. The heavy ions present in galactic cosmic rays produce the effect directly. The proton upsets are caused by the ionization of secondary particles from a nuclear reaction (Chapter 1, Figures 1.1 and 1.2). The upset rates then depend on the likelihood of a proton reaction occurring and leading to reaction products that deposit adequate energy to cause upset. Approximately 1 proton in 10^5 will have a nuclear reaction in the silicon device. Although this seems like a very small number of cases, in space near the Earth the protons in the proton radiation belts are so intense that they will cause many more upsets than the heavy ion cosmic rays in the same environment.

Proton upsets follow a complicated sequence of processes. This has led to two different approaches to upset rate calculations. The basic physics approaches follow the nuclear reaction processes and energetics, and then examine the energy deposition of the reaction products. The semiempirical approaches ignore the details of the reactions, and attempt to describe the energy dependence of the sensitivity of individual parts, so that the rate can be determined from the product of the cross section as a function of energy with the proton spectrum as a function of energy.

Single Event Effects in Aerospace, First Edition. By Edward Petersen.
© 2011 the Institute of Electrical and Electronics Engineers, Inc.
Published 2011 by John Wiley & Sons, Inc.

In 1979, separate papers by Guenzer and McNulty and co-workers introduced experimental evidence that high energy protons, such as those in the proton belts, could cause single event upsets [Guenzer 1979, Wyatt 1979]. The two groups realized that if alpha particles from device contaminants could produce upsets, as shown by May and Woods [May 1979], then alpha particles resulting from proton induced nuclear reactions would also lead to upsets.

McNulty compared upsets due to proton nuclear reactions with those due to dose rate upsets caused by pulsed electron beams [McNulty 1980]. He observed that the dose rate upsets lead to upsets over most or all of the chip after the upset threshold is passed, whereas the proton upset cross section increases gradually with proton energy. The proton upsets are dependent on dose, but not dose rate, indicating a single event mechanism. This paper was the first to examine proton energy deposition in thin silicon detectors as related to proton induced upsets.

8.1 NUCLEAR REACTION ANALYSIS

There has been a long series of papers that examined the upset problem from the basis of the nuclear reactions in silicon that produce the upsets. The first paper was by Petersen in 1980 [Petersen 1980]. This paper presented calculations of nuclear reaction cross sections for protons in various semiconductor materials. The basic reaction processes were presented, and the total reaction cross sections for alpha particle production were obtained. The article also discussed the nuclear breakup into two heavy particles that could deposit large amounts of energy in short distances.

Figure 8-1 shows the total reaction cross section in silicon and the contributions of the individual reactions. The reactions are identified by the particles emitted. The energy of the recoil nuclei depends on the reaction. For example, the typical recoil nucleus from the p, p-alpha reaction will have significantly more energy than the recoil from the p, 2p reaction.

The total reaction cross section peaks near 20 MeV and then decreases to approximately 400 mb at high energies. The continual increase of total energy deposited in silicon from the recoil nuclei and reaction products is not shown in this type of plot. The curve of proton

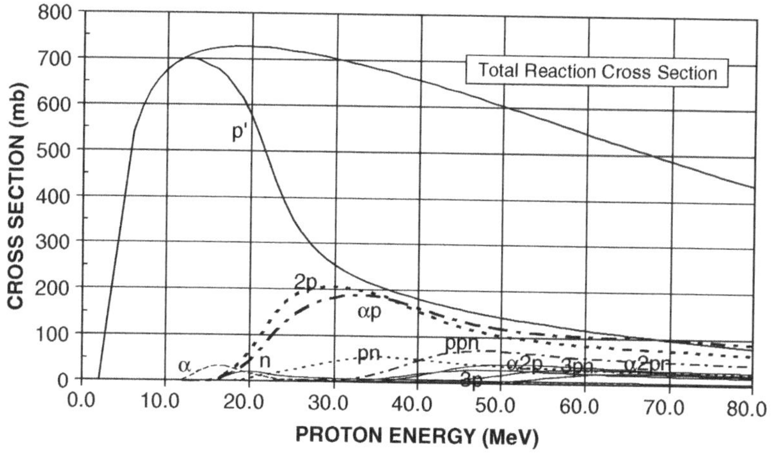

Figure 8-1 Proton reaction cross sections in silicon. The labels indicate the reaction products.

upset cross section as a function of energy rises to a plateau, rather than following the total reaction cross-section curve shown here.

Petersen continued this approach with the first overview of the proton upset problem [Petersen 1981]. The paper discusses the proton intensities in the radiation belt, and the effects on the proton spectrum of the spacecraft shielding. It discusses the energy that is available for upset from the proton elastic scattering and from proton nuclear reactions. The elastic scattering contribution can be important for sensitive devices at low proton energies. The paper presents calculations of the nuclear cross sections and of the resulting upset cross section as a function of device sensitivity and calculates the resulting upset rates in orbit. The last paper in this series was that of Langworthy, who presented an analytic calculation of recoil energy spectra from proton reactions [Langworthy 1982, Petersen 1982]. Figure 8-2 shows his results.

There have been a number of other papers examining the problems from the basic physics aspect. Bradford applied microdosimetric techniques to examine the energy deposition of recoil nuclei from high energy proton reactions [Bradford 1982]. This paper pointed out the importance of nuclear reactions outside the sensitive volume. This is the dominant contribution for very small devices. Bion and Bourrieau developed a Monte Carlo code, incorporating some of the ideas of

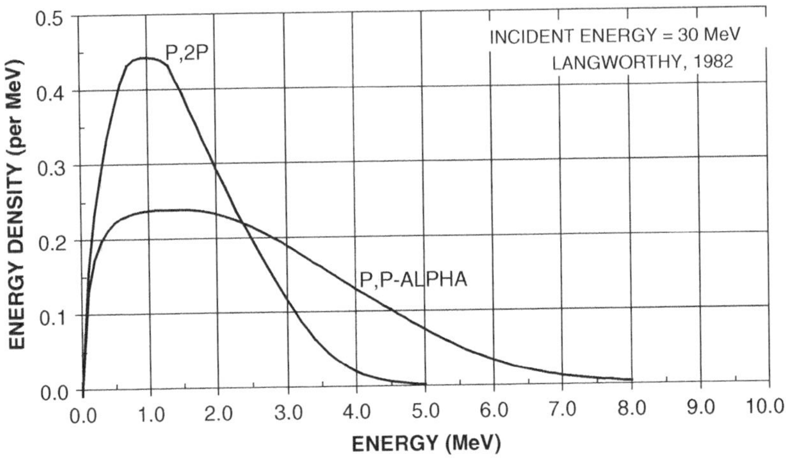

Figure 8-2 Energy spectra of recoiling nucleus from proton reactions in silicon.

Farrell [Farrell 1982, Bion 1989]. They compare their energy dependence with the Bendel one-parameter model and compare calculated upset rates with space results.

Takami and co-workers performed nuclear theory calculations to study the proton reactions in silicon below 50 MeV [Takami 1990]. The intranuclear cascade model used for most Monte Carlo calculations should not be valid below 100 MeV. They used the exciton model and the statistical evaporation model to analyze the nuclear reactions and the optical model to determine the total cross sections. They obtained reasonable agreement with the measured cross sections for the 93L422.

Miroshin and Tverskoy used a nuclear reaction model to develop a two-parameter model for single event upsets [Miroshkin 1994, Miroshkin 1995]. The parameters are device volume and threshold energy. They do not include elastic scattering. The authors present the equation

$$\sigma = n \times V \times \sigma_{in} \times \varepsilon(E_{thr}) \tag{8.1}$$

where

n is number of Si nuclei in a cubic centimeter
V is the sensitive volume of a memory cell

σ_{in} is the cross section for inelastic interactions with silicon
$\varepsilon(E_{thr})$ is a function that expresses the fraction of interactions that deposit an energy greater than the threshold energy E_{thr} in the volume V.

Having measurements of the cross section at two or more different energies, one is able to determine the two unknown parameters V and E_{thr}. The function $\varepsilon(E_{thr})$ is based on the integral spectrum of the deposited energy and can be calculated. The threshold energy is related to the critical charge through the ionization potential of silicon, so that $Q_c = E_{thr}/22.5$ with Q_c in picocoulombs and E in MeV.

The authors did not present either of the functions σ_{in} or $\varepsilon(E_{thr})$, making it impossible for others to use their approach. If these functions were made available, this might be a very valuable method. The total reaction cross section, as shown in Figure 8-1, peaks in the vicinity of 20 MeV and then decreases; whereas the proton upset cross section increases with energy until saturation. The authors' calculated curves all increase with proton energy. There must be an energy term in $\varepsilon(E_{thr})$ to reflect that the total energy available depends on the incident energy. The authors refer to V as the volume of the memory cell. At low incident energies the reaction products have short ranges so the volume should nearly coincide with the device volume. However, Bradford pointed out that nuclear reactions outside the sensitive volume make an important contribution at high energies [Bradford 1982]. Therefore, the volume should also be energy dependent.

The complete equation describing the upset process is then of the form

$$\sigma = n \times V(E) \times \sigma_{in}(E) \times \varepsilon(E, E_{thr}) \tag{8.2}$$

We can make a rough calculation of the proton upset cross section due to nuclear reactions. Consider silicon with 5×10^{22} nuclei/cm^3, a device $10 \times 10 \times 5$ microns^3, and a high energy proton cross section of 400 mb. The cross section of the device is then 10^{-11} cm^2. Notice that the area of the device is 10^{-6} cm^2. The energy deposition function has been measured by McNulty and indicated a nuclear temperature of 2.45 MeV. Figure 8-3 shows the resulting cross-section curve. This calculation ignores the complexities of reaction products leaving or entering the sensitive volume, which overall will raise the

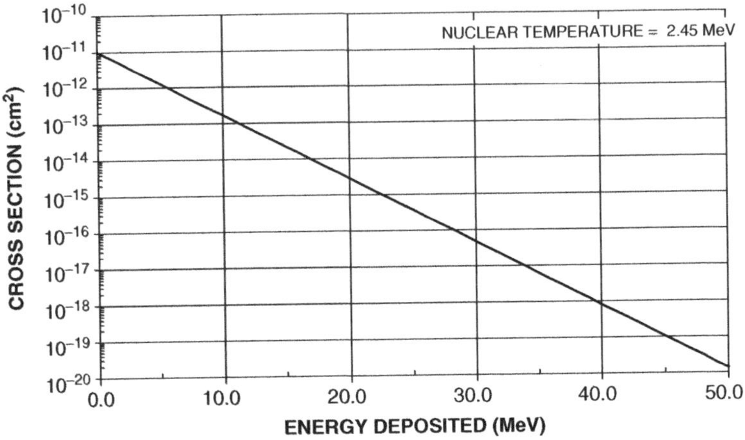

Figure 8-3 Proton upset cross section at high energy calculated for a 10 × 10 × 5 device as a function of device sensitivity.

cross section. It also ignores the fact that the available energy is limited at lower proton energies.

We can make a similar calculation for the contributions due to elastic scattering. This was shown in [Petersen 1996a] and will not be repeated here. It used the cross sections quoted in Figure 6 of [Petersen 1981].

8.1.1 Monte Carlo Calculations

The first presentation of Monte Carlo calculations of recoil energies from proton reactions in silicon was given by Hamm and coworkers [Hamm 1981]. This has been followed by a series of papers by McNulty and co-workers. The first in 1981 presented measurements of the energy deposition of protons in silicon detectors [McNulty 1981]. It then compared the spectra with initial calculations using a Monte Carlo calculation. Farrell and McNulty discussed development of Monte Carlo reaction code specifically for microdosimetric studies of energy deposition from proton reactions in silicon [Farrell 1982]. The Clarkson code (CUPID) was extended to calculate proton upset cross sections [McNulty 1985]. The dimensions of the sensitive volume and the critical energy for upset were estimated from previous heavy ion measurements. They later used the code to estimate the dimensions of the sensitive volume [Abdul-Kader 1987]. Monte Carlo approaches are generally valid above 100 MeV, but CUPID was tailored to fit proton reactions with silicon below 100 MeV.

More recent Monte Carlo calculations have used the GEANT4 code [Truscott 2004]. Truscott discusses the many features of this code that make it a good candidate for detailed examination of proton induced upsets. He also discusses the need for considering nuclear reaction in passivation and other layers above the silicon sensitive volume.

8.1.2 Predictions of Proton Upset Cross Sections Based on Heavy Ion Data

There have been a number of attempts to derive proton cross-section curves from heavy ion data. The simplest is the figure of merit approach discussed below. We will briefly summarize some of the more elegant approaches [Weulersse 2006].

8.1.2.1 SIMPA

Doucin and co-workers developed the SIMPA approach [Doucin 1995]. They convolute the heavy ion cross section $\sigma(E)$ (where E is the energy deposited by the heavy ion) with the probability that a recoil Si ion of energy E is generated by an incident proton. The heavy ion cross section is fitted with exponential function and the probability is deduced from HETC code comparison. The only parameter difficult to evaluate is the sensitive thickness.

8.1.2.2 PROFIT

Calvel developed the PROFIT model [Calvel 1996]. This assumes that all SRAM cells have the same surface, the same depletion depth, but different threshold LETs. It is assumed that all interactions are elastic and all recoils are silicon atoms. It calculates the number of recoil atoms and from this the cross section.

8.1.2.3 BGR Method

The burst generation rate (BGR) was developed for neutrons but can also be used for protons [Ziegler 1979, Letaw 1991, Normand 1998].

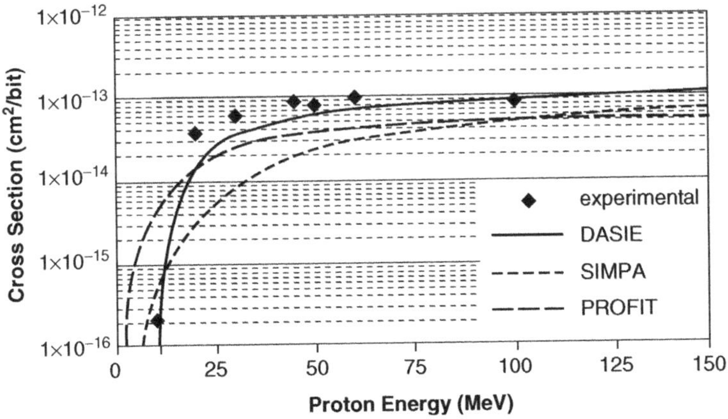

Figure 8-4 SIMPA, PROFIT, and DASIE2 simulations on a 0.6 μm device with a 2 μm sensitive depth [Weulersse 2006]. Copyright © 2006 IEEE.

Upset sensitivity is characterized by critical charge and the sensitive volume. This method considers that the charge induced by secondary ions is deposited locally at the generation point. The BGR approach is discussed in more detail as it is applied to neutrons in Section 9.1.1.

8.1.2.4 Upper Bound Approach

This method was developed by Edmonds [Edmonds 2000]. It develops input parameters from the heavy ion cross-section curve. A charge-collection efficiency function measures the effects of the secondary ion creation localization. It assumes that the secondary ions have short ranges.

8.1.2.5 DASIE

This code attempts to enhance the BGR approach and use a more accurate upset criterion than the upper bound approach [Weulersse 2006]. The tracks crossing the drain are separated from those not crossing the drain, so that the user requires IC layout information. This version of the code is analytic. Previous versions involved Monte Carlo approaches, primarily for neutrons. Figure 8-4 shows a comparison of the results for DASIE, SIMPA, and PROFIT. The approach has been extended to use input from laser mapping [Miller 2006].

8.1.2.6 Barak Approach

Barak and co-workers have attempted to develop simple expressions, developed on analytical and empirical models, for the energy deposited by protons [Barak 1999a, Barak 2001, Akkerman 2002a, Barak 2004, Barak 2006]. They calculate the proton cross-section curves from heavy ion cross sections in devices with submicron sensitive volumes. Figure 8-5 shows a sample of their results.

8.2 SEMIEMPIRICAL APPROACHES AND THE INTEGRAL CROSS-SECTION CALCULATION

The majority of proton upset rate calculations have used the approach based on the work of Bendel and Petersen [Bendel 1983]. Their paper developed a semiempirical equation, based on the nuclear reaction energetics, to describe the energy dependence of the proton upset cross section. The form of the equation was determined by examining the nuclear physics, but the resulting equation is semiempirical with constants to be determined from fits to the experimental data. In Chapter 4 we discussed how the Bendel form of the cross section should be replaced by the lognormal or Weibull forms of the cross section. If we have an exact description of the cross-section curve, the approach is no longer semiempirical.

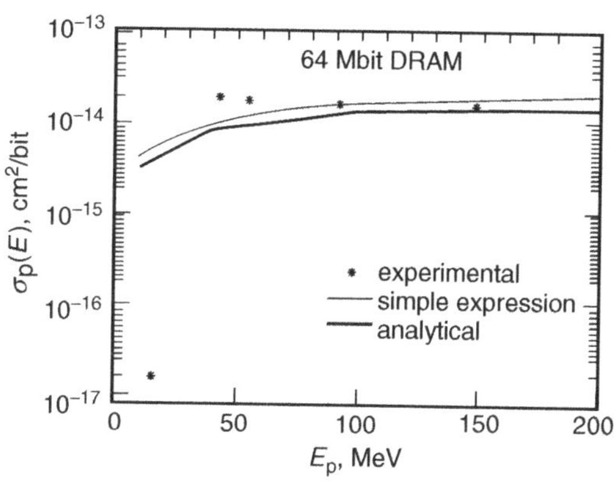

Figure 8-5 Calculated proton cross-section curve for a 0.32 μm 64 Mbit DRAM. The calculations are based on heavy ion data for the same part [Barak 2006]. Copyright © 2006 IEEE.

If we have the proton cross-section curve, the upset rate can be calculated exactly by combining it with the proton spectrum:

$$R = \int_{E_{\min}}^{E_{\max}} \frac{d\Phi}{dE} \sigma_{\mathrm{SEU}}(E) dE \tag{8.3}$$

where $d\Phi/dE$ is the differential proton energy spectrum (protons/(cm^2-MeV-s)) and σ_{SEU} is the SEU cross section at energy E (cm^2/bit). $E_{\min}$ and $E_{\max}$ are the limits of the energy spectrum. The following series of steps will produce the proton upset rates in a given spacecraft:

1. (a) Obtain the experimental proton upset cross section at several energies.
   (b) Determine the lognormal description of the cross-section curve by using a least square analysis.
2. (a) Obtain the average proton spectrum in the particular orbit.
   (b) Obtain the shielding distribution for the devices in the spacecraft. See Figure 8-6.
   (c) Determine the proton spectrum at the devices (see Figures 8-7 and 8-8).
3. Combine the upset cross sections with the spectrum to find the upset rates. See Figure 8-9.

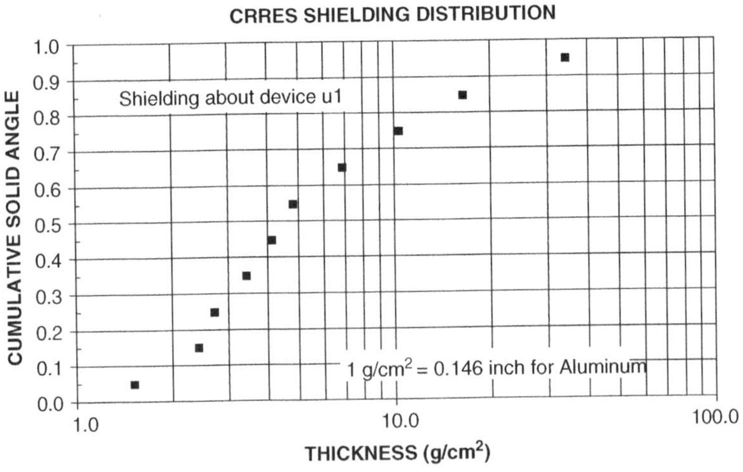

Figure 8-6 Representative satellite shielding distribution. The shielding has been divided into 10% bins.

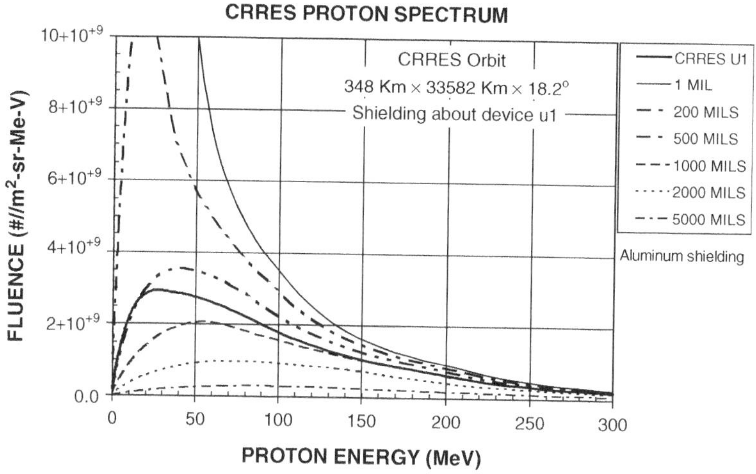

Figure 8-7 Representative proton spectrum inside shielding.

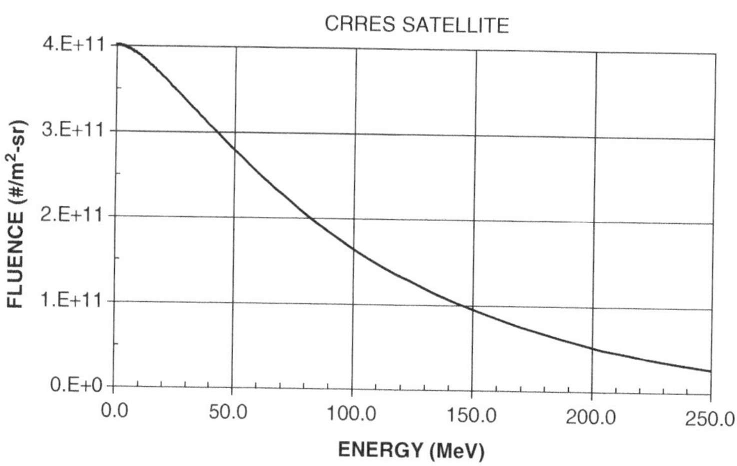

Figure 8-8 Integral version of the proton spectrum behind shielding distribution shown above.

Note that this is not the same procedure used to determine cosmic ray upset rates. The cosmic ray upsets depend on the total energy deposition of each cosmic ray. That means that the path length of the ion in the device is important. Therefore, the IRPP procedure should be used. It is incorrect to directly combine the heavy ion cross section as a function of effective LET and the particle flux as a function of

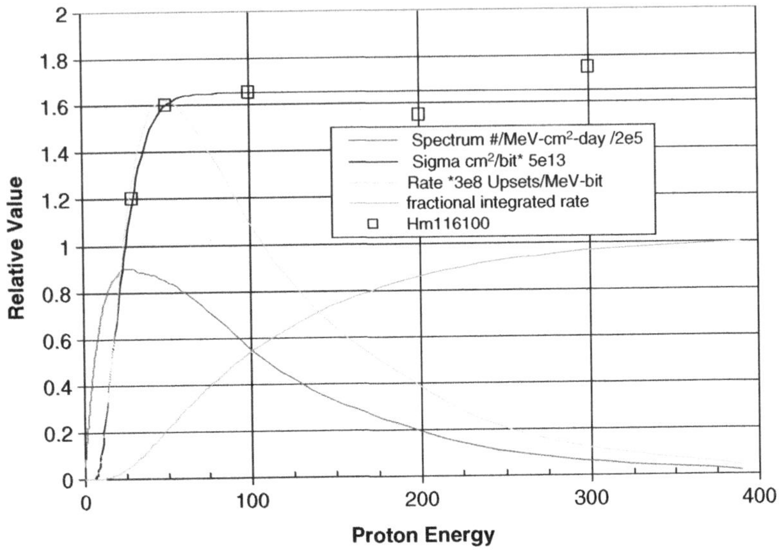

Figure 8-9 Combining the cross-section curve and the environmental spectrum to determine the proton upset rate.

LET. The proton upsets depend on the probability of a nuclear reaction in the device, so it is appropriate to directly combine the cross-section curve and the incident proton spectrum.

The use of the Weibull function to describe the proton cross-section curve has been suggested [Cutchin 1993]. This may be a good choice if adequate good data is available. The problem is that there are rarely many more points than fitting parameters. The parameters of the Weibull curve emphasize the threshold region. Unfortunately, it is the saturation cross-section region that is important. Also, measurements that emphasize the saturation region often do not have enough points in the threshold region to determine all three parameters. This option is available for rate calculations in the CREME96 code [Tylka 1997].

8.3 RELATIONSHIP OF PROTON AND HEAVY ION UPSETS

It initially appeared that no simple relationship existed between upset rates due to heavy ions and upset rates due to protons due to the different mechanisms of energy deposition [Petersen 1983a]. Rollins attempted to develop a method of relating proton upset rates to heavy ion test data [Rollins 1990]. He obtained a relationship between the limiting proton and heavy ion cross sections, the device thickness, and the heavy ion LET threshold. He defined the LET threshold as the value of LET at which the cross section is 10% of its limiting value. The Bendel A parameter could then be determined, enabling the upset rate to be calculated. There was significant scatter in the experimental points used to establish his relationship. Rollins's paper was the first to establish a connection between proton upset rates and heavy ion cross-section measurements.

Petersen presented an alternate approach. He related proton upset thresholds and heavy ion upset thresholds [Petersen 1992c]. This approach allows upset rates in either environment to be estimated using data from either environment. The paper also pointed out that devices with LET thresholds above 10 MeV/mg/cm^2 could have unacceptable proton induced upset rates in the proton radiation belt. The paper reexamined the approach of Rollins and concluded that there was an unsatisfactory amount of spread in the results when fit with that approach. Another approach to relating proton and heavy ion cross sections has been derived by Calvel [Calvel 1996]. The initial result looked good.

It now appears that these approaches should be replaced by the figure of merit, which appears to give a very good correlation of heavy ion and proton upset rates.

8.4 CORRELATION OF THE FOM WITH PROTON UPSET CROSS SECTIONS

If the FOM truly represents a measure of device sensitivity for upset, then it should also be useful for comparing the upset rates due to protons in the proton radiation belts. We can quickly see that there is a simple relationship by examining the limiting proton cross section as a function of the figure of merit for parts for which both proton and heavy ion data are available [Petersen 1998]. This is shown in Figure 8-10. Similar data at cross sections in the range of 10^{-15} to 10^{-14} have been obtained by Liu [Liu 2006b]. We observe that there is a direct relationship between the two quantities. It appears that the figure of merit can be determined directly from the limiting proton cross section σ_{PL}:

$$FOM = 4.5 \times 10^4 \times \sigma_{PL} \tag{8.4}$$

Some people believe that there is some LET threshold above which a device will not be sensitive to proton upsets. The results in Figure 8-10 suggest that proton upset cross sections can be directly related to the figure of merit. A test of this hypothesis, versus the hypothesis that some devices do not upset for incident protons, could be made by looking for cases for which the high energy proton cross section is more than an order of magnitude below the value predicted using Equation (8.4) and a known figure of merit.

8.5 UPSETS DUE TO RARE HIGH ENERGY PROTON REACTIONS

The common upset producing reactions in silicon, such as (p, p-alpha) with a Mg recoil cannot deposit enough energy to have an equivalent

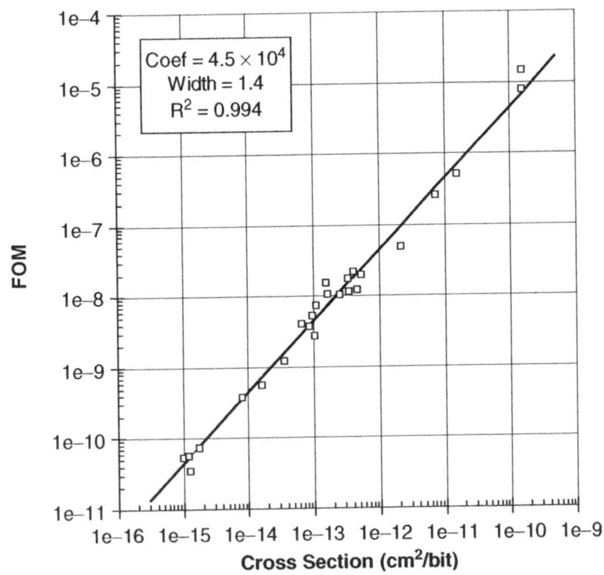

Figure 8-10 The correlation of the figure of merit calculated from heavy ion data with the limiting proton cross section. The results show a direct correlation over five orders of magnitude [Petersen 1998].

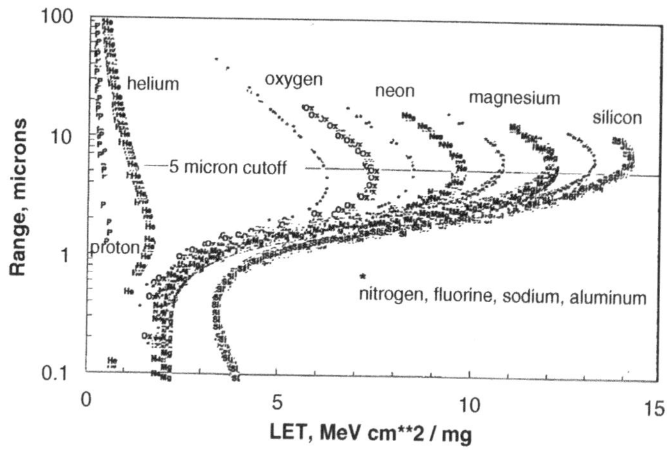

Figure 8-11 Range versus LET for reaction products for 200 MeV protons incident on silicon calculated in a nuclear cascade reaction code [O'Neill 1998]. Copyright © 1998 IEEE.

LET above 12 MeV. Figure 8-11 shows the results of a reaction calculation of the products of 200 MeV protons on silicon using a nuclear cascade code [O'Neill 1998]. This shows the highest LET from proton reactions to be 14 MeV/mg/cm^2.

There may be a few reactions that were not included in these calculations. The reactions that deposit large amounts of energy must be rare nuclear reactions that occur only for high energy protons. Petersen suggested several of these [Petersen 1980]. The prime candidate reaction is the breakup of the silicon nucleus into two nitrogen nuclei, with a Q value of -27.2 MeV and a possible cross section of 10 mb. However, according to Figure 8-11, this would only produce maximum energy deposits equivalent to 18 MeV/mg/cm^2. This type of reaction probably peaks at much lower energies than 100 MeV, as cascade processes dominate at high energies and evaporation dominates at lower energies. However, with the possibility of proton reactions in heavy overlayers, such as tungsten, it is now possible for reaction products to have LETs up to 35 MeV/mg/cm^2. See Chapter 10.

The high energy proton induced upsets have been studied experimentally by Bion [Bion 1989], and by McNulty in a series of papers [Farrell 1982, El Teleaty 1983, Farrell 1984, McNulty 1991b]. They showed rare events that deposited 30–45 MeV. This corresponds to LETs of 30–50 MeV/mg/cm^2 in their detectors. Proton upsets were observed in devices with high LET thresholds by Pickel and co-workers [Pickel 1989]. It is not clear what the LET threshold was for these devices. In practice it appears reasonable to assume that any device that does not upset for heavy ions with a LET of 15 or above will not upset for protons.

8.6 UPSET DUE TO IONIZATION BY STOPPING PROTONS, HELIUM IONS, AND IRON IONS

Some groups have assumed the upsets due to proton ionization should be included in the upset rate calculation. There are three problems with this approach. First, ionizing protons cannot cause upsets in most devices. The second problem is that in low orbits the upset rate should be considered for two different geographical regions. The first region is in the South Atlantic anomaly where the rate for soft devices is dominated by upsets due to proton reactions and the second region is the rest of the orbit where the upset rate is dominated by cosmic rays. The calculated cosmic ray rate should not include any contributions from the trapped protons or from the proton cosmic ray components. The third problem is that soft devices in the South Atlantic anomaly have their rates dominated by the proton reactions so that if this effect existed, it could not be observed.

Upsets due to direct proton ionization can be important for optics systems in the space environment. Marshall has examined a fiber optic data link and shown that the large area detectors have an effective LET threshold of zero, so that they are very sensitive to direct proton ionization [Marshall 1995].

Scaling studies indicate that, as devices scale down and acquire very small critical charges, they will start to upset for direct ionization. This has now been observed in 65 nm devices [Rodbell 2007]. They observed upsets in latches and SRAMs with 40 to 70 degree bombardment with 1.25 MeV protons.

Figure 8-12 shows how the direct proton ionization might show up when protons are included in the rate calculation. As we will see, this probably seriously overestimates the effect. Figure 8-13 shows the LET and range curves for low energy protons. The high LETs occur for very low energy protons near the end of their range. These protons are losing energy very rapidly and do not have the constant

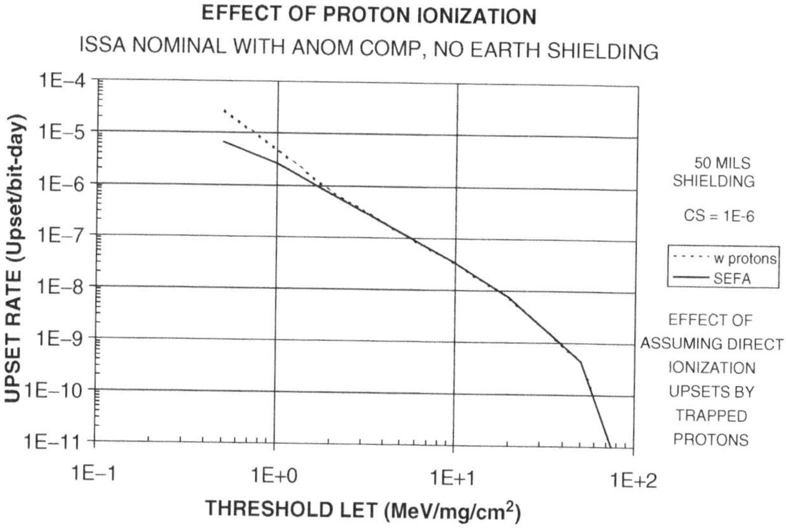

Figure 8-12 Contribution to the upset rate from assumed proton ionization from trapped protons.

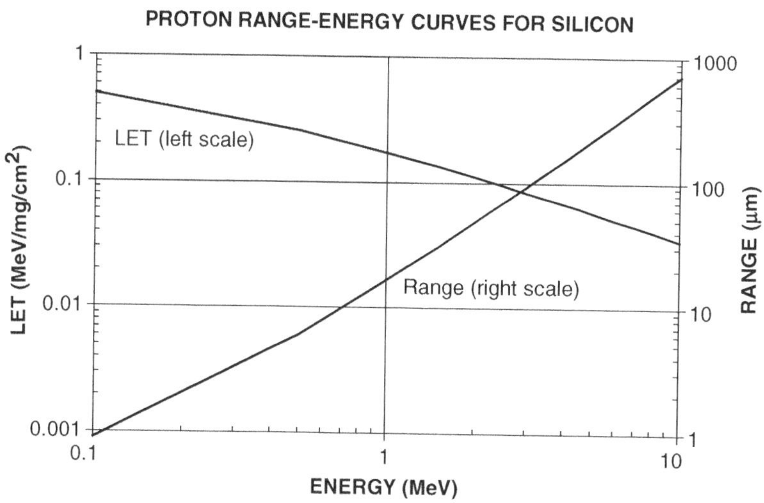

Figure 8-13 The LET and range for low energy protons in silicon.

LETs that are assumed in the upset rate calculations. In the case of low energy protons, the upset probability must be determined from the total energy loss. The effect will be difficult to observe in space, as in the proton belts the rates will still be dominated by the upsets due to nuclear reactions.

Figure 8-14 shows the maximum energy deposition that stopping protons can make in silicon. Here we plot the energy as a function of the stopping distance. Figure 8-15 shows the range as a function of the energy.

The principal problem with the ordinary calculation of upsets due to direct proton ionization is that the calculation assumes that the particle maintains the same LET with which it enters the RPP volume. This is not true for the low energy protons that have very high LETs. We show the actual situation in Figure 8-16. We consider the energy deposited in 10 microns of silicon. The curve shows the energy loss that is calculated using the LET (dashed curve), and the actual energy deposited (solid curve). At very low energies the particles are stopping in the silicon so that the constant LET assumption predicts total energy deposition larger than the total energy available. Even at the peak of the energy deposition curve, the peak values occur for only a very limited range of incident energies. If we refer back to Figure 8-7, we see that very few protons are in this energy range after going through the spacecraft shielding.

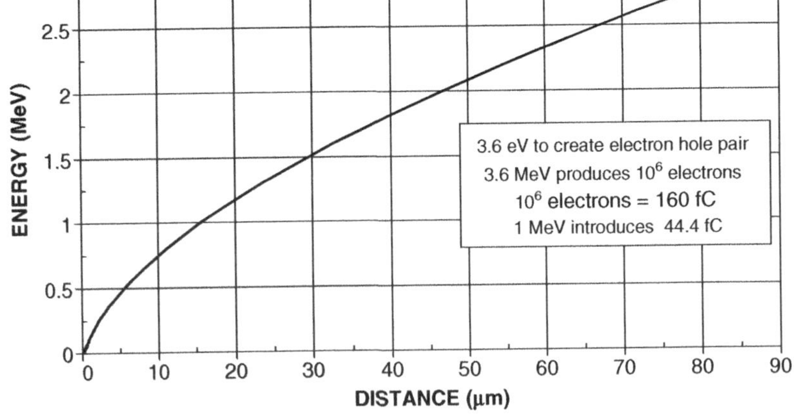

Figure 8-14 Energy deposition by stopping protons.

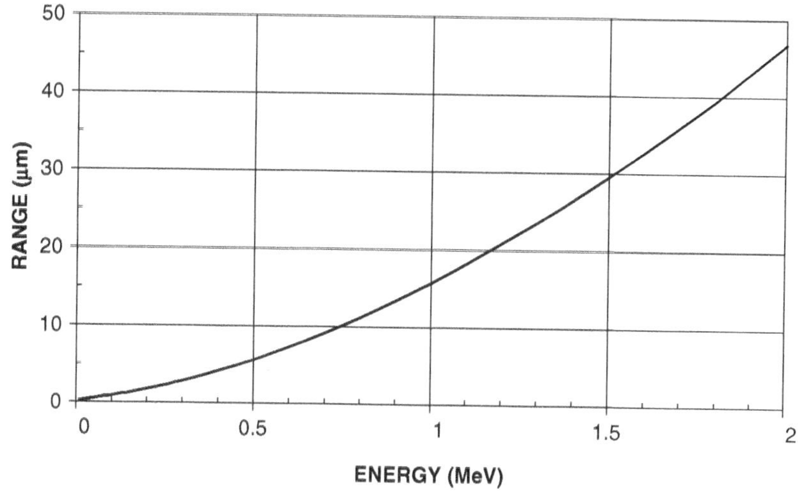

Figure 8-15 Range of low energy protons in silicon.

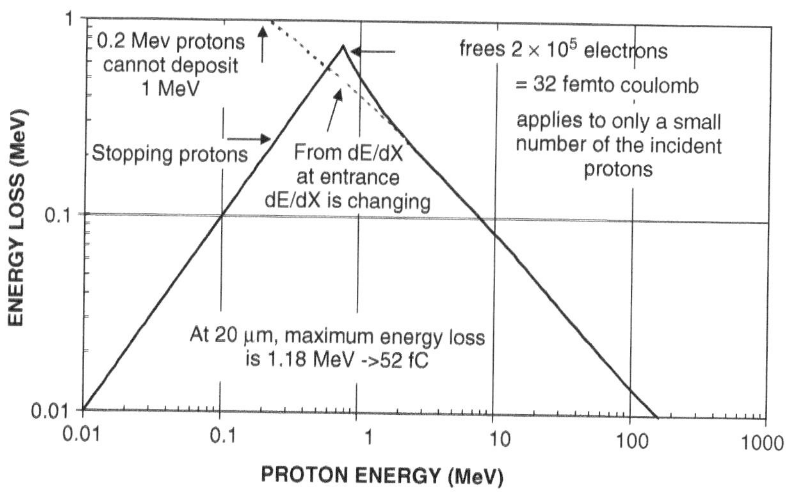

Figure 8-16 Energy loss in 10 microns of silicon as a function of incident proton energy.

Edmonds has examined the effects of solar particle events on an opto-coupler with a large area photodiode that was sensitive to ionizing protons [Johnston 1998, Johnston 1999, Edmonds 2008]. He considered a one week exposure to the CREME96 worst week environment for an estimate of the number of errors accumulated over the duration of the solar event.

We observe that the protons with energies less than 0.1 MeV can be ignored when the environment is the model flare behind a 100 mil shield. For this particular environment, reactions by high energy protons will be of less importance than the direct ionization processes. The importance of the protons relative to heavy ions in the other energy bins depends on the nature of the device.

The photodiode has approximately a 430 μm diameter with a charge collection depth of about 50 μm. The critical charge for a SET is about 0.05 pC, corresponding to a critical energy of 1 MeV. A device with these dimensions will have a significant acceptance angle for which ion path lengths are 100 μm or longer. Figure 8-17 indicates that protons with energies between 1 and 8 MeV will deposit more than 1 MeV over the 100 μm path length. From Table 8-1 we see that the flux from this group is almost two orders of magnitude larger than the heavy ion flux. The direct ionization from protons appears to give a much larger contribution to the SET rate than the heavy ions.

Figures 8-16 and 8-18 show decided differences between the actual deposited energy and the deposited energy calculated assuming

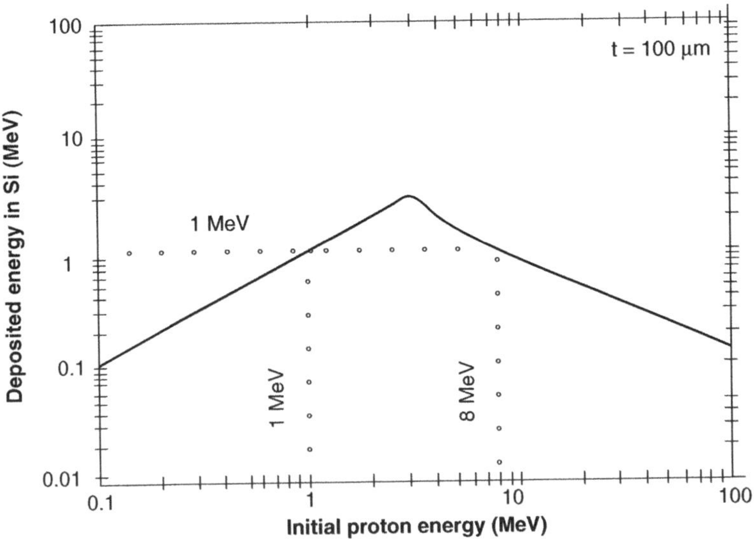

Figure 8-17 Energy deposited by a proton in a 100 μm silicon target plotted against initial proton energy, calculated from range–energy data. Initial proton energies between 1 and 8 MeV will deposit more than 1 MeV in the target [Edmonds 2008]. Copyright © 2008 IEEE.

Table 8-1
Worst Week Fluxes for Selected Particles in Interplanetary Space with 100 mils of Aluminum Shielding

Particle	Flux $1/m^2$-s-sr
Protons with $0 < E < 0.1$ MeV	2.29×10^3
Protons with 0.1 MeV $< E <$ 1.6 MeV	9.42×10^4
Protons with 0.1 MeV $< E <$ 8 MeV	1.23×10^6
Protons with 1 MeV $< E <$ 8 MeV	1.19×10^6
Protons with $E >$ 15 MeV	6.23×10^6
Heavy ions with LET > 01 MeV-cm^2/ mg	1.80×10^4

SOURCE: [Edmonds 2008]. Copyright © 2008 IEEE.

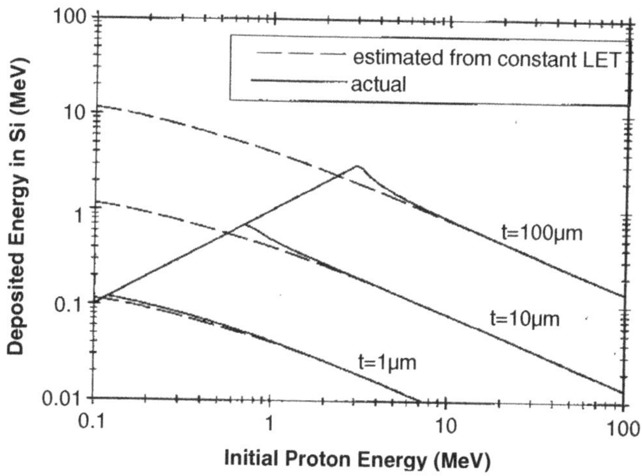

Figure 8-18 Energy deposited by protons in each of several silicon target thicknesses plotted against initial proton energy. The solid curves were calculated from range–energy tables while the dashed curves multiply target thickness by incident LET [Edmonds 2008]. Copyright © 2008 IEEE.

constant LET. The CREME96 code assumes that the LET is constant when it calculated upset rates. We need to examine when the constant LET approximation is useful and when it leads to error. The adequacy of the constant LET approximation can be determined from figures such as that of Figure 8-18. For path lengths of 1 μm, the constant LET approximation is very good (the dashed curve agrees with the solid curve) at all energies that are important [Edmonds 2008].

Therefore, the standard RPP calculations can be used for ionizing proton upset rates for very small devices.

Now let us consider the case shown in Figure 8-17. The critical energy was 1 MeV so that we have SET contributions for protons of energies between 1 and 8 MeV. If we use the constant LET approximation, we consider the flux between 0.1 and 8 MeV. Table 8-1 shows that the flux between 0.1 and 8 MeV is nearly the same as the flux between 1 and 8 MeV, so the constant LET approximation works well for this example. However, consider modifying the example so that the critical energy is 3 MeV. The solid curve in Figure 8-17 shows that no incident particle can deposit an energy exceeding 3 MeV, so the flux of protons that can deposit this energy by direct ionization is zero. In contrast, the constant LET approximation predicts that this energy can be deposited by incident energies up to 1.6 MeV. The corresponding flux in Table 8-1 is about five times the corresponding heavy ion flux. For this example the constant LET approximation predicts the error rate from protons by direct ionization to be five times the error rate from heavy ions, while a more correct calculation predicts no errors from protons by direct ionization [Edmonds 2008].

If the constant LET approximation cannot be used in the upset rate calculation, the common RPP calculations cannot be used. Edmonds has developed an approach that allows for the energy loss of the proton, and you should examine this approach for these cases [Edmonds 2008]. The relative contributions of GCRs, proton reactions, and ionizing protons will depend on the device and the environment. This example was chosen to emphasize the direct ionization by protons.

There is another possible contribution to proton upsets by direct ionization besides the energy deposition of stopping protons. This has been pointed out by Barak [Barak 1996]. We have previously mentioned energy straggling effects for heavy ions. They can also occur for protons, so that low probability events deposit more than average energy. Figure 8-19 shows a calculation of the probability of charge deposition for 50 MeV protons. The figure corresponds to Figures 3-20 and 3-21 in Chapter 3 and was calculated using the Xapsos approach [Xapsos 1992, Xapsos 1993]. Barak has extended his study of straggling and extreme cases in an additional publication and compared his results with those of Weller [Barak 2005, Weller 2003].

Upset rate calculation for stopping helium may have an effect similar to that for stopping protons. Figure 8-20 shows the energy loss

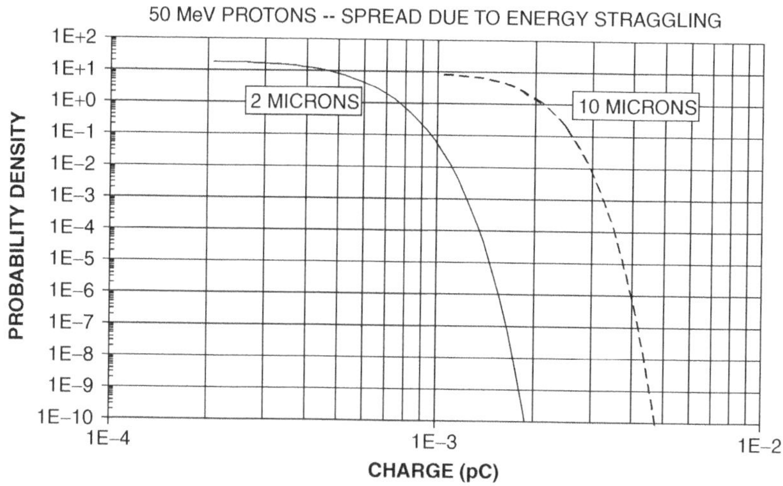

Figure 8-19 Energy deposition probabilities in 2 microns and 10 microns of silicon.

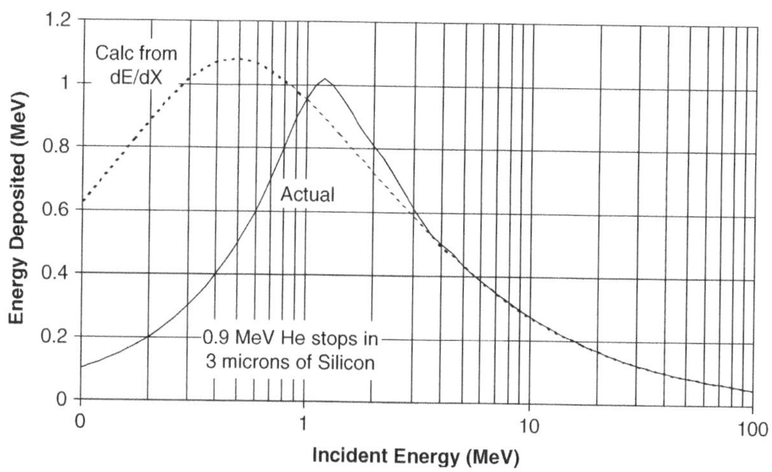

Figure 8-20 Energy loss of stopping helium in 3 μm of silicon.

of helium in silicon. For very low energy particles and very sensitive devices, there may be an overestimate of the upset rates. This may enter calculations involving solar energetic particles where there is a large helium population. This effect is not likely to enter, but rates due to solar particle events should be examined with and without protons and helium ions. If there is an effect, then you should carefully

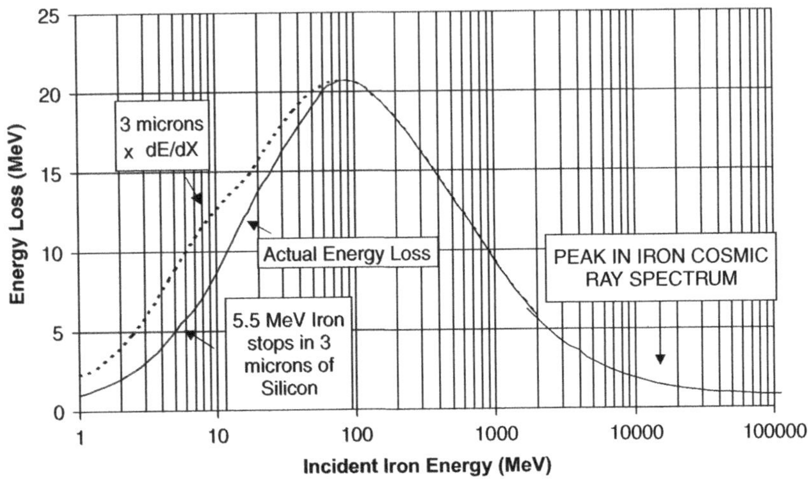

Figure 8-21 Energy loss of stopping iron in 3 μm of silicon.

examine both the environment (shielding effects) and the part size and sensitivity.

For comparison we can examine the effect of iron ions stopping in silicon. This is shown in Figure 8-21. Here we see that stopping iron ions are basically considered properly.

CHAPTER 9

Neutron Induced Upset

After May and Woods [May 1979] identified radio active contamination as a cause of upsets in electronic circuits, electronic manufacturers initiated major efforts to avoid the problem. IBM put together a large, well-funded, task force in 1978, as preprints of the May paper were widely distributed. The IBM efforts were very successful, the task force was disbanded, and the company felt confident in putting mean time between failures (MTBF) guarantees in computer purchase contracts.

In the early 1980s, the common practice, when there was a computer failure identified in a memory module, was to replace the module and set it aside for later diagnostic testing. If the diagnostic testing found no failures, the module was labeled NDF for "no defects found." IBM had sold a major computer system to a large bank in Denver with strict MTBF guarantees in the contract. The system did not meet the requirements and there was a significant financial impact. An initial investigation examined the NDF statistics for computers across the country. For the United States as a whole (mean altitude 770 ft) the NDF result accounted for less than 10% of the modules. For the mountain states (mean altitude 3200 ft) it was five times this level, and accounted for 50% of the modules. For the modules installed in Denver (altitude 5280 ft) the NDF rate was ten times the rate for the country as a whole [Ziegler 1996a].

IBM assembled another large task force and investigated the problems at a number of locations. The investigators convinced themselves

Single Event Effects in Aerospace, First Edition. By Edward Petersen.
© 2011 the Institute of Electrical and Electronics Engineers, Inc.
Published 2011 by John Wiley & Sons, Inc.

and, with some difficulty, management, that the problem was the result of neutron bombardment resulting from cosmic rays hitting the environment, as had been predicted by Ziegler in 1979 [Ziegler 1979, Ziegler 1981, Ziegler 1996a, Ziegler 1996b, Ziegler 2004].

The IBM task force undertook extensive modeling of the problem and developed a family of solutions so that they improved from the rate in a 288 kb memory chip in 1983 to a rate nearly four orders of magnitude less for a 16 Mb chip in 1991 [Ziegler 1996a]. The efforts are described in a series of articles in a special issue of the IMB *Journal of Research and Development* published as Vol. 40, No. 1, January 1996 [Ziegler 1996a, Ziegler 1996b, O'Gorman 1996, Ziegler 1996c, Ziegler 1996d, Srinivasan 1996, Tang 1996, Murley 1996, Freeman 1996].

9.1 NEUTRON UPSETS IN AVIONICS

The study of neutron induced upsets was extended to avionics altitudes by Tabor and Normand [Tabor 1993]. They established that the increase in SEUs for electronic devices scaled with the cosmic ray flux up to altitudes of 20 km (65,000 ft). They also demonstrated that the airborne device and system SER increases with the increasing cosmic ray flux at higher latitudes (up to 70°N). Normand has followed this study with a series of studies to better understand the environment and the upset modeling [Normand 1993, Normand 1996, Normand 1998].

9.1.1 BGR Calculation

The burst generation rate (BGR) approach was developed by Ziegler and Lanford to calculate SEU rate induced by protons or neutrons [Ziegler 1979]. Normand has used and extended this approach [Normand 1993, Normand 1996, Normand 1998, Chadwick 1999, Normand 2001]. The energy deposited by the recoils from the nuclear reactions is what causes the upset, and the SEU rate is given by

$$Rate = Ct \int \frac{d\sigma}{dL} \int_{E_\mathrm{p}} BGR(E_\mathrm{p}, E_\mathrm{r}) \frac{dJ}{dE} dE\, dL \qquad (9.1)$$

where
C is the collection efficiency
t is the collection depth (μm)
$d\sigma/dL$ is the differential form of the Weibull function describing device sensitivity
BGR is the burst generation rate ($cm^2/\mu m^3$): the probability that particles of energy E will produce recoils of energy $\geq E_r$
E_p is the energy of the incoming particle
E_r is the recoil energy ($E_r = t \times 0.23 \times$ LET in MeV)
dJ/dE is the particle differential flux (particles/cm^2-s-MeV)

The collection depth can be considered to consist of three parts: the depletion region, a funnel region, and a diffusion region, with the depletion region generally the largest contributor. However, as Bradford pointed out, reactions also have contributions from outside the sensitive volume, so you do not expect an exact fit to device dimensions [Bradford 1982]. t has been seen to vary between 1 and 6 μm with recent parts having $t = 2$ μm. With these values of t, C has been found to work well at a value of 0.5 [Normand 1996, Normand 1998].

The key part of the method is the BGR function, which expresses the probability that an incident particle will produce recoil particles of appropriate energies. These are calculated using high energy nuclear reaction codes and lower energy cross-section databases [Letaw 1991]. Figure 9-1 shows the values used [Normand 1998].

9.1.2 Integral Cross-Section Calculation

If we have information about the neutron cross-section curve, the upset rate can be calculated exactly by combining it with the neutron spectrum:

$$R = \int_{E_{min}}^{E_{max}} \frac{d\Phi}{dE} \sigma_{SEU}(E)\, dE \tag{9.2}$$

where $d\Phi/dE$ is the differential neutron energy spectrum (neutrons/(cm^2-MeV-s)) and σ_{SEU} is the SEU cross section at energy E (cm^2/bit). E_{min} and E_{max} are the limits of the energy spectrum.

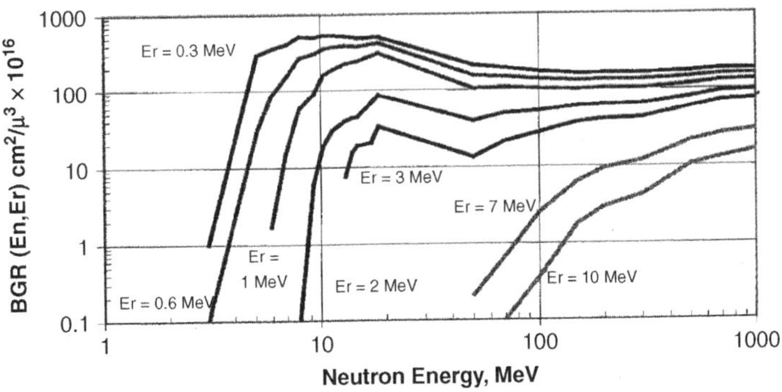

Figure 9-1 The BGR function for neutrons of energy E_p creating recoils of energy E_r. [Normand 1998]. Copyright © 1998 IEEE.

9.1.3 Figure of Merit Calculation

Normand extended his analysis by examining the figure of merit (FOM) approach. In this approach the Rate $= C \times FOM$, where C is a constant that varies with orbit and environment. There is a very good correlation between the FOM approach and the BGR approach. Figure 9-2 shows the results.

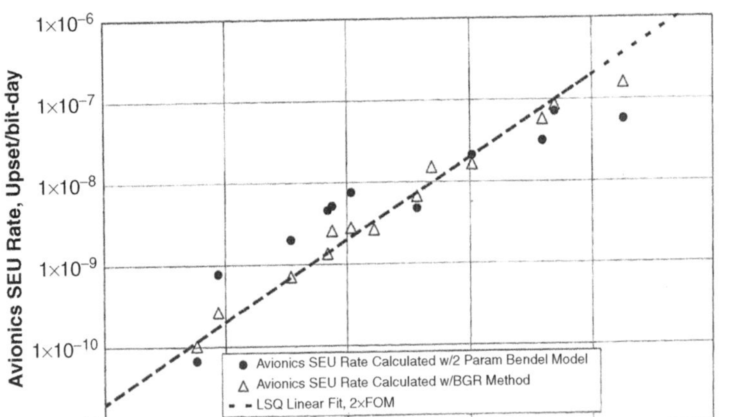

Figure 9-2 Atmospheric upset rates for a variety of devices as a function of the FOM. The calculations are for 40,000 feet and 45°. The rates calculated by the BGR method vary directly with the device figure of merit [Normand 2004]. Copyright © 2004 IEEE.

Figures 9-3 and 9-4 show the variation of the FOM coefficient *C* for various altitudes and latitudes, based on the Boeing atmospheric model [Normand 1993]. It appears that the limiting SEU cross sections for high energy protons and high energy neutrons are the same [Dyer 2004]. Therefore, the same expression, Equation (8.4), can be used for proton and neutron determination of the FOM and we can obtain the figure of merit from measurements with heavy ions, protons, or neutrons.

9.1.4 Upper Bound Approach

For avionics purposes, a simple and conservative approach uses an upper bound value for the integral neutron flux for $E > 10$ MeV of 6000 n/cm^2-h (40,000 ft) and the asymptotic value of the high energy neutron cross section σ_{SEU} (high energy) [Edwards 2004, Normand 2006]. We have previously discussed that the proton cross section at high energies is basically constant (Section 4.3). Recent measurements have shown that this is also true for high energy neutrons, and that the values for neutron and proton cross sections basically agree [Dyer 2004, Baggio 2004].

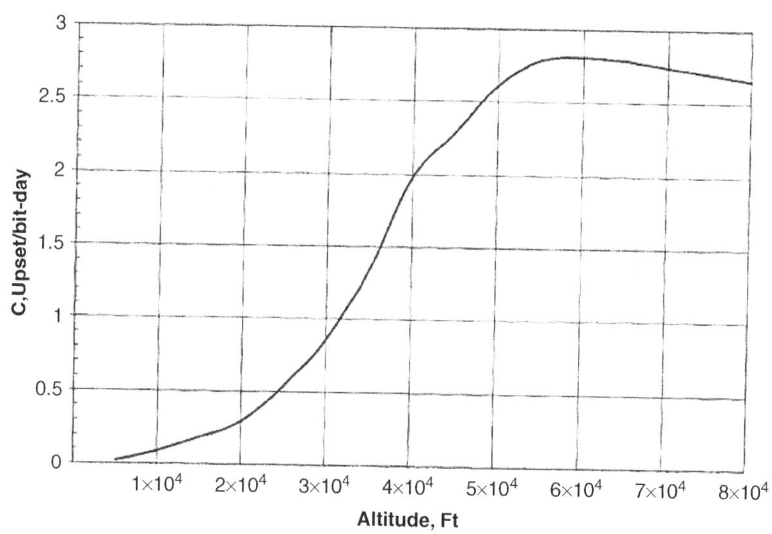

Figure 9-3 The FOM coefficient *C* for atmospheric neutrons as a function of altitude at 45° [Normand 2004]. Copyright © 2004 IEEE.

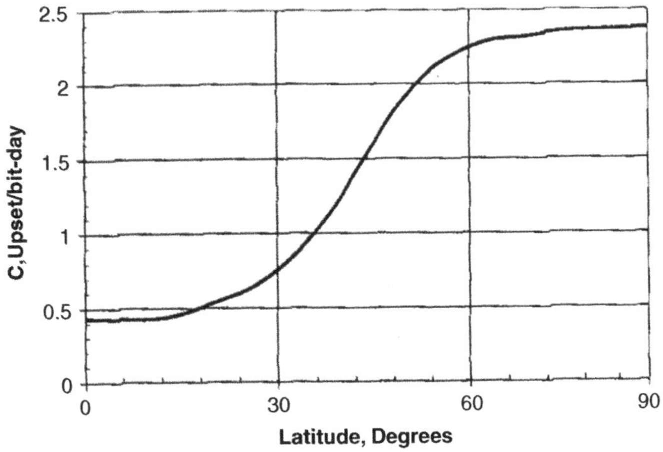

Figure 9-4 The FOM coefficient C for atmospheric neutrons as a function of latitude at 40,000 ft [Normand 2004]. Copyright © 2004 IEEE.

If we were to express this relationship in terms of the FOM, we obtain for 40,000 ft and 45 degrees:

$$\text{SEU Rate} = 3.2 \times \text{FOM (upsets/bit-day)} \tag{9.3}$$

This is consistent with the value in Figure 9-2, although slightly conservative [Normand 2004].

9.1.5 Exposure During Flights

The neutron environment is strongly dependent on both latitude and altitude. Therefore, rout-specific information is required to characterize the neutron environment, and hence the predicted error rate in electronics, on a given flight. Various computer codes exist to model the radiation environments in the atmosphere for the purpose of calculating air-crew dosimetry during routine flights [Hands 2009, Dyer 2009]. Some of these are also capable of determining neutron and other particle species in the atmosphere. Figure 9-5 shows the calculated and measured data for two aircraft flights. The calculation was performed using the QARM model (QinetiQ Atmospheric Radiation Model) [Lei 2006]. You can observe the variation of dose with rigidity and altitude.

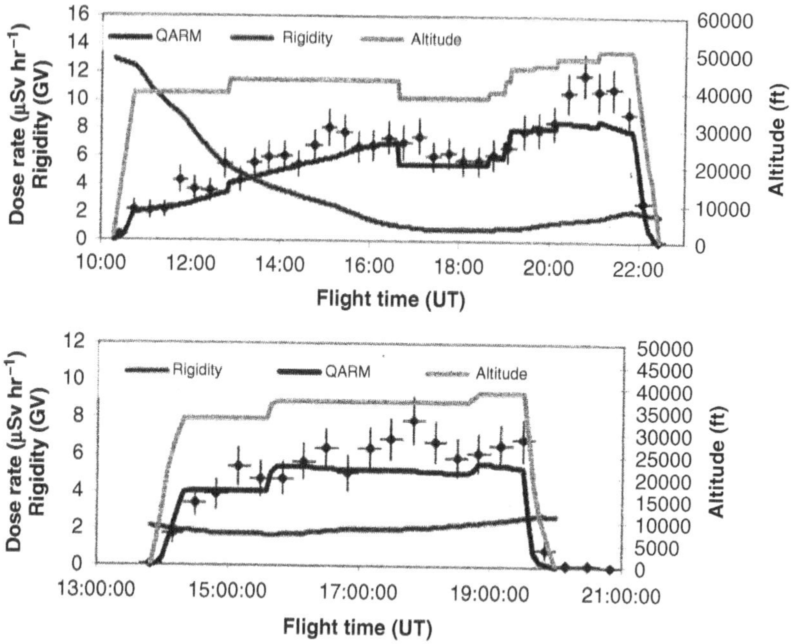

Figure 9-5 Measured flight path dose rates versus the QARM calculated dose rate. Top: An executive jet flight from Shanghai to the United Kingdom in December 2003. Bottom: A commercial flight from New York (JFK) to London (LHR) in June 2001 [Lei 2006]. Copyright © 2006 IEEE.

9.2 UPSETS AT GROUND LEVEL

The IBM group made an extensive study of upsets at ground level [Ziegler 1996a, Ziegler 1996b]. We can use their numbers to calculate the single event rate coefficient at ground level, Denver, and Tombstone. Knowing that the neutron intensity follows an exponential distribution, we can then calculate the altitude dependence of ground level events as shown in Figure 9-6. Gordon and co-workers measured neutron flux at five altitudes and show the exponential behavior as a function of atmospheric depth [Gordon 2004]. They present an analytic model that can be used to determine the flux for a range of altitudes and locations. This is available at www.seutest.com.

The numbers in Figure 9-6 can be used to give us rough estimates of probable ground level SEE rates. The ground level rate coefficients are in the range of 6×10^{-3} to 2.2×10^{-2} upsets/bit-day. The

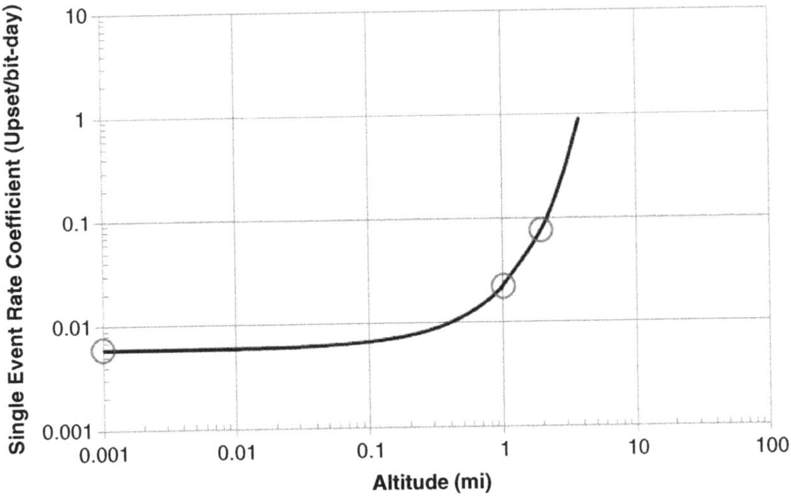

Figure 9-6 The variation of single event rates with altitude at ground level (after Ziegler) [Ziegler 1996b]. The data points indicate IBM results at sea level (New York), Denver, and Tombstone.

probable range of the figure of merit for modern unhardened parts is 2×10^{-9} to 2×10^{-8}. This suggests that ground level SEE rates will be in the range 1.2×10^{-11} to 4.4×10^{-10} upsets/bit day. These might be in very many bits in one computer, or in fewer bits in very many computers.

In many situations, we are examining SEE in nonmemory parts such as controllers, microprocessors, or FPGAs. In these cases we obtain a cross section per device, which is some unknown number times the cross section per bit. In these cases we can calculate the FOM per device using the cross section per device. The SEE rate per device is calculated using the same orbit-specific rate coefficient.

CHAPTER 10

Upsets Produced by Heavy Ion Nuclear Reactions

10.1 HEAVY ION NUCLEAR REACTIONS

Several papers pointed out the contributions to upsets from reactions in passivation and other layers above the active silicon layer [Gasiot 2002, Wrobel 2003, Truscott 2004]. Truscott demonstrated the usefulness of the Monte Carlo code GEANT4 for the analysis of these events.

There have been a number of observations of upsets produced below the normal cross-section curve when testing with low LET heavy ions. Figure 10-1 shows an example. There have been several explanations proposed [Liu 2006a], but the one that has received the most attention is that the upsets are due to nuclear reaction of the incoming ions with the tungsten overlayer in the chip [Warren 2005].

The Warren paper was well received. However, the accompanying paper by Howe pointed out that the reactions could make a major contribution to the upset rates in space for hard devices [Howe 2005]. Many found this difficult to believe as they know that only 1 in 10^4 ions has reactions, while all ions produce ionization. Figure 10-2 shows the relative cross sections for the direct ionization and for the indirect ionization from nuclear reactions [Warren 2005]. So the

Single Event Effects in Aerospace, First Edition. By Edward Petersen.
© 2011 the Institute of Electrical and Electronics Engineers, Inc.
Published 2011 by John Wiley & Sons, Inc.

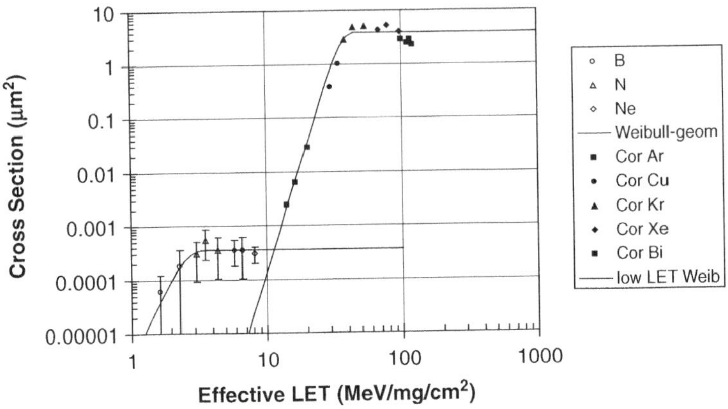

Figure 10-1 BAE data for "Chiplet 1 RAM0 SPRAM" taken at LBL. The data shows a definite low LET shoulder. The data has been fit with two different Weibull functions. The upper set is fitted making Connell geometrical corrections to the original data.

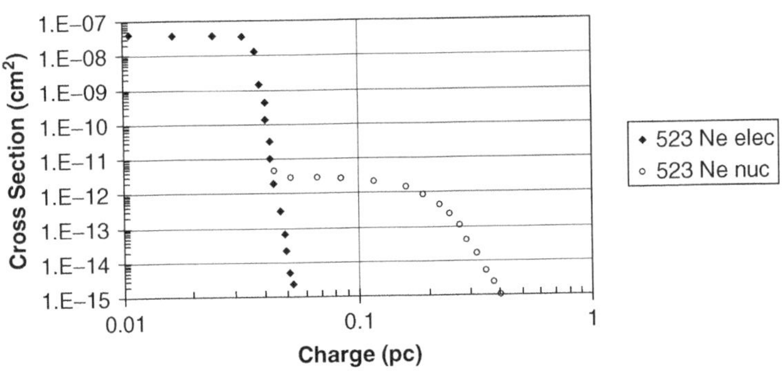

Figure 10-2 Integral cross-section curve for 523 MeV neon on silicon. The cross section corresponds to the probability that the deposited charge in a $2 \times 2 \times 2\ \mu m^3$ cube exceeds the given value. The lower curve extends beyond 1 pC if a tungsten layer is introduced [Warren 2005]. Copyright © 2005 IEEE.

question is: How can the tail compete with the dominant ionization component?

Figure 10-3 shows how the reactions contribute. The small open circles show CREME96 calculations of upset rate versus charge deposition for a $2 \times 2 \times 2\ \mu m^3$ device in geo-min conditions with 100 mils of shielding. The small open squares are Howe's similar results. These curves include all ions in the space environment. The small crosses show the ionization upset rate due to oxygen only. Note that

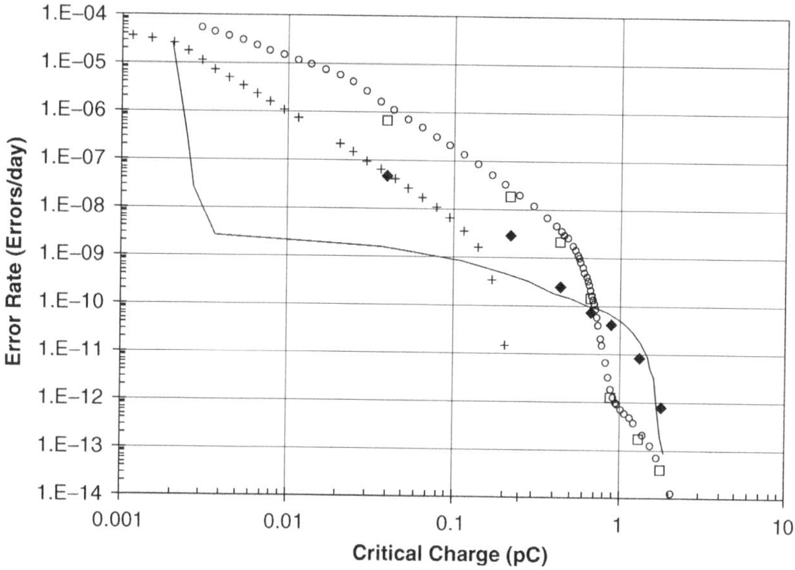

Figure 10-3 SEU error rates showing the ionization contribution and the contribution for oxygen and alpha particle reactions. Based on a figure by Howe [Howe 2005]. The rates indicated by small circles are the standard ionization RPP upset rates for a $2 \times 2 \times 2\ \mu m^3$ device in the space environment. The rates indicated by the solid curve and the solid squares are from nuclear reactions and can contribute more upsets than direct ionization for devices that upset only with very large charge deposition.

high energy particles contribute at low LET while low energy particles contribute at high LET and that high energy particles dominate in space. The solid curve corresponds to Figure 10-2 except it is the reaction contribution for high energy oxygen on tungsten [Howe 2005]. The solid diamonds indicate the corresponding rate calculation including both oxygen and helium as calculated by Howe. We observe that the oxygen reactions can contribute upsets for very hard devices that are not sensitive below the iron ionization peak.

Note from Figure 10-3 that there is basically a window of vulnerability. Devices with critical charges below 0.8 pC will be dominated by the ionization, while devices with critical charges above 2 pC will not be vulnerable to any upsets [Haddad 2006]. The width of the window depends on the width of the cross-section curve. Devices with very wide curves may not be appreciably affected even if their critical charges are above 1.0 pC. The very rapid decrease in the ionization upset rate curve between 0.6 and 1.0 pC indicates that the relative contributions of the reactions and ionization may change rapidly with critical charge, even as the rate itself decreases rapidly, for devices with very narrow cross-section curves.

10.2 UPSET RATE CALCULATIONS FOR COMBINED IONIZATION AND REACTIONS

Next, we want to calculate the upset rate for the device in Figure 10-1. We will use a spreadsheet calculation and a CREME96 calculation. The spreadsheet calculation, shown in Figure 10-2, folds together the upset rate as a function of LET, as calculated in CREME96 with the cross section as shown by the Weibull curves in Figure 10-1. We plot the differential total rate and the differential rate due only to ionization. We see that the nuclear reactions increase the rate by 15%. As the tungsten producing the reactions is on one side of the device, only one-half of the calculated rate due to reactions is included.

The question arises about whether we can use the lower Weibull curve to accurately represent the device response in space. The experimental points represent the results of bombardments with the light elements boron, neon, and nitrogen, presumably due to their hitting tungsten in the overlayer. The reaction cross sections are basically geometrical for high energy particles, and the geometry is dominated by the large tungsten nucleus. Therefore, all light, high energy particles will have similar behavior and the plot of number of reactions versus LET of the test light, high energy particles will be representative of all of the light, high energy particles in the space environment. The similarity is apparent when one examines the cross-section curves in Warren [Warren 2005].

We use the following parameters for the Weibull calculations using the CREME96 website. The depth is assumed to be 2.25 μm consistent with the value used by Warren [Warren 2005]. The rates have to be calculated using the full volume of the device, and then the reaction result scaled down to represent the actual probability and the geometrical factor. The scaling factor is $(\frac{1}{4} \times 0.00036 \times 1/2)$ The 1/4 scales from 4 μm^2 to 1 μm^2, the 0.00036 scales down to the measured cross section, and 1/2 allows for the fact that only particles from one side of the plane contribute. Note that the depth here is different from the depth used in Figure 10-4. An increased depth for the same LET

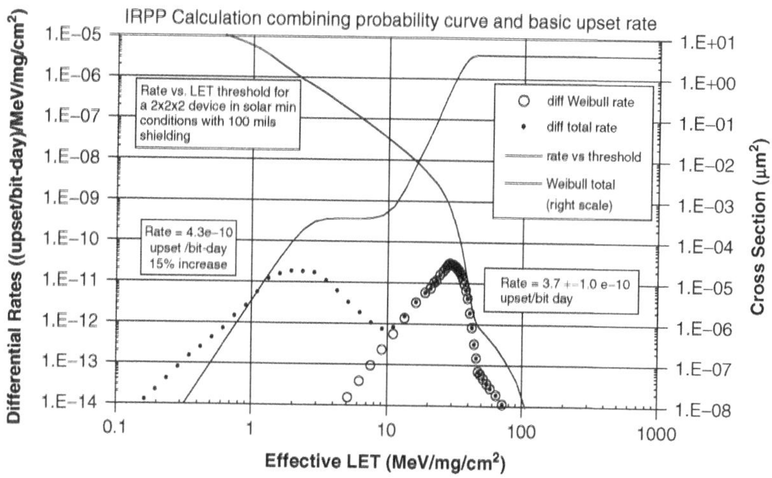

Figure 10-4 Spreadsheet calculation of the ionization and total upset rates in geosynchronous solar minimum environment with 100 mils of shielding for the device characteristics shown in Figure 10-1.

Table 10-1
Contributions of Ionization and Reactions to Total Upset Rate

LO	W	S	CS	CREME	Scaling CS	Result SEU/bit/day
0	36	7.95	4.0	2.5e−10		2.5e−10
0	2.52	5.13	4.0	1.1e−6	$\frac{1}{4} \times 0.00036 \times \frac{1}{2}$	5.0e−11
						3.0e−10 total

implies a higher critical charge in CREME96, so the calculated upset rate decreases for the thicker device, especially when the device is on the edge of the iron falloff. In this calculation the reaction component is 20% of the ionization component or 17% of the total. It may be more realistic to assume that the reactions contribute over only one-third of the total solid angle as the reaction products outside this range have long path lengths to reach the sensitive volume. In this situation the reactions contribute 13% of the ionization or 11% of the total. Table 10-1 shows the contributions.

Reed has presented the MRED code result for this case. He compared the result with space results, shown in Figure 10-5 [Reed 2007]. MRED stands for Monte Carlo Radiative Energy Deposition, developed by researchers at Vanderbilt University [Reed 2006]. The MRED

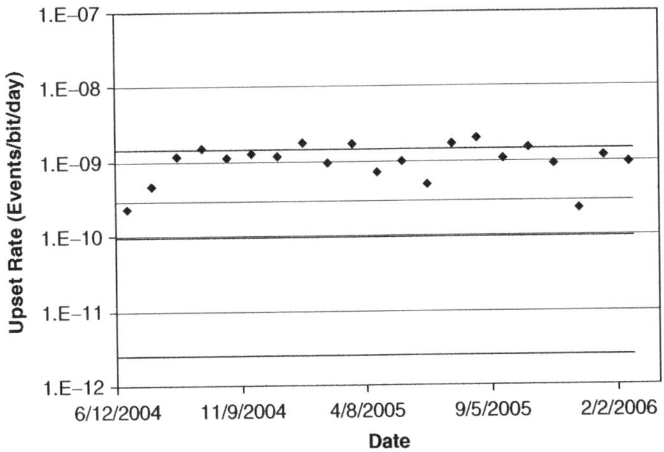

Figure 10-5 Comparison of IRPP calculation of upset rates with measured values on a NASA space mission and with results of the code MRED. The dashed line indicates IRPP two term calculations using CREME96 and the solid lines indicates the upper and lower limits of the MRED predictions. Data and MRED predictions from [Reed 2007]. Copyright © 2007 IEEE. The lower small dashed curve indicates the CREME96 results quoted in the paper, which are either the result of a RPP calculation rather than an IRPP calculation or the assumption of a harder device than used in the present and MRED calculations.

results agree very well with the IRPP results using CREME96, shown as a dashed line in the figure.

Reed quotes a CREME 96 result of 2.5e−12 errors/bit/day from his reference 23. This value appears to be from a RPP calculation, corresponding to one of the hollow points shown in Figure 10-3. For a very SEU hard device, such as this, there is a factor of 100 difference between the RPP rate and the IRPP rate. The rapid change in the environment with LET leads to large contributions from the low LET portion of the cross-section curve. The quoted difference could also arise if their CREME96 calculation assumed a slightly harder device then used in our calculations or in the MRED calculation.

10.3 HEAVY NUCLEAR ION REACTIONS SUMMARY

Nuclear reaction produced upsets are conceptually very important. They will be evident only for very hard devices with heavy metal overlayers. Dodd has shown these effects for a very hard device, but not for several soft devices [Dodd 2007b]. However, even for a very SEU hard device such as the BAE memory they contribute less than 17% of the total upsets. The rate is very well predicted by the traditional IRPP CREME96 calculation and, with slightly larger error bars, the Monte Carlo code MRED.

CHAPTER 11

Samples of Heavy Ion Rate Prediction

In this chapter we want to give a number of examples of heavy ion rate predictions. Initially, we will examine special conditions and will perform spreadsheet calculations. We will then show that onset thresholds should not be used for estimating the severity of SEE rates. Finally, we will lead the reader through the various steps involved in using the common codes such as CREME96.

11.1 LOW THRESHOLD STUDIES

Our first case will examine the influence of the threshold region on the calculated upset rates. As the Weibull and lognormal curves sometimes have different threshold behavior, we will compare them to examine the differences and the resulting impact on the calculated rates. Petersen pointed out that the three-parameter lognormal curve sometimes does a poorer job of fitting the threshold than the four-parameter Weibull and suggested that it was occasionally useful to insert a low LET cutoff on the lognormal function [Petersen 2005]. The argument is based on the fact that charge collected at the node of interest depends on the location of the ion hit. Some locations allow a small charge deposition (low effective LET) to produce a large collected charge and upset the device. That is, there appears to be an internal gain,

Single Event Effects in Aerospace, First Edition. By Edward Petersen.
© 2011 the Institute of Electrical and Electronics Engineers, Inc.
Published 2011 by John Wiley & Sons, Inc.

called efficacy, in that paper. As you move toward the threshold, the lognormal curve is representing higher and higher efficacy. Higher efficacy must correspond to higher internal electric field, corresponding to drift or the funnel effect. However, the device geometry can affect this charge collection. Therefore, there is a limit in the efficacy and a corresponding cutoff in the lognormal cross-section curve. The four-parameter Weibull function can describe this directly, or a low LET cutoff in the lognormal function can describe this effect.

In order to compare cross-section curves representing several different functions, it is necessary to do spreadsheet integral rectangular parallelepiped calculations (IRPP). These are performed in two steps. First, one uses a code such as CREME96 to do the RPP calculation of upset rate versus LET for the geosynchronous solar minimum case [Tylka 1997]. This is then combined with the cross-section curve for the integral result. The geosynchronous solar minimum case is conservative and representative. Similar calculations can be used for any orbit [Petersen 1998]. Figure 11-1 shows two representative curves of RPP upset rates versus effective LET. CREME calculates the basic curve by using the data for the environment as a function of LET and calculating the upset rate for the given geometry as a function of critical charge. You then change critical charge to effective LET by combining it with the depth of the device. Note that for thin devices, the cos θ effects give contributions at large effective LETs. You can have

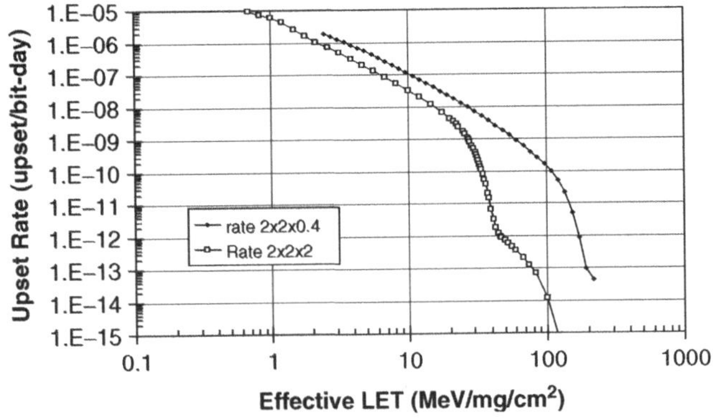

Figure 11-1 RPP upset rates calculated using CREME96. These rates are for solar minimum conditions at geosynchronous orbit with 100 mils of shielding. The rates are for a bulk $2 \times 2 \times 2\ \mu m^3$ device (left) and a $2 \times 2 \times 0.4\ \mu m^3$ EPI device (right).

contributions for LETs to 120 MeV/mg/cm^2 for near cubic devices, while the contributions go to 200 MeV/mg/cm^2 for thin devices. The total upset rate is calculated by integrating over this curve using the weighting defined by the cross-section curve. You occasionally see statements that if the threshold is above 100 MeV/mg/cm^2, no rate analysis is required. This is wrong. If the possible SEE could be fatal, the rate certainly should be examined.

11.2 COMPARISON OF UPSET RATES FOR WEIBULL AND LOGNORMAL FUNCTIONS

Figure 11-2 shows representative cross-section curves for CMOS/bulk and CMOS/SOI devices. The solid curves are Weibull curves, while the dashed curves are lognormal functions that match the Weibull curves at the one standard deviation points. We are interested in the differences in the predicted upset rates introduced by the differences in the two curves in the threshold region.

In Figures 11-3 and 11-4 we show the input and output curves for the integral upset rate calculations. The environment curves on the left

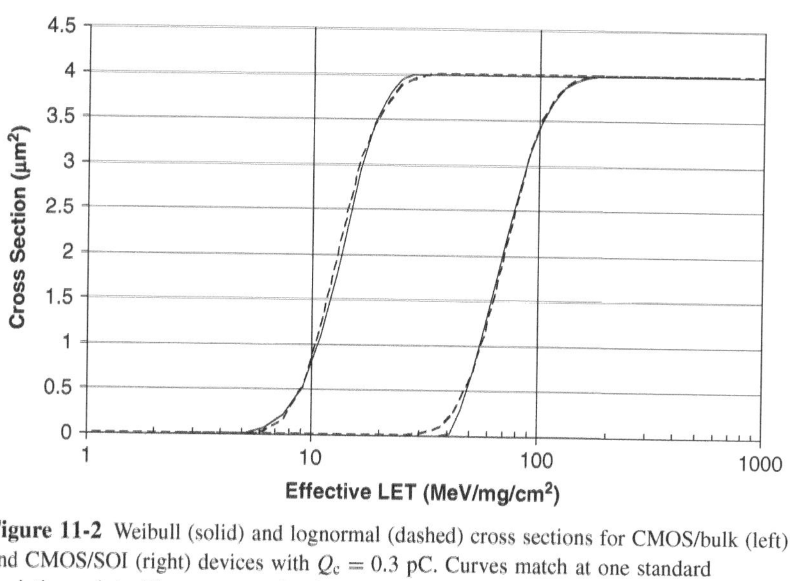

Figure 11-2 Weibull (solid) and lognormal (dashed) cross sections for CMOS/bulk (left) and CMOS/SOI (right) devices with $Q_c = 0.3$ pC. Curves match at one standard deviation points. The curves on the right can also represent the cross-section curves of circuit hardened CMOS/bulk $2 \times 2 \times 2$ μm^3 devices with $Q_c = 1.5$ pC.

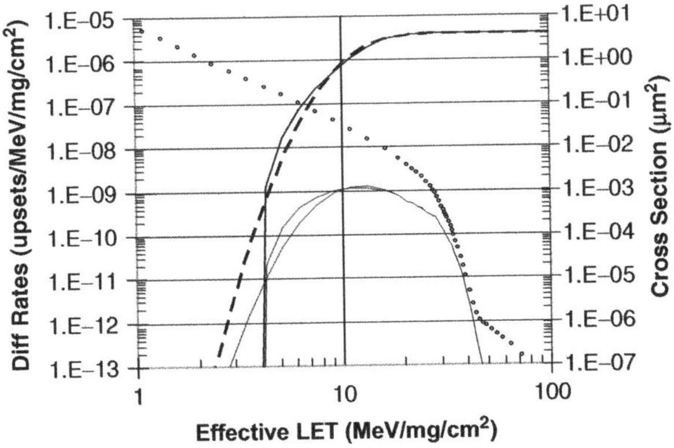

Figure 11-3 Upset rate calculation for representative bulk CMOS device $2 \times 2 \times 2\ \mu m^3, Q_c = 0.3$ pC. The space upset rate designated by the small circles is weighted with the cross-section curves (upper curves on right scale). The two lower curves show the resulting differential upset rates. There is very little difference in the resulting total upset rates.

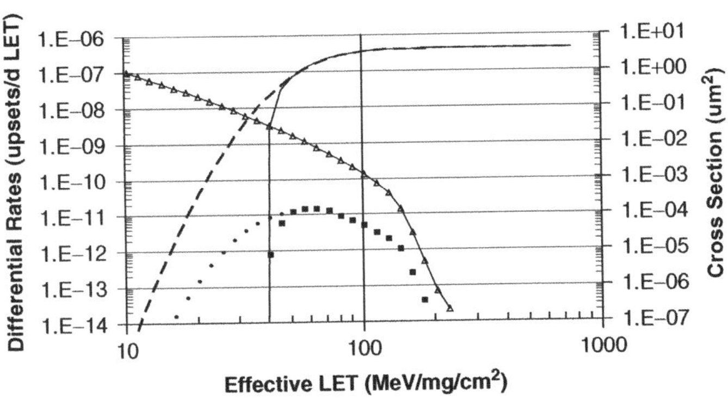

Figure 11-4 IRPP calculation for representative CMOS/SOI device $2 \times 2 \times 0.4\ \mu m^3, Q_c = 0.3$ pC. The space upset rate, represented by small circles, is combined with the cross-section curves (Figure 11-2) to produce the differential upset rates (lower curves), which are summed for the total upset rate for the device.

are folded together with the devices cross-section curves on the right to produce the differential upset rates as a function of effective LET. The cross-section curves are the same as in Figure 11-2, except plotted on a log scale. The two lower curves show the calculated differential upset rates. Figures 11-5 and 11-6 show the cumulative values of the differential upset rates.

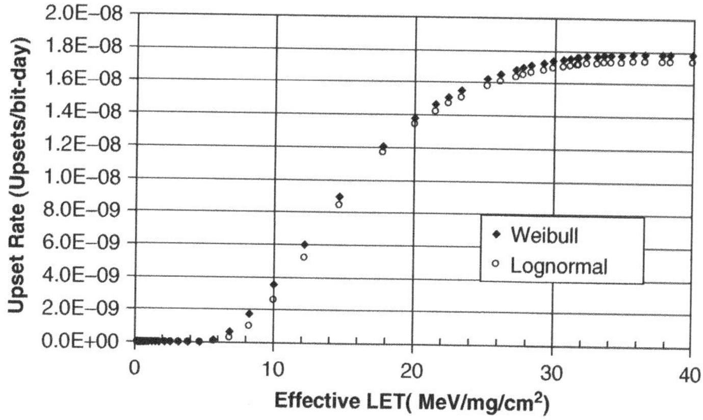

Figure 11-5 Running total upset rate for bulk CMOS rate calculation for device with $Q_c = 0.3$ pC. These curves summarize the lower curves in Figure 11-3.

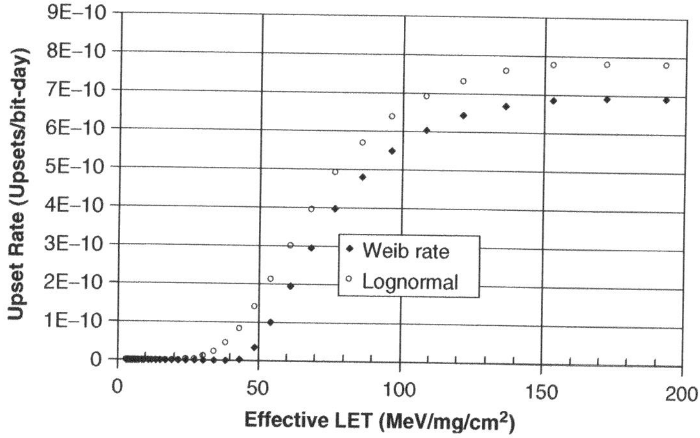

Figure 11-6 Running total upset rate for the CMOS/SOI device shown in Figure 11-4.

The Weibull function for the bulk device has $L_0 = 4, w = 12.4$, and $s = 2.7$. The corresponding lognormal curves have $m = 2.66$ and $s = 0.31$. Figure 11-3 shows that there appears to be a significant difference between the two cross-section curves in the threshold region. The Weibull function is again the solid curve and the lognormal function is shown dashed. The total rates are 1.78×10^{-8} upsets/bit-day for the Weibull curve and 1.74×10^{-8} upsets/bit-day for the lognormal curve. The differences in the threshold region have little impact on the calculated upset rates.

The Weibull function for the SOI device has $L_0 = 42$, $w = 40$, and $s = 1.4$. The corresponding lognormal curves have $m = 4.3$ and $s = 0.33$. Figure 11-4 shows that there appears to be a significant difference between the two cross-section curves in the threshold region. The Weibull function is again the solid curve. The two lower curves show the calculated differential upset rates.

The total rates are 6.9×10^{-10} upsets/bit-day for the Weibull curve and 7.8×10^{-10} upsets/bit-day for the lognormal curve. The lognormal curve with a low LET cutoff has an upset rate of 7.2×10^{-10} upsets/bit-day. The differences in the threshold region have little impact on the calculated upset rates.

Note that very thin devices such as this can have upsets at effective LETs above 100 MeV/mg/cm^2. Therefore, SEE survivability needs to be considered for circuit hardened devices with onset thresholds between 100 and 200 MeV/mg/cm^2.

The plots of differential upset rates in this form are difficult to interpret visually. Figures 11-5 and 11-6 show the upset rates on a cumulative basis. For the bulk device, the major contributions to the upset rate come between LETs of 10 and 25 MeV/mg/cm^2. For the SOI device, the major contributions come between LETs of 50 and 100 MeV/mg/cm^2. There is little contribution to the upset rate below 10% of the limiting cross section.

We have been emphasizing that the threshold region has little impact on the total upset rate. However, there is one situation where the threshold region is very important, at least conceptually. Consider the right-hand cross-section curves in Figure 11-2 as if they are for a hardened CMOS bulk device. The critical charge would be 1.5 pC for this device. In this case the critical LET is well above the iron drop-off in the differential RPP upset rate curve, but the lower portion of the threshold region may be below the peak. The situation is shown in Figure 11-7. Our representative lognormal cross-section curve generates a relatively large number of upsets, but the Weibull curve or the cutoff lognormal curve has about a factor of 10 less upsets. Figure 11-8 shows the cumulative upset rates derived from the differential rates shown in Figure 11-7. The total upset rates are 6×10^{-12} upsets/bit-day for the lognormal curve, 3.5×10^{-13} and 2.7×10^{-13} upset/bit-day for the lognormal with cutoff and the Weibull curves. These rates are very small, but you do have to include measurements in the threshold region to distinguish between them.

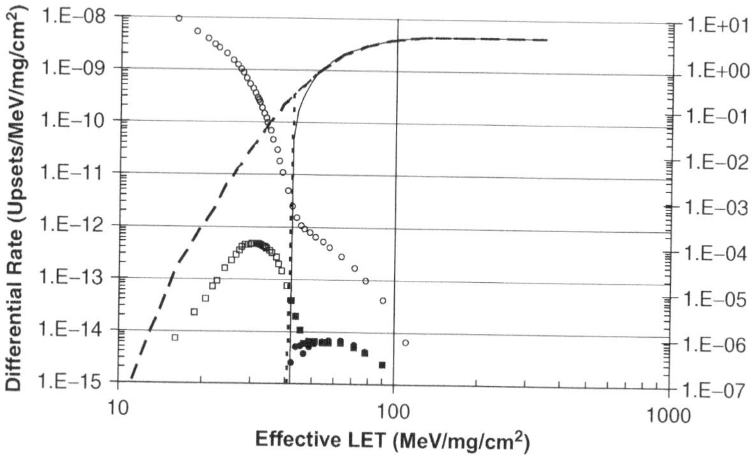

Figure 11-7 Upset rate calculation for representative circuit hardened bulk CMOS device $2 \times 2 \times 2\ \mu m^3$, $Q_c = 1.5$ pC The space upset rate designated by the small circles is weighted with the cross-section curves (upper curves on right scale). The Weibull (solid) and lognormal (long dash) cross-section curves are the right curves in Figure 11-4 now plotted on a logarithmic scale. We have added a short dashed curve for the lognormal with cutoff. The three lower curves show the resulting differential upset rates. In this case there can be large differences in the calculated upset rates depending on the threshold contribution in the region of the iron LET peak.

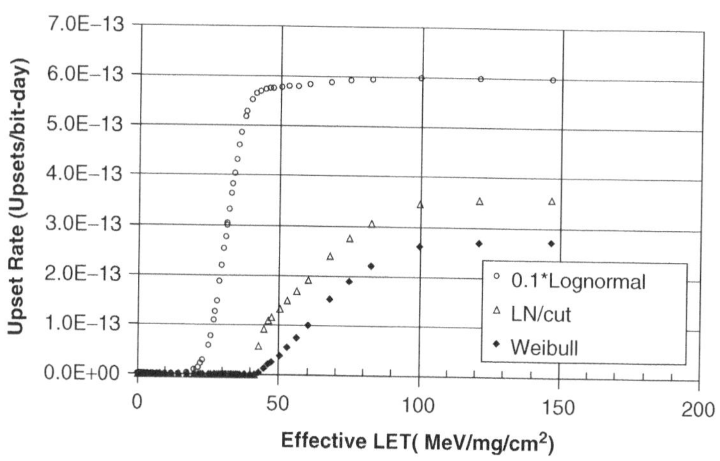

Figure 11-8 Running total for the sum of the differential upset rates shown in Figure 11-7. The contribution can vary greatly depending on the relative position of the threshold and the falloff of the iron region of the LET spectrum.

In general, the onset threshold is not a useful parameter for estimating upset rates, except as it reflects changes in L_C. It is tempting to examine the effect of threshold changes by changing the Weibull threshold parameter. An apparently obvious way is to change L_0 and keep w and s constant. This shows significant changes in the predicted rates. However, this has the effect of changing the width (lognormal σ) and the peak (L_c) of the distribution, so the Weibull threshold is not an independent parameter. Also, previous research has shown that the shape of the cross-section curve is nearly constant for a given technology [Petersen 1996a, Petersen 1996b]. Therefore, a given w and s correspond to a very narrow range of thresholds. As will be seen below the rates vary primarily with the limiting cross section and $L_{0.25}$.

11.3 LOW THRESHOLD–MEDIUM L_C DATA

Figure 11-9 shows an example of data for a part with a low threshold and medium L_c [Morris 2007]. This is an unhardened 0.18 μm bulk CMOS device. A linear Weibull fit finds a negative threshold. The logarithmic fit indicates a threshold of 0.4 MeV/mg/cm^2. A linear fit forcing the threshold of 0.4 MeV/mg/cm^2 produces $cs = 5.34\ \mu m^2, w = 35.7, s = 1.18$.

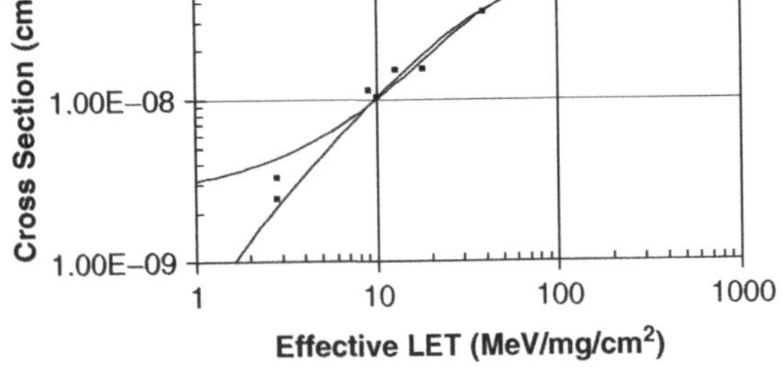

Figure 11-9 Data for an unhardened 0.18 μm bulk CMOS device by Silicon Space Technology. The dashed curve indicates the linear Weibull fit with a negative threshold and the solid curve a logarithmic Weibull fit with a threshold of 0.4 MeV/mg/cm^2. $R^2 = 0.99$. Despite the very low threshold, the upset rate in the solar minimum geosynchronous environment with 100 mils of shielding is only 2.51×10^{-7} upsets/bit-day.

Table 11-1
Upset Rate for Different CREME96 Calculations

	$Z_{min} = 0$	$Z_{min} = 3$
Linear fit, 0 threshold	7.87×10^{-7}	4.7×10^{-7}
Linear fit, 0.4 threshold	2.55×10^{-7}	2.51×10^{-7}

It is possible when doing upset rate calculations to use a default environment that includes all ions, including hydrogen and helium. This is a mistake for low threshold devices that have not been shown to upset for proton direct ionization. If one uses the linear Weibull fit, forcing the threshold to be zero, one is including an unrealistic proton contribution to the upset rate. Table 11-1 shows the differences in the calculated upset rates using CREME for geosynchronous solar minimum conditions with 100 of mils shielding (upsets/bit-day).

Although the proton ionization contributes at LETs above 1 MeV/mg/cm^2, it is probably satisfactory when forcing a threshold to use values of 0.5 MeV/mg/cm^2 or above. One should not accept thresholds less than this unless there is demonstrable upset by ionizing protons as in electro-optical devices. Often, as in this case, one can use the logarithmic fit to obtain thresholds above zero.

With an upset rate of 2.51×10^{-7} upsets/bit-day and a figure of merit of 3.24×10^{-10}, this is one of the hardest unhardened parts. This can be seen by examining later figures that show compilations of upset rates versus onset threshold and versus figure of merit. (This data is not included in those figures.) The low threshold is not indicative of a sensitive part with a high upset rate.

11.4 SEE SENSITIVITY AND LET THRESHOLDS

We have to be clear what we mean by device upset sensitivity. There are two approaches to device sensitivity. They are both useful in particular applications, but there is little correlation between them. In most cases, we are interested in relative upset rates in given space environments. A device is sensitive if it has a relatively high upset rate, and is less sensitive, or SEE harder, if it has a relatively low upset rate. Therefore, operationally, sensitivity should be directly related to space upset rates.

The space upset rates depend both on the device and on the orbit. However, the relative upset rates for various parts remain the same nearly independent of orbit. Petersen demonstrated that the relative rates follow the FOM for a variety of orbits, whether proton intensive, heavy ion intensive, or a combination of the two [Petersen 1998]. Furthermore, Normand demonstrated that the FOM also described the device sensitivity to atmospheric neutrons [Normand 2004]. Therefore, the device upset sensitivity is a fundamental property of the device and does not depend on the device's application usage in space.

There are a few specialized situations in which the concept of uniform relative sensitivity may break down. Petersen indicated that there was more spread in the plots of upset rate versus the FOM for orbits in the heart of the proton belts due to more low energy protons [Petersen 1998]. Something similar may happen for solar flares with their abundance of low energy protons. Devices with large amounts of boron will have more upsets in thermal neutron environments.

The second approach to device sensitivity is based on the onset threshold for upset. It appears clear that devices with low LET onset thresholds are more sensitive than devices with higher thresholds. Petersen demonstrated that there is a small very sensitive region in a device that corresponds to the onset threshold [Petersen 2005]. It is sometimes valuable to determine this location and/or the corresponding LET either experimentally or theoretically. In general, this value will scale with the critical charge or critical LET as changes in the device circuit parameters do not change the shape of the cross-section curve. It is not clear that it is practical to use this threshold for hardness assurance purposes. If the threshold is extremely low, the device may be sensitive to direct proton ionization. That is true of electro-optical devices and a few commercial memories.

As the onset threshold corresponds to a measure of device sensitivity, it appears intuitively obvious that it should be a measure of relative upset sensitivity in space. The rest of this section will demonstrate that this assumption is wrong and that there is very little correspondence between space upset rates (relative device sensitivity) and onset thresholds. Therefore, operationally, the onset threshold should not be used as a measure of device sensitivity. We really need two different names for the two measures of device sensitivity.

It has been demonstrated that the principal contributions to upset rate come from the center of the upset cross-section distribution; see

[Petersen 1992b] and discussion above. Despite this, a number of authors still regard the threshold region as important and the onset threshold as a valid indication of relative upset rates in space [O'Bryan 2005, Irom 2005]. These references are representative of a long series of articles.

There are several operational definitions for the SEU threshold. The one used commonly is the LET such that a fluence of 10^7 ions/cm^2 does not produce an upset. If we assume one upset for 10^7 ions, we can use the number of bits in the device to determine the corresponding cross section. Then with this cross section, or the device Weibull fits, it is possible to determine the corresponding LET threshold values. We will construct a scatter plot of the space upset rate as a function of device threshold using the 1e^7 thresholds. Figure 11-10 shows that there is little correlation between the space upset rates and the device threshold. A given threshold may correspond to a range of five orders of magnitude in upset rate. Factors other than the onset threshold

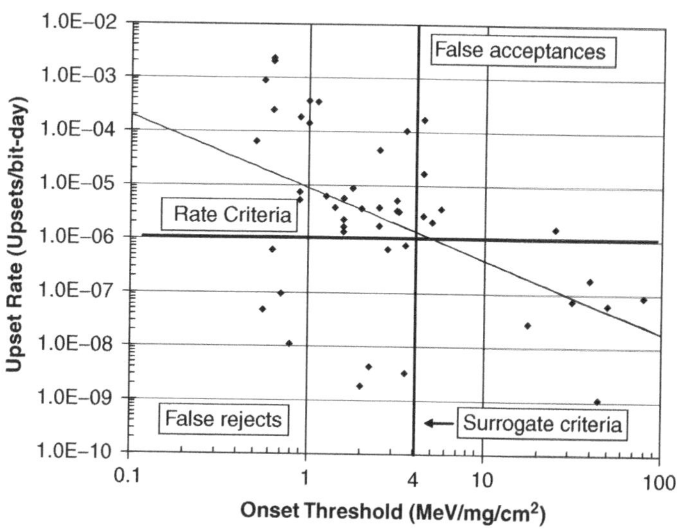

Figure 11-10 Scatter plot of space upset versus the device threshold, where the threshold is defined as the point at which there is one upset for a fluence of 10^7 particles per square centimeter. The rates are calculated for the geosynchronous orbit with 100 mils of shielding and solar minimum conditions. The curve is a best fit leading to Rate $\sim T^{-1.3}$. R^2 for the fit is 0.27 so there is only 27% correlation between the upset rate and the device threshold. The horizontal and vertical lines correspond to desired space upset rate criteria and corresponding surrogate criteria based on the onset threshold.

determine 73% of the upset rate. The use of the LET threshold as a direct indication of device upset sensitivity is wrong.

If data is fitted using the Weibull function, one of the resulting parameters is a threshold. Some might believe that this is a meaningful measure of device sensitivity. Figure 11-11 shows the same correspondence using the Weibull threshold. There again can be a range of more than four orders of magnitude in upset rate for a given threshold value. If the data is fit with the lognormal function, there is not a number identified as a threshold, although many other communities would use the median of the differential lognormal function as the threshold.

The reason for the scatter is that the upset rate depends not only on the critical charge (indirectly related to the threshold) but on the area of the device, the depth of the sensitive volume of the device, and the width of the cross-section curve.

The LET threshold is sometimes used as a guide for estimating the severity of the SEE problems. The severity should be determined from the actual upset rates, but the LET threshold is used as a surrogate. Some programs use it as a cutoff for acceptable devices. Other programs use it as a guide to suggest when they should pay special attention to graceful error recovery. An example is a guideline

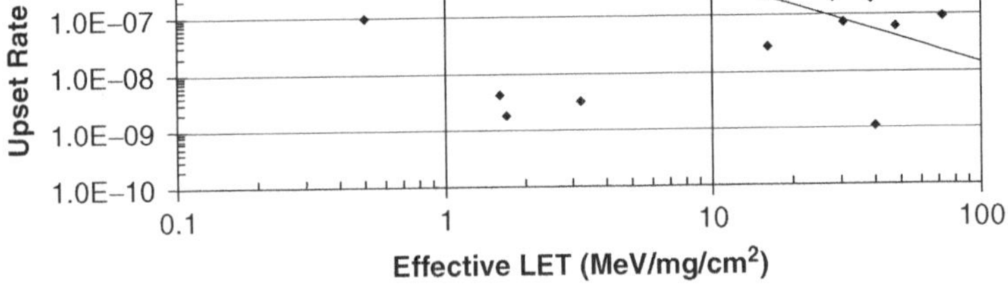

Figure 11-11 Scatter plot of space upset rate versus the device threshold, where the threshold is the threshold of the Weibull function fit to experimental data. The curve is a calculated fit assuming that the upset rate is inversely proportional to the device threshold. R^2 for the fit is 0.36 so there is only 36% correlation between the upset rate and the device threshold. The rates are calculated for the geosynchronous orbit with 100 mils of shielding and solar minimum conditions.

that says: if the threshold LET is less than 4 MeV/mg/cm^2, the program needs to expend special effort on graceful error recovery. This may correspond to a more specific guideline of less than 1×10^{-6} errors/bit-day.

Now consider Figure 11-10, and determine how many false rejects and false acceptances are obtained for various combinations of upset rate and threshold. For the example of a 4 MeV/mg/cm^2 threshold used to approximate a cutoff rate of 1×10^{-6} errors/bit-day, there are 9 rejections of satisfactory parts, while 6 parts are accepted that need further study. The intuitively obvious use of the onset threshold as a filter is wrong 30% of the time. We have both false acceptances and false rejections. It is not obvious which is worse. Tabulated thresholds such as those in [O'Bryan 2005, Irom 2005] serve as a very poor general screening tool.

A different type of threshold is obtained by laser testing. The laser threshold corresponds to the minimum energy deposition so that only one pixel is upset. As the pixel size is approximately 1 square micron, this corresponds to a 1 micron cross section. We will use the device parameters tabulated in Petersen [Petersen 1998] for comparison. The results are shown in Figure 11-12. This threshold does not correspond to the onset threshold, and there is some correspondence with the

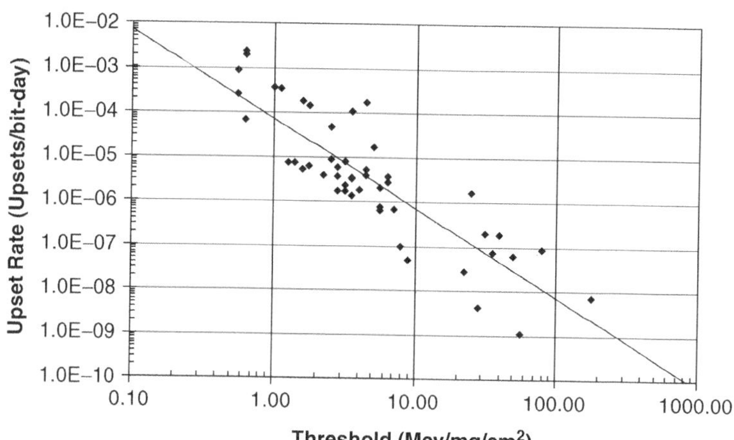

Figure 11-12 The geosynchronous upset rate plotted as a function of LET threshold determined by a cross section of 1 μm^2. This corresponds to upsetting one pixel in laser experiments. The line varies as the square of the threshold, and $R^2 = 0.74$.

upset rate. Due to the large amounts of scatter, the laser threshold is not useful for comparing device sensitivities. It might be very suitable for comparing the sensitivities of various generations of a given part for hardness assurance purposes.

Proton upset rates also have very little correlation with the onset threshold. Figure 11-13 shows upsets rates for a representative orbit as a function of the LET threshold of the device. The data is from [Petersen 1998]. The upset rates can vary over four orders of magnitude for a given threshold value.

The LET onset threshold is sometimes used to estimate the importance of proton SEE effects for orbits in the proton belts. However, this is unnecessary. As several authors have pointed out, if *any* upsets are observed at a LET less than 15 MeV/mg/cm^2, the device will be susceptible to proton induced upsets [O'Neill 1998, Lum 2004]. If upsets are observed at lower LETs, the program should perform detailed upset rate measurements, rate calculations in orbit, and evaluation of SEE mitigation effectiveness. An alternative requirement is the observation of upsets with Si or Ne beams. Of course, the upset rates can be estimated using the FOM approach.

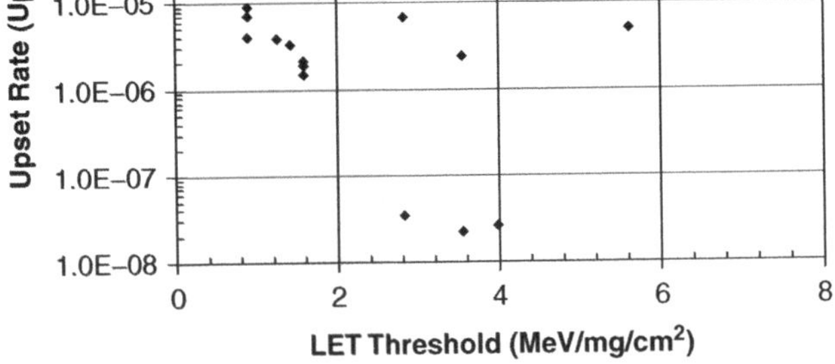

Figure 11-13 Proton upset rates for an 1111 km, 63 degree, solar minimum orbit with 100 mils of shielding as a function of the heavy ion LET threshold for a group of representative devices. The scatter plot shows little correlation between the upset rate and the device threshold, and any value of the threshold can correspond to a large range of upset rates.

If any upsets are observed below a LET of 120 MeV/mg/cm^2 (200 for CMOS/SOI), the program needs to examine the survivability probabilities. It is not necessary to examine the onset threshold to see these upsets. The SEE analyst must be prepared to evaluate failure probability for any level of threshold. It has happened that the analyst determines an apparently trivial error probability, but that probability turned out to be the largest single item in the mission error budget, so that the computer system had to be completely redesigned using harder parts.

Figure 11-14 shows the scatter plot of the upset rate versus the figure of merit (FOM). This shows a much better correlation than the correlation with the device threshold. The slight scatter in the FOM rates arises as it treats all devices as if they had the same geometry. In lieu of complete upset rate calculations, the figure of merit is appropriate as an approximate indication of device upset sensitivity (upset rate) for either heavy ions or protons [Petersen 1998]. Papers that summarize device measurements and thresholds should also include either the device FOM or upset rates in a standard orbit. Either can be used to determine the SEE rate in any other particular orbit [Petersen 1998].

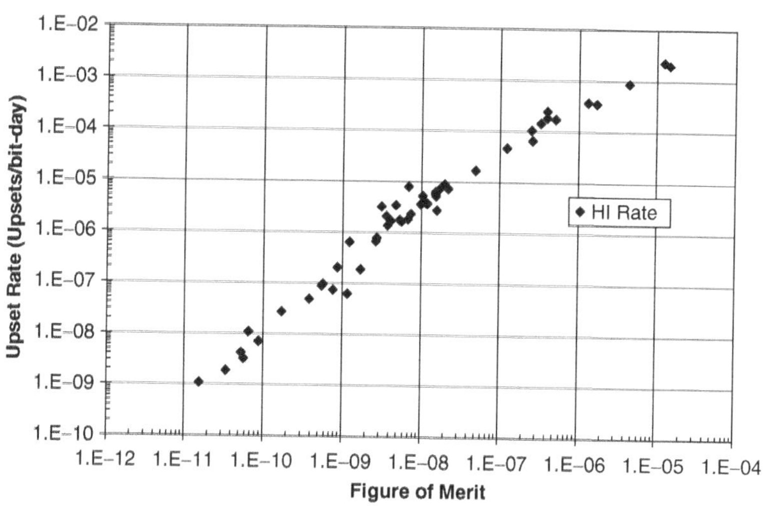

Figure 11-14 Scatter plot of space upset rates versus the device figure of merit. There is a much better correlation than for the device threshold. The upset rates are calculated in CREME96 for the solar minimum geosynchronous environment with 100 mils of shielding.

11.5 CHOOSING AREA AND DEPTH FOR RATE CALCULATIONS

In some cases information on the depth is not known. This is an important factor in the rate calculations, especially for smaller devices. It now appears that depth information can be obtained from details of the cross-section curve [Petersen 1997b]. This was discussed in Section 5.3. There we demonstrated how the angular data at high LET gives a direct indication, even without detailed curve fitting, of the relative depth and width of the device.

Similarly, the cross-section curve gives a direct measure of the area of the sensitive region. Therefore, the cross-section data should always be extended to high LET to obtain both the area and depth of the device.

A way around the problem of depth is to always assume that the depth is 1 micron. This was originally proposed by Kolasinski [unpublished]. For large soft devices, this is a valid approximation, as the use of the same depth to calculate critical charge from critical LET and to calculate the upset rate, using that critical charge and depth, leads to nearly compensating errors. As devices become smaller and harder, the approximation becomes worse. Except for devices that are intentionally built with vertical structures, the depth is in general 10% to 20% of the width or length.

Edmonds has pointed out that instead of asking for X, Y, and Z; codes such as CREME should ask for the RPP dimension ratios Z/X and Z/Y [Edmonds 2005]. The default values are 0.2. These ratios can easily be determined from studies of the angular cross-section variations.

11.5.1 SOI Devices

It was originally assumed that the sensitive area was the gate (body) with dimensions determined by the gate width (W) and length (L) and silicon film thickness (t_S). It has now become clear that the area extends over the entire transistor. Musseau and co-workers performed a charge-collection experiment in which the cross section for the peak was much larger than the gate area [Musseau 1997]. This and the wide spread in charge collected are consistent with later studies. If only the gate was involved, the size of the track would be important in submicron devices. Therefore, for all types of devices, one should use the cross-section area indicated by the cross-section curve, but not counting multiple upsets.

11.5.2 Inclusion of Funnel in CREME Calculation

The funnel needs to be an input into the CREME only if there is clear evidence that a funnel is present. This can be determined from the analysis of the cross-section curve discussed in Section 5.3. Some analyses have attempted to make a conservative rate calculation by including the funnel when there is no evidence for it. ***This is not valid***. CREME calculates the critical charge corresponding to a given LET by combining the LET with the total charge-collection depth, including the funnel. The longer path length with the funnel leads to a larger calculated critical charge. Therefore, the device appears harder and the corresponding rate lower. It is a mistake to include a funnel length in CREME unless there is clear evidence for it. Conservative rate calculations are performed by assuming no funnel and Z/X and Z/Y ratios approaching 0.01.

11.6 RUNNING CREME96 TYPE CODES

The NRL site for CREME96 is being shut down as this book goes to press. However, the entire CREME96 code is being transferred to CREME-MC at Vanderbilt University. This is both good and bad, as none of the necessary upgrades are being made during the transfer. They are not adding the integral lognormal description of the cross-section curves. They are not removing the Bendel–Petersen description of the proton cross-section curves, which should be replaced by the lognormal. However, they do have the Weibull description of the proton curves, which is an adequate substitute.

The instructions for the CREME96 portion of the code continue to be very complete. We will copy a few of the initial instructions that give an overview of how the program works. We will then give an example of the input and output of a CREME96 heavy ion upset rate calculation.

Before you can access the CREME96 codes, you must follow the registration and login procedures. Before you start CREME96 calculations, you should know the answers to these questions:

> What is the orbit of your satellite?
>
> What kind of radiation environment are you designing for? The typical "quiescent" environment that prevails most of the time? Or do you want to consider extreme cases, such as solar particle ("flare") events and (if your orbit is inside the Earth's magnetic field) geomagnetic storms?
>
> How much shielding does the satellite provide between your system and the external space radiation environment?

In addition, if you want to calculate single-event effect (SEE) rates:

> What are your devices' characteristics? What are the ground-test data on your devices' single event effect (SEE) cross sections, due to both proton induced recoils and direct ionization?

If you have answers to these questions, you're ready to make calculations.

The CREME96 program suite consists of eight primary modules, not all of which are needed for each calculation. The names and functions of these modules (as well as the input/output files—explained in the complete set of instructions) are summarized in Table 11-2.

For example, if you wanted to calculate the direct ionization induced SEE rate on an interplanetary mission, you would run the CREME96 programs in this order:

> FLUX then TRANS then LETSPEC then HUP
>
> If you wanted to calculate the direct ionization SEE rate in a low Earth orbit (LEO), you would run:
>
> GTRN then FLUX then TRANS then LETSPEC then HUP
>
> If you wanted to calculate the trapped proton induced SEE rate in a low Earth orbit, you would run:

Table 11-2
CREME96 Primary Program Modules

Module	What It Does	Input Files	Output Files
GTRN	Evaluates geomagnetic shielding	None	*.gtf; *.gt#, # = 1-X
TRP	Evaluates trapped proton fluxes	None	*.trp; *.tr#, # = 1-X
FLUX	Evaluates the ionizing radiation environment at the external surface of the spacecraft.	*.gtf or *.gt#;	*.trp; all optional *.flx
TRANS	Transports nuclear fluxes through the spacecraft shielding	*.flx (or *.tfx);	*.shd(optional) *.tfx
LETSPEC	Calculates linear energy transfer (LET) spectra	*.tfx (or *.flx)	*.let; (*.dlt optional)
HUP	Evaluates direct ionization ("heavy ion") induced SEEs	*.let	*.hup
PUP	Evaluates proton induced SEEs	*.tfx (or *.flx)	*.pup
DOSE	Evaluates dose due to nontrapped particles (GCRs, SEPs)	*.tfx (or *.flx)	*.dse

Note: If you are interested in geosynchronous or interplanetary orbits, you do not need to run GTRN or TRP. (See GTRN for more advice.) In these cases, begin your calculations with FLUX.

TRP then TRANS then PUP
(However, you should always evaluate both proton induced and direct ionization induced SEE rates for Earth orbits.)

The following section discusses some of the input and output files for heavy ion upset rate calculations. Note that at step three (LETSPEC) the mass range may be changed so that hydrogen and helium are not included. Table 11-1 gave an example of why this may be necessary. Also note that in step four, the input format sometimes expects decimal points; otherwise there may be arithmetic problems.

11.6.1 CREME96/FLUX

External Space Ionizing-Radiation Environment: User-Supplied Parameters

1. Atomic number of lightest element to be included: Choose 1. You may want to increase this to 2 or 3 at later stages.
2. Atomic number of heaviest element to be included: Choose 92 even though this increases computational time. This is a basic reference file and you may want to perform calculations of hard devices that have contributions from elements above 28. Many submicron devices have low thresholds, but wide widths (large Weibull W) so that major contributions are made from elements heavier than 28.
3. Environment model:
   Solar-Quiet (no "flare") Conditions:
   Solar Minimum (Cosmic Ray Maximum)
   Solar Maximum (Cosmic Ray Minimum)
   Enter year (e.g. 1996.80)
   Solar-Energetic Particle ("flare") Conditions:
   Worst Week
   Worst Day
   Peak 5-minute Averaged Fluxes
4. Spacecraft Location:
   Near-Earth Interplanetary/Geosynchronous Orbit
   Inside Earth's Magnetosphere

11.6.2 CREME96/TRANS

Nuclear Transport Module: User-Supplied Parameters

1. Shielding material: Aluminum.
2. Shield thickness: 100 mils, or use an existing file.

11.6.3 CREME96/LETSPEC

Linear Energy Transfer (LET) Spectra User-Supplied Parameters

1. Atomic number of lightest element included: 3 unless you have a good reason to go lower.

2. Atomic number of heaviest element included: 92 unless you have a soft device.
3. Minimum energy value: 0.1 MeV/nuc.
4. Device material: silicon.

11.6.4 CREME96/HUP

Direct Ionization Induced Single Event Upset (SEU) Rate User-Supplied Parameters

Label:	BAE	shoulder1	RPP not IR	bae2x2x2
Comment				
Comment				
Bit RPP:				
X	0.	0.	2.	0.
Y	0.	0.	2.	0.
Z	2.25	2.25	2.25	2.0
Funnel	0	0	0	0
Bits/Device:	1	1	1	1
1. Weibull:	(•)	(•)	( )	(•)
Onset	0.0	0.0		0.0
Width	36.0	2.52		36.0
Exponent	7.95	5.13		7.95
Limiting X-sctn	4.0	4.0		4.0
2. Crit Charge:	( )	( )	(•)	(•)
Qcrit			0.907	
XS/bit:			4.0	
3. Table	( )	( )	(•)	(•)
filename:				

11.6.5 CREME96 Results

```
15 bDb0ZF.HUP 21010
%Created by CREME96:HI_UPSET_DRIVER Version 210
  on 20100726 at 130445.2
%Input Integral LET Spectrum File: input.let
%Created by CREME96:LETSPEC_DRIVER Version 210
  on 20100105 at 140150.7
%ZMIN = 2 ZMAX = 92 LETMIN = 1.00E+00 LETMAX
      = 1.10E+05 MeV-cm2/g LBINS = 1002
%EMINCUT = 1.00E-01 MeV/nuc
%TARGET MATERIAL = SILICON
%Input File to LETSPEC_DRIVER: geomin09_100.tfx
%Created by CREME96:TRANSPORT_DRIVER Version 210
  on 20100105 at 140017.5
%ZMIN = 1 ZMAX = 92 EMIN = 1.0000E-01
  EMAX = 1.0000E+05 MeV/nuc MBINS = 1002
%Thickness =   100.0000 mils    ALUMINUM
%Input File to TRANSPORT_DRIVER: input.flx
%Created by the External Environment Module,
  Version 104 on 20100105 at 135901
%ZMIN = 1 ZMAX = 92
%IMODE  =   0 SOLAR-QUIET MODE: YEAR = 1977.0000
%ITRANS =   0 GEOSYNCH/NEAR-EARTH INTERPLANETARY
  FLUXES
REPORT NO.            1:       BAE T O D O
RPP Dimensions: X  =   2.00000 Y  =   2.00000
  Z  =   2.25000 microns.
Funnel length =   0.00000 microns.
CROSS-SECTION INPUT    4   WEIBULL FIT:
   ONSET    =     0.000 MeV-cm2/milligram
   WIDTH    =    36.000 MeV-cm2/milligram
   POWER    =     7.950 (dimensionless)
   PLATEAU  =     4.000 square microns/bit
Number of bits =   1.00000E+00
Rates:     SEEs/bit/second/    bit/day/
  device/second/      device/day
***** 1 2.98161E-15   2.57612E-10
  2.98161E-15    2.57612E-10
REPORT NO.    2:      shoulder 1 T O D O
RPP Dimensions: X =    2.00000    Y =  2.00000
  Z = 2.25000 microns.
Funnel length = 0.00000 microns.
CROSS-SECTION INPUT    4    WEIBULL FIT:
    ONSET    =    0.000 MeV-cm2/milligram
```

```
    WIDTH   =   2.520 MeV-cm2/milligram
    POWER   =   5.130 (dimensionless)
    PLATEAU =   4.000 square microns/bit
Number of bits =    1.00000E+00
Rates:    SEEs/bit/second/   bit/day/
  device/second/         device/day
***** 2 1.22468E-11   1.05812E-06
  1.22468E-11   1.05812E-06
REPORT NO. 3: RPP T O D O
RPP Dimensions: X =   2.00000 Y =    2.00000
  Z =   2.25000 microns.
Funnel length = 0.00000 microns.
CROSS-SECTION INPUT 5
  Critical charge       = 9.07000E-01 picocoloumbs
  Cross-Section = 4.00000E+00 square microns/bit
Number of bits = 1.00000E+00
Rates:    SEEs/bit/second/   bit/day/
  device/second/        device/day
***** 3   2.98227E-17   2.57668E-12   2.98227E-17
  2.57668E-12
REPORT NO. 4: BAE 2x2x2 T O D O
RPP Dimensions: X = 2.00000 Y = 2.00000
  Z = 2.00000 microns.
Funnel length = 0.00000 microns.
CROSS-SECTION INPUT 4 WEIBULL FIT:
   ONSET   = 0.000 MeV-cm2/milligram
   WIDTH   = 36.000 MeV-cm2/milligram
   POWER   = 7.950 (dimensionless)
   PLATEAU = 4.000 square microns/bit
Number of bits = 1.00000E+00
Rates:   SEEs/bit/second/     bit/day/
   device/second/        device/day
***** 4  4.69534E-15   4.05678E-10  4.69534E-15
  4.05678E-10
```

11.7 CREME-MC AND SPENVIS

CREME-MC is a new CREME site at Vanderbilt University: https://crème-mc.isde.vanderbilt.edu/CRÈME-MC. This includes the old CREME96 capabilities plus additional codes for Monte Carlo and physical modeling. It will not be finalized until after this book is published, so we can only give an indication of its total capabilities.

The code adds multilayer planar structures to evaluate the effects of semiconductor material near the sensitive volume [Warren 2005]. It also adds an updated Galactic Cosmic Ray Model. It presumably will add calculation of efficacy curves [Warren 1999, Warren 2008].

There are a variety of Monte Carlo radiation transport modules to simulate protons, alpha particles, and heavy ions:

Capture the effects of high energy particles including nuclear reactions.

Capture energy loss variation, straggle, and stopping of particles.

Reproduce ground-based test data with a monodirectional flux including at angle irradiation.

Produce error rate predictions for isotropic space environments.

Evaluate coincident upsets in multiple devices.

Model devices with multiple weighted sensitive volumes, both rectangular parallelepiped and ellipsoidal volumes.

Furnish probabilistic solar proton models.

CREME-MC and CREME96 are updates of CREME86 [Adams 1986]. Another update is SPENVIS (www.spenvis.oma.be) developed by ESA in Belgium. It has limitations on user's Internet domain names and affiliations so may not be accessible to everyone.

11.8 EFFECT OF UNCERTAINTIES IN CROSS SECTION ON UPSET RATES

In Figure 5.10 we showed data that had large statistical uncertainties. We can estimate the effect of those uncertainties on the calculated space upset rates by replacing the calculated cross-section curve with curves one standard deviation of each side of it. Figure 11-15 shows the original data and the additional curves.

The original data is fit with the parameters $CS\,8\ \mu m^2, L_0\ 3.2, w$ 35.8, and s 0.74. The outer curves have the same curve parameters and cross sections of 5.44 μm^2 and 10.56 μm^2. The corresponding upset rates calculated in CREME are $7.0 \times 10^{-8}, 1.2 \times 10^{-7}$, and 1.8×10^{-7} upsets/bit-day. The one standard deviation curves have

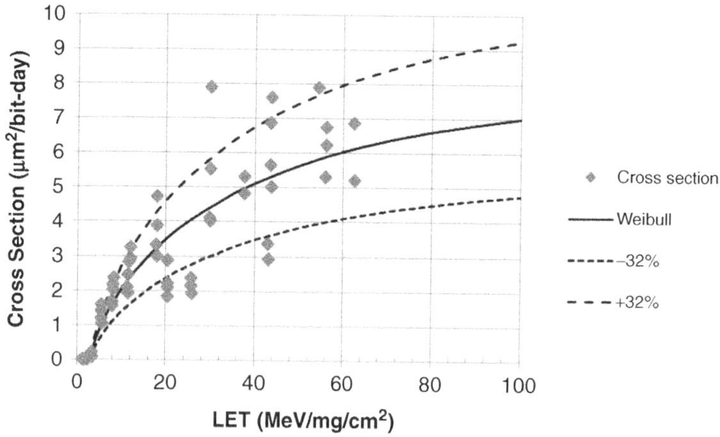

Figure 11-15 The data of Hansen (Figure 5.10), the Weibull fit to that data, and curves one standard deviation (±32%) on each side of it.

upsets rates reduced by 42% or increased by 46%. There is a factor of 2.6 between the low value and the high value. The calculations were for geosynchronous orbit with solar minimum conditions and 100 mils of shielding.

The upset rate can be determined using the figure of merit, Equation 7.38,

$$FOM = \frac{\sigma_{\mathrm{HL}}}{L_{0.25}^2}\left[\frac{(\mathrm{MeV/mg/cm^2})^2}{\mathrm{cm^2}}\right],$$

and Equation (7.39)

$$L_{0.25} = L_0 + w \times 0.288^{(1/s)}$$

and Equation (7.40)

$$R = C \times FOM$$

The error propagation equation for this situation is [Beers 1987]

$$S_R^2 = S_\sigma^2 + 4S_{L_{0.25}}^2 \qquad (11.1)$$

If we assume that $L_{0.25}$ is determined primarily by L_0, we can substitute S_{L_0}. In principle this is valuable. In practice, there has been virtually no work examining the uncertainty in the threshold. We also observe a contradiction with the last example, where the uncertainty in the calculated upset rate was greater than the uncertainty in the cross section. Sophisticated least squares fitting programs sometimes produce the uncertainties in the fitting parameters, but they have not been applied to the single event problem. Uncertainties in cross-section measurements and analysis will lead to uncertainties in calculated upset rates, but the details await further investigation.

CHAPTER 12

Samples of Proton Rate Predictions

12.1 TRAPPED PROTONS

A primary source of single event effects for low altitude spacecraft are the protons in the trapped radiation belts [Barth 1997, Xapsos 2006b]. We have shown samples of the environment in Section 2.3.2.

12.2 CORRELATION OF THE FOM WITH PROTON UPSET RATES

We can now examine the upset rates for parts for which we have both Weibull and Bendel parameters. There will be little difference in the upset rates determined with this function or using the lognormal function. Figure 12-1 shows the correlation of calculated upset rates as a function of the FOM for a series of circular polar orbits.

We observe that the upset rates and the FOM correlate over six orders of magnitude. There are a few devices that have relatively poor agreement. This suggests that we now go back and check the initial data and data fitting for these devices. In the past there have not been any criteria that we could use to examine the reasonableness of data and data fitting. It appears that a comparison of predicted rates using fit parameters and predicted rates using the FOM with the

Single Event Effects in Aerospace, First Edition. By Edward Petersen.
© 2011 the Institute of Electrical and Electronics Engineers, Inc.
Published 2011 by John Wiley & Sons, Inc.

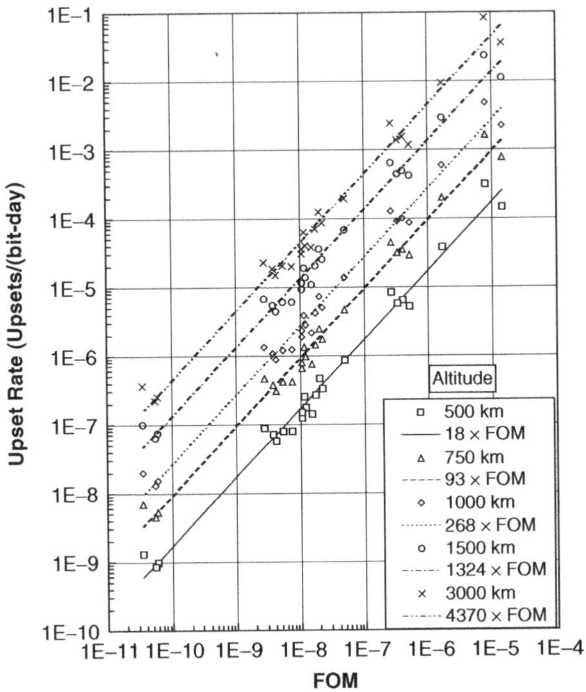

Figure 12-1 Proton upset rates for low polar orbits. The numbers (18, 93, 268, 1324, and 4370) are the rate coefficients for these orbits.

appropriate rate coefficient gives a good check on the quality of the fit parameters.

Figure 12-2 shows similar comparisons for equatorial orbits. There is again very good correlation of the upset rates with the FOM. Note in both Figures 12-1 and 12-2 that results for given parts remain consistent so that deviations from the fitted line result from the part parameters and are not due to changes in the environment.

Figure 12-3 shows the results for high altitude equatorial orbits. The fits become somewhat worse as we move higher. The width factor corresponding to one standard deviation increases from a factor of 1.5 to a factor of 3.5. However, this uncertainty inside a range of 10^5 is often acceptable. The poorer correlations are due to the presence of many more low energy protons in the high altitude regions of the belts.

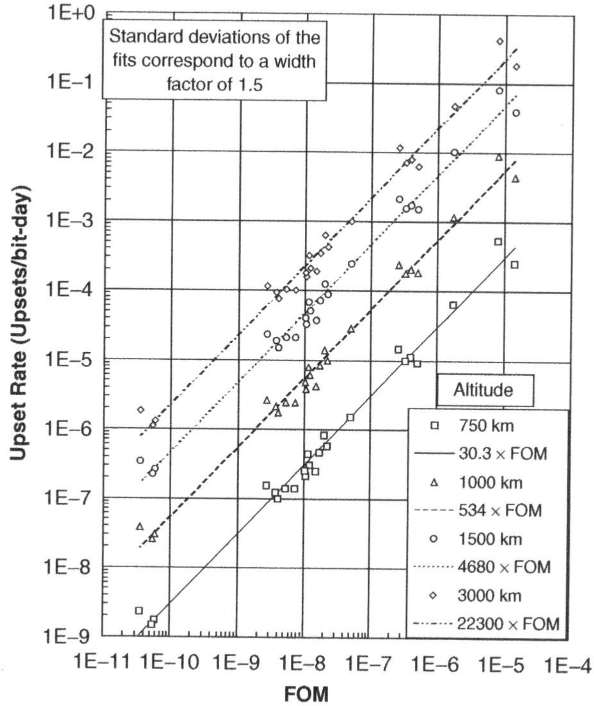

Figure 12-2 Proton upset rates for low equatorial circular orbits.

These results demonstrate that we can accurately calculate proton upset rates using the figure of merit. The figure of merit in turn can be obtained either from heavy ion measurements or from proton measurements. We now have a single parameter that completely describes SEU sensitivity. We have previously tried to use the LET threshold or critical energy as the single SEU parameter, but they did not work.

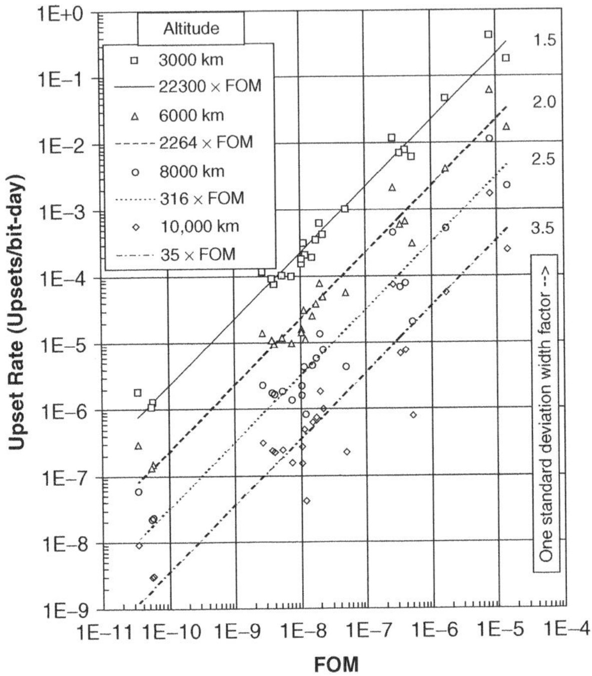

Figure 12-3 Proton upset rates for high equatorial circular orbits.

CHAPTER 13

Combined Environments

13.1 RELATIVE PROTON AND COSMIC RAY UPSET RATES

The upset rates at low Earth orbits are dominated by the trapped proton contributions. We can examine the various components. Figure 13-1 shows the rates for the 93L422 for cosmic rays. The rates decrease for circular orbits as the altitude decreases. There is much more of an effect at the equator than there is over the poles. Figure 13-2 shows the rates due to the trapped protons. There is a strong peak with altitude at the middle of the belts. There is slightly more intensity near the equator, but the rates are still high for polar orbits through the heart of the proton belt. Figure 13-3 shows the combined upset rates for the 93L422 as a function of altitude and inclination for circular orbits.

Remember that the relative upset rate for a device depends only on its figure of merit and not on the particular environment. Therefore, this set of curves can be used to estimate the rate for any device by scaling the results with the relative figures of merit of the devices.

13.2 CALCULATION OF COMBINED RATES USING THE FIGURE OF MERIT

Now that we have shown that the FOM works for the calculation of heavy ion upset rates and for the calculation of proton upset rates, we

Single Event Effects in Aerospace, First Edition. By Edward Petersen.
© 2011 the Institute of Electrical and Electronics Engineers, Inc.
Published 2011 by John Wiley & Sons, Inc.

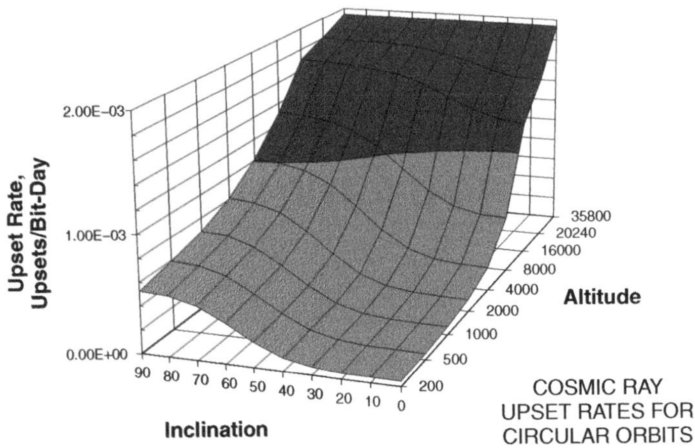

Figure 13-1 Upset rates for 93L422 due to cosmic rays at various inclinations and altitudes near earth. The figure of merit for this device is 1.4e−5.

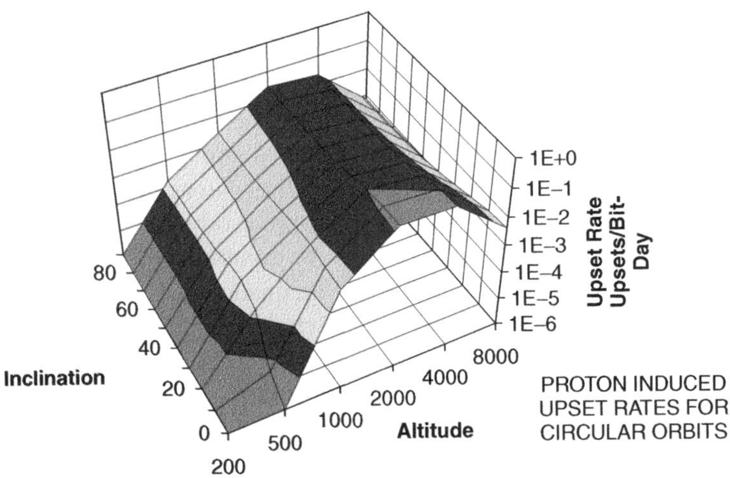

Figure 13-2 Upset rates for 93L422 due to protons at various altitudes and inclinations near earth. Figure of merit is 1.4e−5.

can combine the calculations to obtain the total upset rate in any orbit. Figure 13-4 shows the correlation between the upset rates and the FOM in a relatively high altitude low Earth orbit. We see that the total rate for an individual part in a particular orbit can be obtained by using a single part-dependent number, the FOM, and two orbit-dependent numbers, one for heavy ions and one for protons.

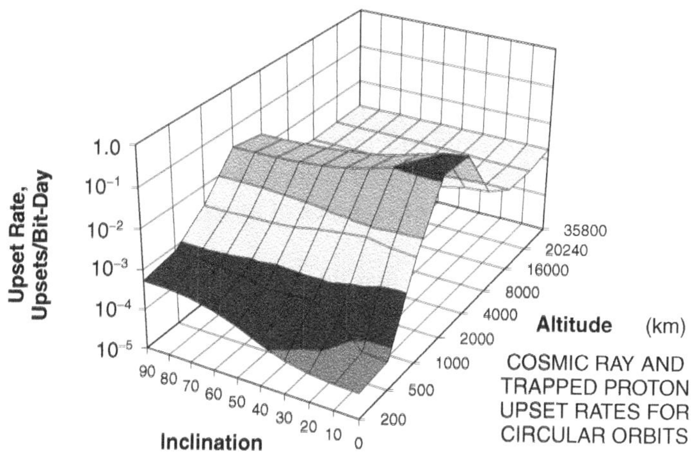

Figure 13-3 Upset rate for 93L422 as a function of orbit in the proton radiation belt for the combined environments. Figure of merit is 1.4e−5. You can scale for other devices. They will have much lower figurers of merit and many more bits.

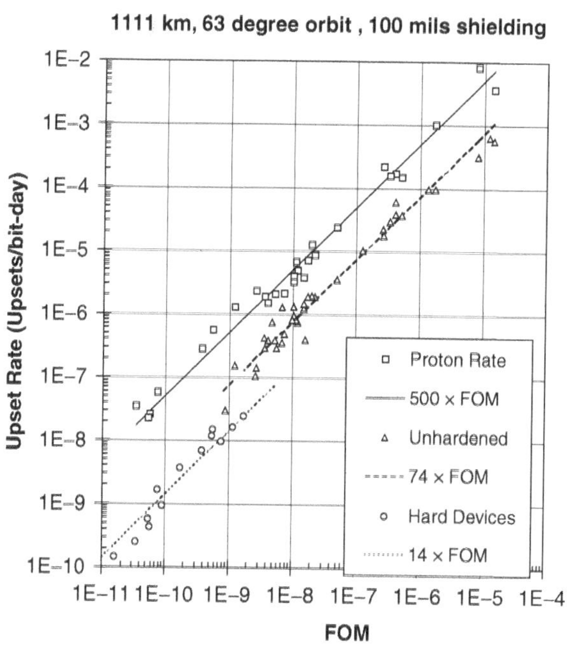

Figure 13-4 FOM calculation of both heavy ion and proton upset rates for a high altitude low Earth orbit. The rate coefficients for this orbit are 500, 74, and 14.

Figure 13-5 shows the results for the SEU rate calculations for a low Earth orbit. The width factors corresponding to one standard deviation are 1.5 for protons and 1.7 for heavy ions. For this orbit we can say that the average upset rate is 62 times the figure of merit, with 45% of the upsets due to protons and 55% due to heavy ions. Of course, in practice, the instantaneous upset rates depend on the location in the orbit, with the two types of upsets occurring in different locations.

Petersen also pointed out that this approach can be used for non-circular orbits. Figure 13-6 shows the results for an elliptical orbit [Petersen 1998]. Note that we not only specify the inclination, apogee, and perigee, but also the argument of the perigee and the initial

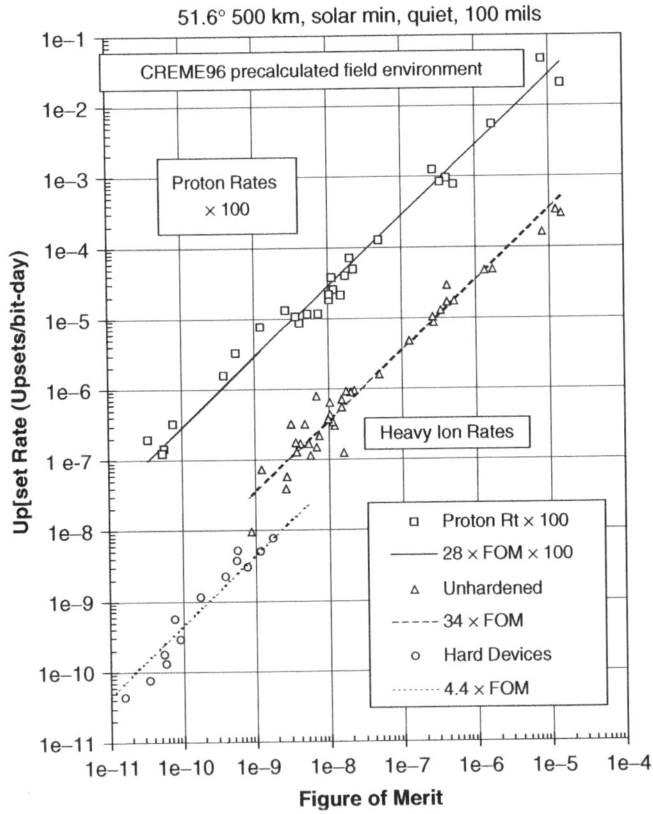

Figure 13-5 FOM calculation proton and heavy ion upset rates for a low Earth orbit. The proton data has been displaced upward by two decades.

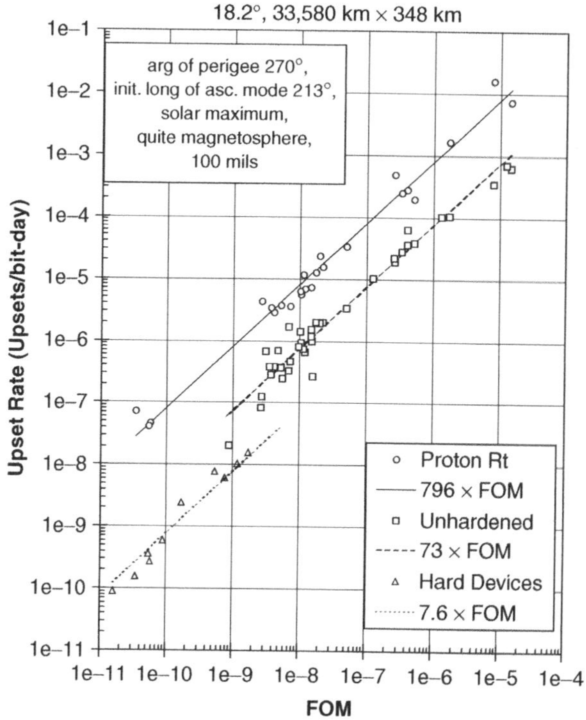

Figure 13-6 FOM calculation of both heavy ion and proton upset rates for an elliptical orbit.

longitude of the ascending mode. These last two parameters are necessary to specify the actual path through the radiation belt. The primary three parameters describe a family of paths. The satellite will be launched into a particular part of this orbit space, and then slowly precess through the entire set of orbits. Figure 13-7 shows the effect of precession.

One question that was not addressed above is why the upset rate is sensitive to the FOM and the limiting proton cross section, but apparently not to the details of the threshold region of the proton cross-section curve. Typical thresholds are in the range 23±6 MeV [Petersen 2008]. If we examine the integral proton spectrum behind 100 mils for the orbit used in Figure 13-6, we find that the integral number above these energies varies by ±16% in this range. This is

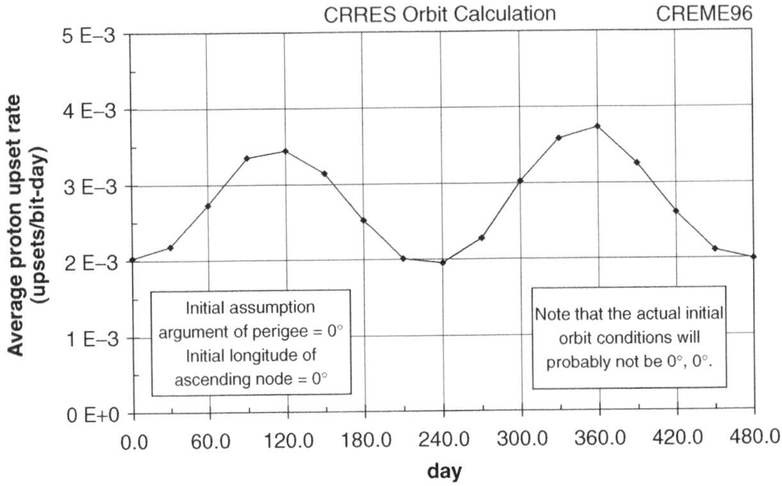

Figure 13-7 Rate changes due to precession in an elliptical orbit as a function of time in orbit.

much less variation than that obtained by varying the FOM or the orbit. It is therefore reasonable to approximate the upset rate using the limiting cross section and not including the proton cross-section threshold parameters. The ±16% seen here is part of the 1.5 width factor seen in the various proton upset rate calculations.

13.3 RATE COEFFICIENTS FOR A PARTICULAR NEW ORBIT

It is straightforward to calculate the single event rate coefficient for your particular orbit. You obtain good measurements and corresponding rate calculations for a few devices. Knowing the rates and the figures of merit for the devices, you can determine the rate coefficients. These rate coefficients can then be used for other devices, without the full rate calculations.

As a first step, you can use some of our results. The previous calculations presented results using a large number of devices. Now that we have results for a number of devices, we can choose a few particular devices, calculate the rates in an orbit of interest for those parts, and use that result to determine the rate coefficients for that orbit. Rate predictions for any other device can then be made without using the full arsenal of rate prediction tools for all of the parts of interest.

The devices in Table 13-1 appear to be suitable for the general determination of rate coefficients. You calculate the expected upset rates for your particular orbit with these devices using the IRPP or lognormal approaches (or Weibull, or as a last resort Bendel). You then use their figures of merit to calculate the corresponding individual rate coefficients and average those to obtain a rate coefficient for that orbit. The parameters are the Weibull heavy ion and proton parameters.

13.4 RATE COEFFICIENTS FOR ANY CIRCULAR ORBIT ABOUT THE EARTH

We can use the approach of Section 13.2 to calculate the rate coefficients for the large family of circular orbits about the Earth. We choose particular orbits on a grid of inclinations and altitudes. Figures 13-8–13-11 show these results for heavy ion and proton induced upsets. The rate coefficients for any particular orbit can easily be obtained by interpolation along and between the curves. These curves give a nice graphical view of the upset rate environment near earth.

13.5 RATIO OF PROTON TO HEAVY ION UPSETS FOR NEAR EARTH CIRCULAR ORBITS

It is sometimes useful to know if a given orbit is dominated by protons or by heavy ions. With this knowledge, you can choose to do experimental upset rate measurements corresponding to the dominant environment. Figure 13-12 shows approximate contours of the ratio of the proton to the heavy ion rate coefficients for the family of circular Earth orbits. The fine structure is an artifact of the graphing routine, as data is only available at the grid points. Protons are of interest below the $R = 1$ line due to their concentration in the South Atlantic anomaly.

Table 13-1
Devices and Figures of Merit for Rate Coefficient Calculations

Device	L_0 (MeV/mg/cm^2)	W	s	$x = y$ (μm)	z (μm)	Sigma (μm^2)	FOM
Heavy ion unhardened							
2901B	4.2	10	1.5	61.23	1	3750	5.11×10^{-7}
HM-6116	4.2	7.9	2.5	20	1	400	4.94×10^{-8}
MT4C1004	1.54	14.5	1.45	5.57	1	31	5.25×10^{-9}
Heavy ion hardened devices							
HM-6508	48	50	1.15	22	2	490	1.16×10^{-9}
6504RH	30.75	40	1.4	13	2	170	7.63×10^{-10}
01G9274	1.6	28	3.25	1.52	0.8	2.3	5.37×10^{-11}
Protons	L_0 (MeV)	W (MeV)	s			Sigma (cm^2)	FOM
MI5C2568	5.62	16.97	4.28			3.73×10^{-13}	1.68×10^{-8}
Hitachi 4 Mb	10.40	19.25	1.93			7.72×10^{-14}	3.48×10^{-9}
Mitsubshi 1	5.0	10.31	1.62			1.22×10^{-14}	5.49×10^{-10}
Sony 1 Mb	10.16	12.91	2.09			2.03×10^{-14}	9.13×10^{-10}
IBM 16 Mb	0.00	24.53	3.07			1.3×10^{-15}	6.18×10^{-11}

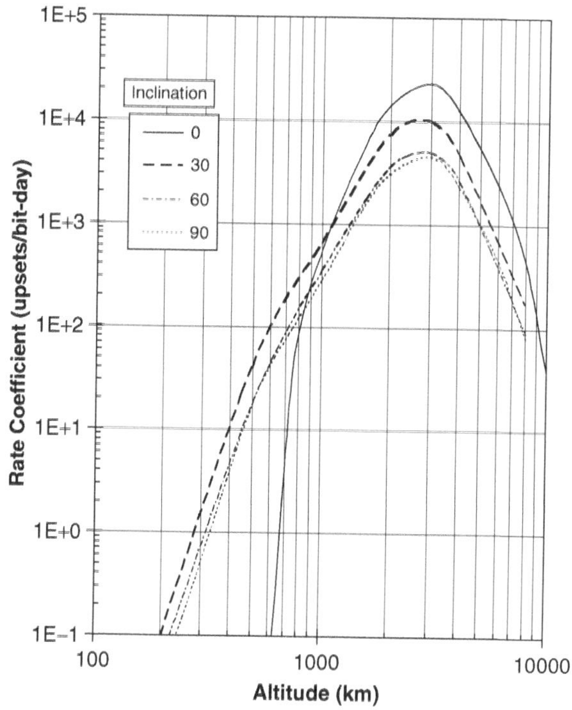

Figure 13-8 The proton rate coefficient for FOM calculations as a function of altitude for several inclination circular orbits. These curves show the geographical dependence of proton upset rates.

13.6 SINGLE EVENTS FROM GROUND TO OUTER SPACE

We have studied upset rates with very high altitude cosmic rays, as the cosmic rays encounter the Earth's magnetic field, in the proton belts, as the ions attenuate in the upper atmosphere, as proton reactions increase the rate, and the attenuation down to ground level. These results can be combined in a single figure as shown in Figure 13-13.

These rates are calculated for 45° situations. If we examine Figure 7.23, Figure 9.4, Figure 13-8, and Figure 13-9, we observe that the rates are generally within factors of 2, extending to factors of 3 over the poles or near the equator. The curve is wrong, on the high side, for uninteresting low altitudes near the equator. As the rate coefficient varies by six orders of magnitude, and the figure of merit varies by

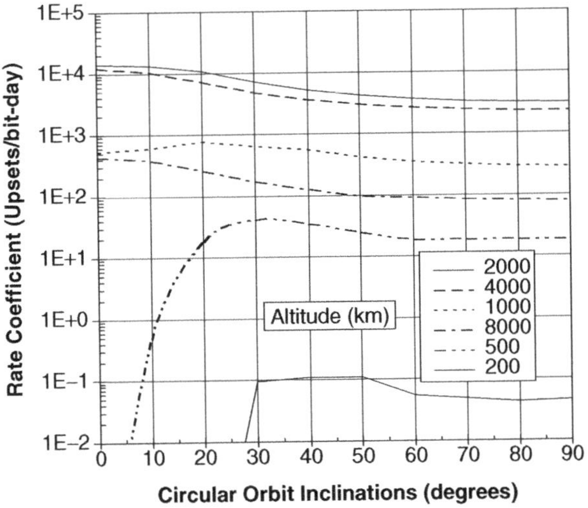

Figure 13-9 The proton rate coefficient as a function of inclination for circular orbits at several altitudes. The proton upset rate in any circular orbit can be determined from these plots.

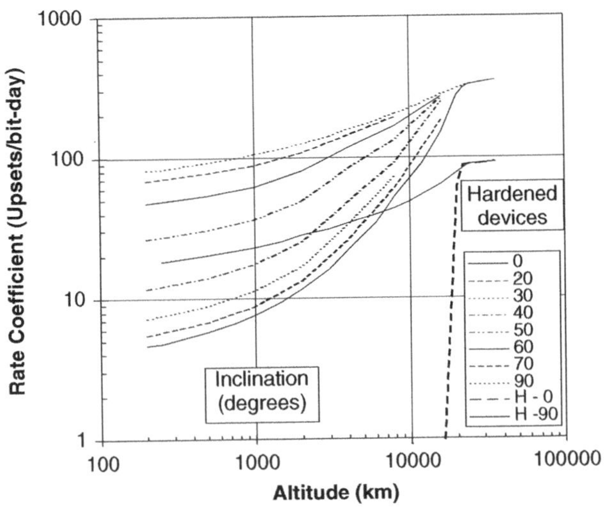

Figure 13-10 The heavy ion rate coefficient as a function of altitude for circular orbits at several inclinations. These curves show the geographical dependence of heavy ion upset rates.

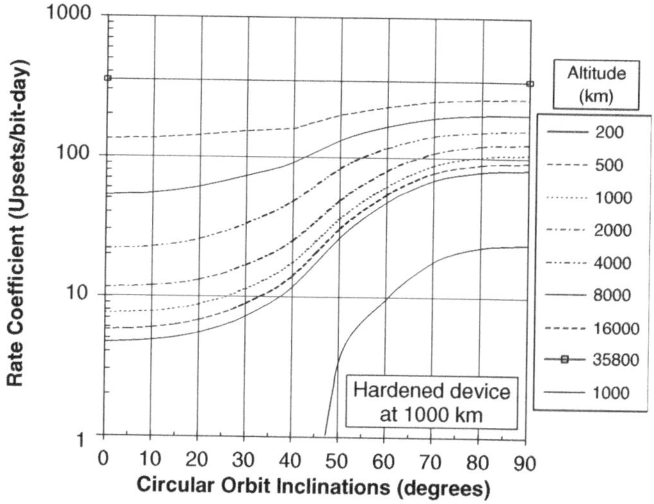

Figure 13-11 Heavy ion rate coefficients as a function of inclination for circular orbits at several altitudes. The heavy ion upset rate in any circular orbit can be determined from these plots.

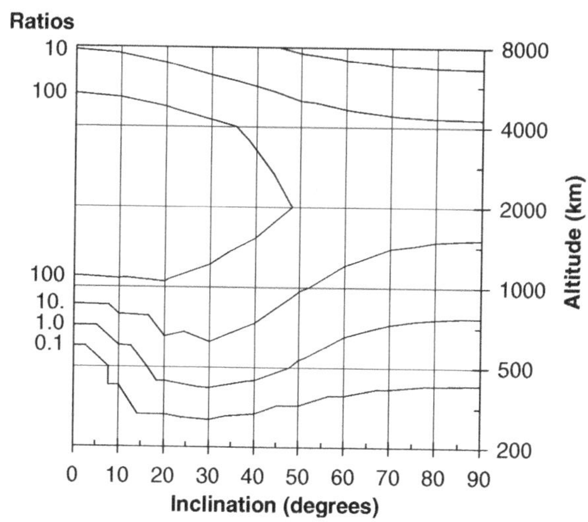

Figure 13-12 Approximate contours of the ratio of proton to heavy ions upsets for circular orbits near Earth. The fine structure is an artifact of the plotting routine.

six orders of magnitude, the total variation of upset rates is twelve orders of magnitude. In this situation a factor of 2 or 3 is normally unimportant, and you have the information to make the corrections if necessary. For example, Figure 13-4 for 1111 km, 63° indicates rates a factor of 2 below those indicated by this curve. Therefore, this curve would be sufficient for a first approximation. There is little variation of event rates as the inclination varies from 45° to 30° or 45° to 65°, and only slightly more for extensions to 0° or 90°, so this chart can be used to estimate rates for all common orbits, including geosynchronous, as well as ground level and avionics altitudes.

The figure of merit varies from 10^{-11} to 10^{-5}. The lowest two orders of magnitude are for rare hardened devices. The highest two orders of magnitude are primarily for rare old large devices. Therefore, with no knowledge of a part, one can make a guess that its figure of merit is 10^{-8}. Combine this number with an appropriate rate coefficient from Figure 13-13 and you have a zeroth order approximation for its upset rate.

The orbital calculations are performed for the devices listed in Table 13-1. The rates in the proton belts are orbit averaged rates. You should also calculate the peak rates, as they are significantly higher.

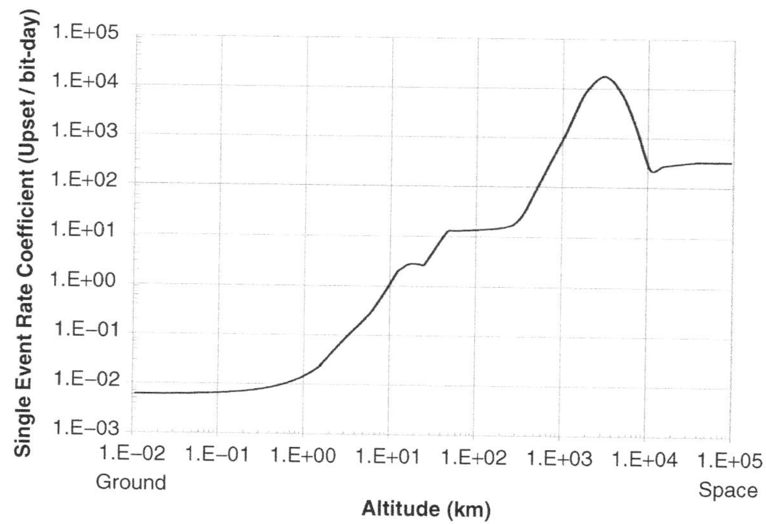

Figure 13-13 The altitude variation of single event effects at 45° from ground to outer space. For satellites it is 45° inclination and approaching ground it is 45° latitude. There is slight variation from 30° to 60°, and a larger but still small variation from 0° to 90° so this curve can be used for all common circular orbits. At high altitudes, there is no variation with inclination, so this chart can also be used for geosynchronous orbits.

CHAPTER 14

Samples of Solar Events and Extreme Situations

Solar particle events (SPEs) generally constitute the most severe environments in which space systems must operate [Adams 1984]. Large events produce enhanced particle fluxes in near-Earth space that typically persist for several days. The particles are called solar energetic particles (SEPs). These events are often referred to as "flares." Solar flares often occur coincidentally. However, flares are not the causative agent of the SEP fluxes that are important for space system design. We now know that these events are produced by interplanetary shocks driven by coronal mass ejections. Solar events were discussed in more detail in Section 2.3.5 and references therein.

Figure 14-1 shows the energy spectra of representative particles in a solar particle event. We notice the very large fluxes at low energies, and how the flux falls off rapidly with energy. The spectrum is much softer than the cosmic ray spectrum. Figure 14-1 also shows cosmic ray helium for comparison. Figure 14-2 shows the LET spectra that may be present in worst case events and compares these with the solar minimum cosmic ray spectra [Tylka 1996, Tylka 1997].

Because the SEP spectra are so soft, shielding does have an effect. Figure 14-3 shows the SEP LET spectra behind various levels of shielding. If it is absolutely necessary to reduce the rate by three orders of magnitude to that expected from cosmic rays, 1 inch of shielding can be added. It is rarely necessary to be this extreme.

Single Event Effects in Aerospace, First Edition. By Edward Petersen.
© 2011 the Institute of Electrical and Electronics Engineers, Inc.
Published 2011 by John Wiley & Sons, Inc.

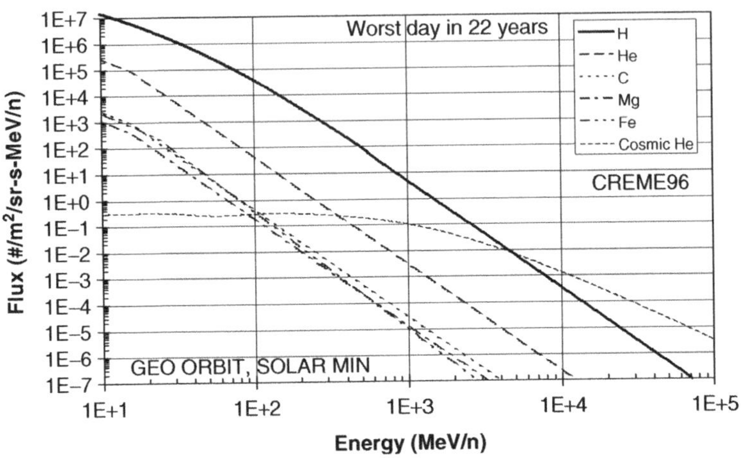

Figure 14-1 The energy spectra of solar energetic particles. The spectrum of cosmic ray helium is shown for contrast.

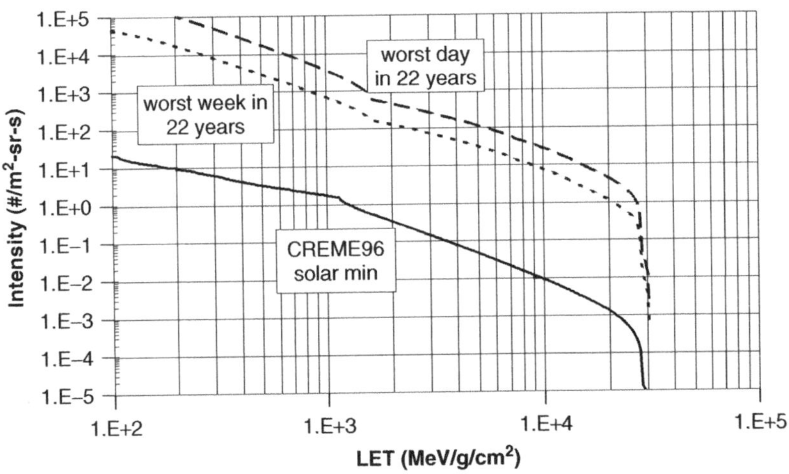

Figure 14-2 Geosynchronous integral LET spectra from the worst case solar flares, obtained from CREME96.

Figure 14-3 shows some structure at low LET due to the contributions of the hydrogen and helium ions (protons and alpha particles). The high intensity of relatively high energy protons means that there can be a significant contribution to the error rate due to proton

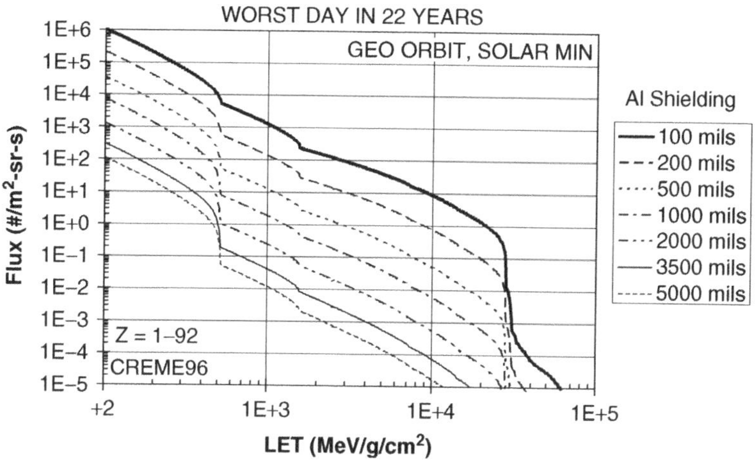

Figure 14-3 Solar energetic particle LET spectra behind shielding.

reactions. There may be nearly equivalent amounts of upsets due to ionization and reactions [Tylka 1996]. Ionization rate calculations may indicate upsets due to both protons and alpha particles. These may be due to stopping particles and may or may not be real. They need to be carefully examined.

Xapsos has published updated versions of these charts showing the 50% and 99% confidence level spectra [Xapsos 2007]. For long-term missions, the Solar Particle Events exceed the Galactic Cosmic Ray fluxes during solar maximum at 100 mil shielding levels. Solar heavy ions are more easily shielded than the galactic cosmic rays, so shielding analysis is significant.

Xapsos has also demonstrated the solar particle events are completely random in time and that it is not possible to predict the time of occurrence and magnitude [Xapsos 2006a].

The particles from a solar event penetrate through the proton belts. Figure 14-4 shows the spectrum from the worst day event as calculated using CREME96. Figure 14-5 demonstrates surprisingly that, even though the spectrum is soft, the upset rate still scales with the figure of merit. The thing that does not scale is the rate coefficient. The low energy spectrum differs by nearly two orders of magnitude, but the rate coefficient changes from 500 (Figure 13-4) to 1192.

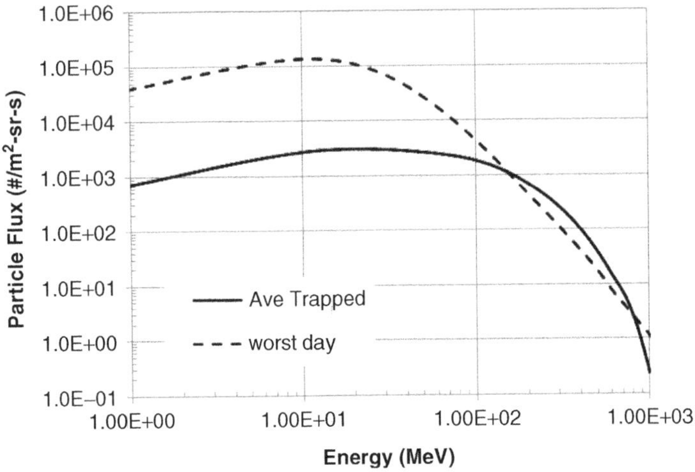

Figure 14-4 The solar event worst day spectrum for a 63°, 1111 km orbit compared with the average trapped proton spectrum with 100 mils of shielding.

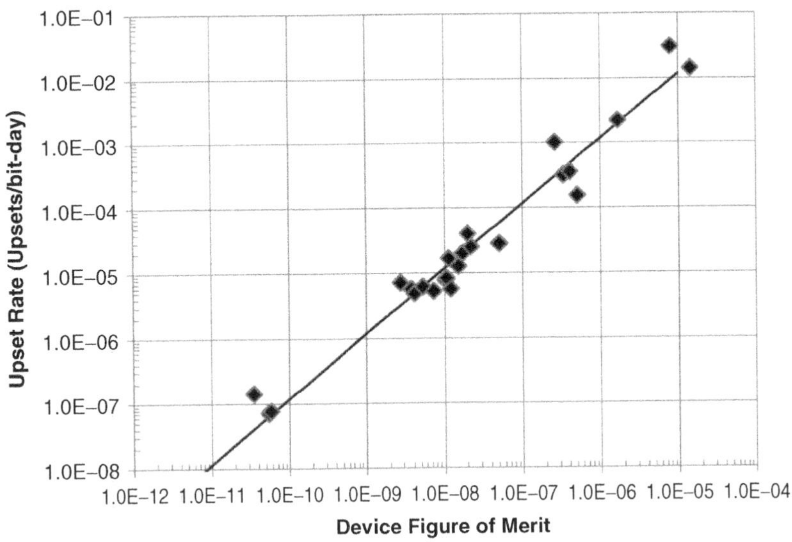

Figure 14-5 The upset rate as a function of the device figure of merit for the worst day spectrum for a 63°, 1111 km spectrum. The rate coefficient is 1193 for this case, compared with the average rate coefficient of 500 for this orbit as shown in Figure 13-4.

Another environment of interest is that of the Tactical ABM. We will not discuss that here, but a good discussion is given by Solin [Solin 2005]. The rates can be calculated in CREME and the authors also demonstrate the use of the figure of merit.

CHAPTER 15

Upset Rates in Neutral Particle Beam (NPB) Environments

15.1 CHARACTERISTICS OF NPB WEAPONS

Charged particles cannot be sent over large distances toward an object in space because they will be deflected by the Earth's magnetic field into a complex trajectory with a relatively short radius of curvature. Electrically neutral atoms, however, are not deflected by a magnetic field but travel through space in a straight line. Neutral particle beams can, therefore, be aimed accurately at a target object. As soon as the beam particles strike the surface of the target, they are ionized and they then produce ionization as they penetrate into the target. The most likely NPB particles will be hydrogen atoms (protons when ionized) or deuterium atoms (deuterons). Energies between 75 and 250 MeV may be taken as the range of energies that would be useful for a NPB weapon to produce. At such energies the NPBs have great penetrating power. Thus, for example, at the low end, 75 MeV deuterons can penetrate 0.5 inch of aluminum and, at the high end, 250 MeV protons can penetrate 7.0 inches of aluminum. The fact that such beams are difficult to shield against is one of the important advantages that a NPB weapon has over other types of space weapons. Heavy atom beams are

Single Event Effects in Aerospace, First Edition. By Edward Petersen.
© 2011 the Institute of Electrical and Electronics Engineers, Inc.
Published 2011 by John Wiley & Sons, Inc.

not considered usable because the energies required to achieve similar penetrations are so high as to be impractical.

The neutral particle beam weapon, as considered for development in the United States, consisted of a source of negative ions, a radiofrequency quadrupole (RFQ) focusing and accelerating section, a linear accelerator to accelerate the beam to high energies, beam sensing and steering optics, and, at the last, a neutralizing section that turns the high energy negative ions into neutral atoms for the flight through space. The American Physical Society made a preliminary unclassified feasibility study that was published in the *Reviews of Modern Physics* [Bloembergen 1987]. We will use their numbers.

Table 15-1 shows representative beam characteristics. For a sample calculation, let us assume a 100 MeV beam of protons with 100 mA current, on target for 1 second. The energy is not important for our considerations; it primarily determines the depth of penetration, and secondarily determines the beam divergence. One hundred MeV protons have a range of 3.7 centimeters in aluminum, adequate to penetrate most targets. The current and time play together, as the dominant effects in electronics depend on the total fluence, so a 0.25 mA beam on for 400 seconds will have much the same effect as a 100 mA exposure for 1 second.

The range to the target is an essential parameter in these calculations because a NPB has an angular divergence and the flux reaching a target will therefore decrease with the square of the range. The angular divergence of a NPB is expected to be very small so the fraction of the NPB output energy that can be put on target will be high; that is,

Table 15-1
Representative NPB Characteristics at 1000 km

Beam current, protons	100 mA
Beam energy	100 MeV
Beam penetration	3.7 cm
Beam divergence	2.26 microradians
Dose rate on target	3.6E5 rads/s
Particle flux	3.9E12 P/cm^2/s
Equivalent 1 MeV neutron flux	5.7E12 N/cm^2/s
Power density on target	62 watts/cm^2
Temperature rise in silicon	5°C/s

it will be highly "directed," but the divergence must necessarily be taken into account [Bloembergen 1987].

The flux of NPB particles reaching a target is then given by the equation

$$F = \frac{I \times 6.24 \times 10^{15}}{\pi \times (0.1 \times R \times D)^2} \left(\frac{p \times \text{km}^2 \times \ \mu\text{rad}^2 \times \text{s}}{\text{cm}^2 \times \text{mA}} \right) \qquad (15.1)$$

where

F = number of NPB particles per cm^2 per second
I = output beam in milliamperes (number of neutral particles/second)
R = range in kilometers
D = half-angle beam divergence in microradians (μrad)

In view of the fact that many unhardened electronic devices fail at ionizing radiation doses of less than 100,000 rads and some fail even below 10,000 rads, the tables show that at target ranges of 100 kilometers or less, even a NPB with a beam current of only 0.25 milliamperes (mA) can produce a very significant threat environment. On the other hand, the huskier weapons can present a threat at 35,000 km, given adequate exposure time and the cumulative nature of the damage.

As an aside, note that a 100 mA proton beam would deliver a dose lethal to humans at 1,000,000 km in slightly over a minute, or from low Earth orbit to the Moon's surface in 10 seconds.

15.2 UPSETS IN THE NPB BEAM

In Chapter 4 we observed that representative limiting cross sections for proton induced upsets were constant for proton energies in the range of 75–200 MeV, the range of interest for NPB exposures. We also observed that the cross section ranged from 10^{-15} to 10^{-12} cm^2-upset/bit. In Table 15-1 we indicated that a representative proton flux was $4 \times 10^{+12}$ $\text{p/cm}^2\text{/s}$. The upset rate will be the product of these two terms:

$$Rate = 10^{-15} \times \frac{\text{cm}^2 - \text{upset}}{\text{bit}} \times 4 \times 10^{12} \frac{p}{\text{cm}^2 - \text{s}}$$

$$Rate = 4 \times 10^{-3} \text{ to } 4 \frac{\text{upsets}}{\text{bit} - \text{s}} \tag{15.2}$$

Devices that upset are normally used in conjunction with error detection and correction circuitry, such as Hamming codes. With modern devices having thousands to millions of bits, the operation of the equipment will depend on the robustness of the error handling system. Error correction schemes usually slow down the memory operation somewhat and use up part of the memory capacity but, in general, they are widely used and considered an indispensable part of making memories reliable. If there is a possibility that a SEU can occur in a microprocessor, then other types of fault-tolerant design may need to be used. Typically, fault-tolerant designs for microprocessor systems involve redundant units and voting or comparisons of repeated processing steps.

The system sensitivity to SEU depends on the amount of fault tolerance used. Even very small proton upset cross sections can lead to system failure in a NPB environment if carefully designed fault tolerance is not used. Figure 15-1 shows the interplay between upset cross section, NPB particle flux in particles per cm^2 per second, and

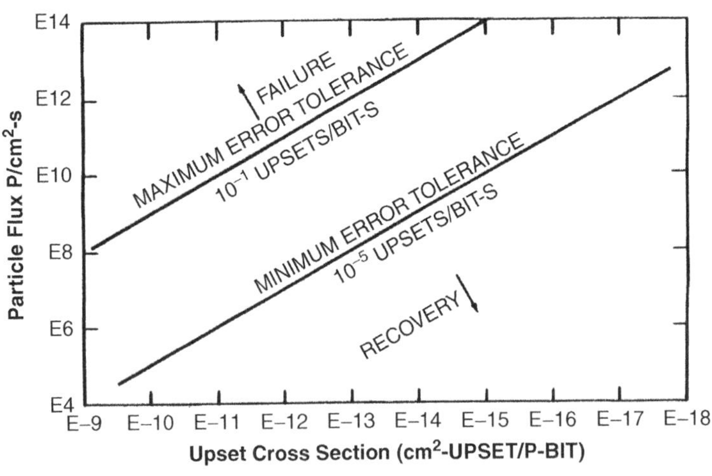

Figure 15-1 Upset rate contour For combined flux and upset cross section for different system error tolerances. The very approximate maximum error tolerance assumes double bit errors being introduced in a triply redundant memory. The minimum error tolerance assumes recovery times on the order of seconds for recovery from a single bit error in the memory.

the amount of fault tolerance used in a small memory system. The figure shows that even high levels of fault tolerance are not adequate if the parts used in the system are sensitive to SEUs. At the other extreme, this flux will cause failure even if very hard parts are used, but error tolerance schemes are not. The reader may want to consider the numbers for his/her systems.

Although single event effects, in general, are not expected to be an important space system vulnerability mechanism, experiments have shown that, in the case of a NMOS IDT 16K static RAM, a significant number of SEUs were produced by high energy protons at a dose rate an order of magnitude lower than that required to produce a dose rate upset of the circuit [Xapsos 1987]. Such a device, if it were used in a satellite could potentially suffer SEUs below the circumvention level set for the satellite. Representative experimental data based on the IDT 16K RAM and for a 16K CMOS RAM are shown in Figure 15-2. This plot uses generated data as one of the devices was advanced for the time and the data had limited distribution. The enhanced upset sensitivity due to the combination of dose rate upset and SEUs are illustrated by the dashed lines. The interpretation of these results is that the presence of photocurrents in the devices due to ionizing radiation dose rate reduces the additional local charge that is required to upset the device and thus makes the devices more susceptible to SEUs (Xapsos 1989). Thus, for example, for the CMOS device that is exhibiting rail

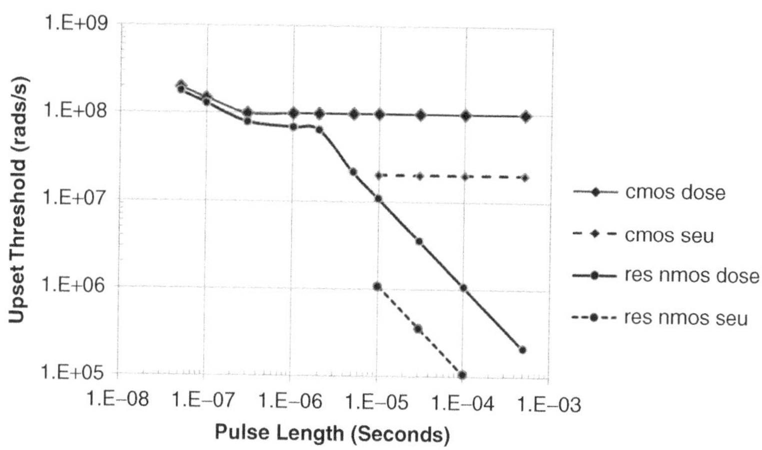

Figure 15-2 Dose rate upset threshold versus pulse length showing lowered threshold due to single event upsets.

span collapse at a dose rate of about 1.0E8 rads per second, it was found that the cross section for producing SEUs is larger than it is when the device is not exposed to a significant ionizing dose rate.

The minus one slope for the NMOS device shows that the dose rate upset is caused by the accumulation of charge (because it has a large resistive load, this device has a time constant on the order of milliseconds). At long pulse lengths, therefore, the device can accumulate enough charge so that the extra local charge required to produce a SEU is greatly reduced. With reference to Figure 15-1, this enhanced SEU sensitivity would appear at particle fluxes of around 1E12 particles per cm^2 per second and would result in the uppermost line moving to the right, that is, toward failure occurring for less sensitive devices.

CHAPTER 16

Predictions and Observations of SEU Rates in Space

Petersen has summarized the available data on the observation and prediction of SEU rates in space [Petersen 1997a]. There are three basic questions:

1. How good can a prediction be?
2. How bad can a prediction be?
3. How does the quality of the prediction depend on the type of orbit?

The paper considered 121 reports of predicted and observed rates. The analysis then excludes solar flares, cases with poor statistics, and cases that are essentially duplicates; leaving 77 comparisons. The heavy ion predictions based on the CREME environments and the proton predictions based on the AP8 environments are both very successful for their basic environments, but are less accurate for low Earth orbits (LEOs). The quality of the results depends strongly on whether the predictions are based on tests with flight parts or with generic parts. The results appear consistent with suggested modifications in these environments based on recent space measurements. The

Single Event Effects in Aerospace, First Edition. By Edward Petersen.
© 2011 the Institute of Electrical and Electronics Engineers, Inc.
Published 2011 by John Wiley & Sons, Inc.

methods that are used for upset rate predictions appear to be adequate for the current generation of devices.

There are a number of cases with large differences between predictions and observations. There are too many of these cases for one to be sanguine about making predictions. The major problem appears to be the variability from part to part. The relatively large deviations observed when predictions are made using generic parts' data indicate that design calculations for future operational spacecraft, when specific parts' data is not available, should allow for a factor of 10 underestimate by the predictions.

16.1 RESULTS OF SPACE OBSERVATIONS

There have been a number of comparisons of space results with ground based predictions. Table 16-1 presents the available results. The table takes the original author's results as given. The results are presented in terms of the ratio of prediction to flight results. There are results from 23 satellites, with 126 comparisons, from 36 articles, with 50 parts (many duplicates of a particular piece part type).

We are interested in the upset predictions for the commonly used proton and heavy ion environments. Therefore, we exclude the papers of flare predictions [Normand 1990, Croley 1995, and Tylka 1996]. The work of Croley, Tylka, and their co-workers gave very good agreement of observations with predictions based on the observed environments, although Tylka and co-workers have some criticisms of the paper by Croley and co-workers. The work was excluded for our purposes as it used measured environments, rather than the standard environmental models. The good agreement does reinforce the observation that the methods of upset calculation are satisfactory.

We excluded the results of single event burnout predictions by protons, even though their agreement was consistent with our results for proton upsets [Adolphsen 1996b, Oberg 1996]. We also exclude results with very poor statistics and results that were replaced by later results. The results of O'Neill and Badhwar were not included as they involve a forced agreement of prediction and observation [O'Neill 1994]. This leaves a total of 91 comparisons of prediction and observation. We then excluded cases in which several different predictions were made for one space measurement. As these predictions were

Table 16-1
Ratio of SEU Rate Predictions to Flight Results

Author	Mission	Orbit	Part	HI	Protons	Notes
Bendel 1983	Solar Max	500 km	93422		1.2	b, p
Blake 1986	Low polar	LEO	HM6508	1		b, m
Shoga 1987	Leasat F2 85	GEO	93L422	0.85		b, j
	Leasat F2 86		93L422	0.84		b, j
	Leasat F3 86	GEO	93L422	0.85		b, j
Binder 1988	Leasat	GEO	93L422	1.4		a, j, n
	GPS		NMOS	1.2		b, n
	GEO		54L78	1.25		b, n
Shimano 1989	MOS-1	909 km	93419		1	b, l, m
Scott 1989a	Orbiter/ CRUXIII		8464-15L	1.4		b, o
			81256	6.1		b, o
			1600s-5v	16.5		b, o
			1600s-3v	9.2		b, o
	CRUX IV		6504	0.49		b, o
	CRUX V		1600	3.7		b, o
			1601 −2.5 v	2.2		b, o
Adams 1989	19		MK4116		1.7	b, n
Bion 1989	UoSAT-2	700 km, 98°	HM6564	0.875		b, p
			HM6516	1.3		b, p
Harboe-Sorensen 1990	UoSAT-2	700 km	TMS4416	0.19	0.29	b, n
			HM6264	3.8	1.57	g, n
			HM6116	7.1	2	g, n
			HM6504	1.25		b, n
			TC5516	1.1		b, n
Normand 1990	TDRS flare		93L422		2	i
Wilkinson 1991	TDRS 1985	GEO	93L422	0.52		b, j
(Shoga 1987 pred)	1986		93L422	0.58		b, j
Koga 1991		700 km, 90°	OW62256	4		b, n
Goka 1991	MOS-1	909 km	93419		0.83	b, m
Adams 1991	Meteosat-3	GEO	EDH8832C	1.7		a, m
			TMP320C25	0.75		a, m
			MB84256	6.8		f, m
	October 1989 flare		EDH8832C	6.7		I, m

(*continued overleaf*)

Table 16-1
(Continued)

Author	Mission	Orbit	Part	HI	Protons	Notes
	October 1989 flare		MB84256	9.8		I, m
McNulty 1991a	CRRES	Transfer	2164	1.07	0.43	h
Langworthy 1991	CRRES		93422		0.31	h
			93L422		1.2	h
compare with			92L44		1.5	h
Campbell 1992			21L47		2.3	h
			71681		2.7	h
			6116		2.6	h
Petersen 1992a	CRRES		93422	1.9		h
			93L422	2.7		h
			82S212	1.3		h
			93L44	2.4		h
			21L47	4.9		h
			2164	4.2		h
Smith 1992	Leasat	GEO, Solar Min	AMD 93L422	0.9		a, n
	Leasat	GEO, Solar Min	FSC 93L422	1.05		a, n
	TDRS1	GEO, Solar Min	FSC 93L422	0.475		b, n
Underwood 1993	S80/T	1320 km	D43256		1.41	a, m
			D43100		0.38	a, m
			58257		0.14	b, m
			581000		0.27	b, m
	UoSAT-5	770 km	HC628128		1.26	b, m
			D43256		0.88	b, m
			581000		2.92	b, m
			D43100		0.21	b, m
	Kitsat-1		581000		0.23	b, m
			58257		25.9	b, m
			581001		0.07	b, m
LaBel 1993	TOMS		256k Hitachi		1.065	b, k, n
	SAMPEX		256k Hitachi		1.023	b, n

Table 16-1
(Continued)

Author	Mission	Orbit	Part	HI	Protons	Notes
Harboe-Sorensen 1993	UoSAT 2,3,5	670 km	HM6116	7.6	1.17	b, n
		800 km	HM6264	6.9	0.93	b, n
		770 km	HM62256	1.8	0.41	b, n
			HM628128	0.8	0.57	b, n
	October 1989 flare		HM6116	50.8		I, n
			HM6264	230		I, n
Smith 1994	TDRS	September 1989 flare	93L422		0.82	I, o
	TDRS	October 1989 flare	93L422		0.85	I, o
	CRRES	Transfer	93L422		0.57	b, h, o
Reed 1994b	CRRES	Transfer	AM93L422		0.47	b, h, o
			AM21L47		0.81	b, h, o
			2164A		0.84	b, h, o
O'Neill 1994	Shuttle		IMS 1601 epi	1	1	d
Vampola 1994	TOMS	1200 km	Hit62256		1.095	b, n
Chenette 1994	CRRES	Transfer	93L422	1.46		b, h
Falguere 1994	MIR	350 km, 52°	HM65756	3.6		b, m
			MC68020	0.26		b, f, m
Croley 1995	TDRS flare		93L422		1.02	I, o
Shinn 1995	STS 51		IMS1601	0.71		a, o
	STS 56		IMS1601	0.97		a, o
Mullen 1995	Apex	362 × 2544, 70°	HM514100		2.06	a, m
Ecoffet 1995	Spot I 1987	800 km	HEF4736	5.4		b, m
	Spot 1 1991			4.6		b, m
	Spot 2 1991			3.9		b, m
Adolphsen 1995	APEX/CRUX		Micron 1M	3.0	0.47	b, e, n
			EDI 1M		2.7	b, e, n
			Hitachi/Elmo 1M	1.46	1.17	b, f, n
			Micron 256K	6.42		b, n
			EDI 256K	72.2	55.5	f, n
			IDT 256K	5.5	5.7	b, f,

(*continued overleaf*)

Table 16-1
(Continued)

Author	Mission	Orbit	Part	HI	Protons	Notes
Petersen 1996c present predictions	CRRES	Transfer	93422	1.08	0.12	a, b, o
			93L422	1.09	0.47	a, b, o
			82S212	0.73	1.51	a, b, o
			92L44	1.59	0.59	b, o
			21L47	2.23	0.91	b, a, o
			4164/2164	0.22	1.02	b, o
			71681		1.06	a, o
			6116	2.8	1.02	b, f;a, o
Johlander 1996	Freja	600–1766 km, 63°	Hitachi 628128		0.88	c, m
Adolphsen 1996a	Crux/Apex	362–2544 km, 70°	MOSFET burnout		0.5	a, n
Tylka 1996	TDRS-1	Geo SPEs	Fairchild 93L422	1	1	I, o
	Leasat	Geo SPEs	AMD 93L422		1	I, o

Notes: **a**, flight parts; **b**, may be generic, not described as flight; **c**, definitely generic; **d**, forced fit, match flight, environment; **f**, statistics poor or confusing; **g**, replaced by Harboe-Sorensen 1993; **h**, replaced by Petersen 1997a; **i**, solar particle events; **j**, replaced by Smith 1992; **k**, replaced by Vampola 1994; **l**, replaced by Goka 1991; **m**, reasonable single thickness shielding; **n**, nominal shielding; **o**, shielding using sector analysis; **p**, sector analysis for similar satellite.

often close to each other and close to the observed value, their inclusion gave an artificial peak at a ratio of one. This leaves 77 cases for our comparisons.

Preliminary reports of these results were reported in other papers [Petersen 1995, Petersen 1996a]. Those results were obtained from analysis that included the overlapping cases, while excluding the data of Adolphsen, Underwood, and their co-workers for which the authors emphasized that they were using generic parts, not flight parts. Therefore, those results showed somewhat better overall agreement than we report here.

Both Harboe-Sorensen and Underwood emphasize the importance of testing parts that are nearly identical with flight parts [Harboe-Sorensen 1993, Underwood 1993]. They both observe large variations in test results when they use a variety of parts. This appears to be the most likely explanation for many of the large deviations observed in the predictions.

Adolphsen and co-workers discussed this aspect of the problem in some detail [Adolphsen 1996a]. There can be large variations from lot to lot, and (surprisingly) large variations within a lot. Unfortunately, the amount of variation may not be apparent in tests on one or two devices in the laboratory, but may be very apparent in tests on forty devices during flight.

The results are separated into two different populations: one for heavy ion upsets, and one for proton upsets. If authors give only combined results, these cases can be placed with the dominate source of upsets. The authors know the location of the upsets in orbit. Plots of these results indicate clearly a region of high upset rates in the proton belts or the South Atlantic anomaly (SAA) and regions of low upset rates over the poles or above the proton belts. The regions with high upset rates correspond to the regions with calculated high intensities of protons. The regions of low rates correspond to regions of galactic cosmic rays (GCRs), where the heavy ion upsets dominate the upset rates. The authors seem consistent in their separation of the two regimes.

The data are examined as a function of the logarithm of the ratio of the predicted value to the observed value. The data are characterized in terms of their average and width. The width factor is the ratio of prediction to observation corresponding to the logarithmic width (standard deviation) of the data. A width factor of 1.4 indicates that ratios within a standard deviation are within factors of 0.71 to 1.4 of the average ratio.

The variations of the results of a number of independent experiments will have a random distribution about the mean. Even though individual experiments may have systematic errors, the ensemble should reflect the population being measured. If there are systematic errors in the environments or prediction methods, those will be evident from the shift of the average values away from a ratio of 1.0.

We will first examine the quality of the predictions for the cases where the parts are identified as flight parts. There are very few of these cases, and many of the papers did not clearly address this issue.

Figures 16-1 and 16-2 plot the results. The data have been placed in bins according to the number of observations of the ratio of the predicted value to the observed value. The bins are a factor of 1.26 wide, and the center bin covers a range from 0.89 to 1.12. A frequency of 3 at a ratio of 1.0 indicates that there were three results that had a

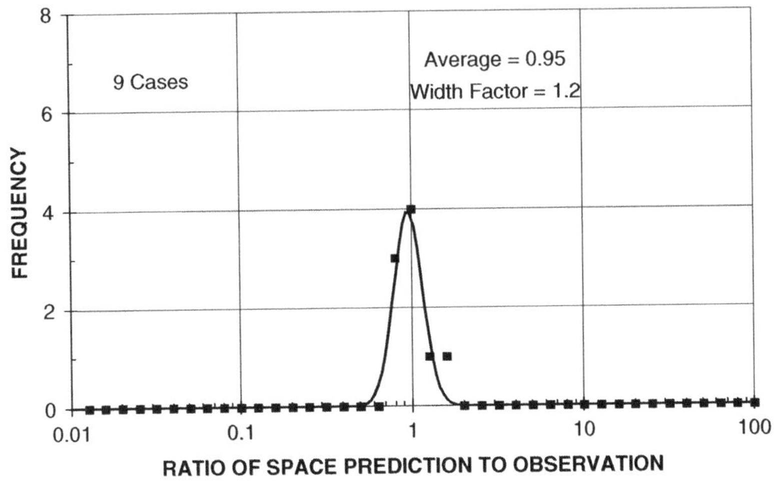

Figure 16-1 Frequency distribution of heavy ion predictions based on data from flight parts.

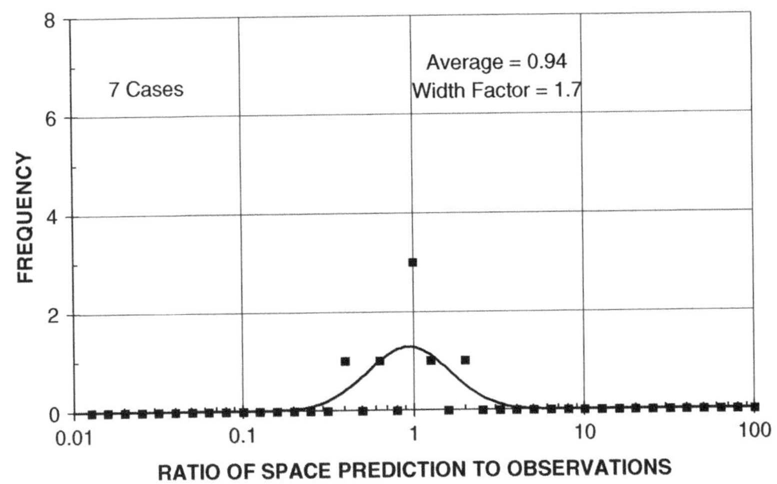

Figure 16-2 Frequency distribution of proton predictions based on data from flight parts.

ratio of prediction to observation in the range from 0.89 to 1.12. Later figures that have wider distributions will be shown with bin widths of 1.39 and 1.78. Note that the bins are used only to help visualization of the data and that the bin centers and widths do not enter the calculation of the data characteristics discussed above.

It appears that the distributions are symmetric, so that the basic form of the distribution is Gaussian (normal). The Gaussian curve is determined by the average and standard deviation of the points in the central peak. The characteristics of a Gaussian distribution can be obtained with good confidence from very few points. All of the plots show the corresponding Gaussian curve.

Figures 16-1 and 16-2 show relatively narrow distributions for both protons and heavy ions. The width factor of 1.2 for the heavy ion cases, based on 9 data points, leads to the conclusion that we can say with 95% confidence that the true mean of predictions for flight parts will be between 0.82 and 1.10 [Natrella 1963]. This in turn suggests that experimenters should attempt to obtain results good to 5% to 10% for future tests of space predictions. The experimenter wants to do a good job with the factors under his control if he is to test the factors not under his control. That means he should obtain high quality experimental measurements on flight parts and make his predictions using a careful analysis of the shielding surrounding his part.

The results in Table 16-1 are labeled with note "a" for identified flight parts, note "b" for cases that were not clearly identified as flight parts, and note "c" for cases that clearly are not flight parts. We get a clearer view of the actual flight distributions if we view all of the "a *or* b *or* c" results in one figure.

Dyer has pointed out that the comparisons should be made for the various types of environments [Dyer 1997]. For heavy ions, for example, we should consider the results separately for low Earth orbit (LEO), elliptical orbits (perhaps distinguishing geostationary transfer orbits (GTOs)), and geostationary orbits (GEOs). For protons we should separate LEOs and elliptical orbits. The GEO results should represent the basic cosmic ray environment described in CREME, whereas the other two heavy ion regimes involve modification by the Earth's magnetic field. The elliptic orbits pass through the heart of the proton belts, so we would expect the results to compare to the AP8 proton environments, whereas the proton LEOs are primarily sampling the South Atlantic anomaly (SAA), which may not be well described by AP8. As more data become available, the LEO results might be classified according to the altitude and inclination.

There is not enough data to separate the data in Figures 16-1 and 16-2 according to the individual environments. The widths will probably be narrower when adequate data are available to analyze flight parts according to the environment.

Figure 16-3 shows the results for heavy ion induced upsets at geosynchronous orbit. The center of the distribution occurs at a ratio of 0.98, which appears to indicate that the environments are sufficiently accurate in this case. Figure 16-4 shows the heavy ion results for LEOs. In this case the center of the distribution indicates that the predictions are high by a factor of 2. Figure 16-5 shows the results for elliptic orbits. Here we have combined the Apex orbit (362–2544 km, 70°) and the CRRES orbit (348–33,582 km, 18.2°). The average in this case is 1.67, perhaps indicating the mixing of the two distributions shown in Figures 16-3 and 16-4. Future results for elliptical orbits should attempt to trace the upset results as a function of position in orbit. In principle, the results at apogee and perigee should agree with the corresponding other experiments.

Figures 16-6 and 16-7 show the results for comparison of predictions and observations for proton induced upsets. The predictions appear to work well for the elliptic orbits that penetrate the heart of the proton radiation belts. The LEO proton upsets are produced in the SAA. It appears that the number of protons in this region is a factor of 1.4 times the number predicted by AP8.

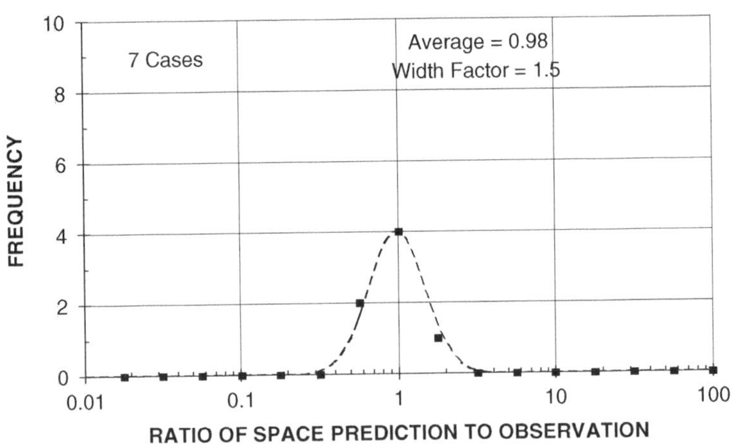

Figure 16-3 Quality of space predictions for heavy ion induced upsets at geosynchronous orbit. These occur for the primary cosmic ray environment, unmodified by the Earth's magnetic field. The good agreement supports use of the basic CREME environments.

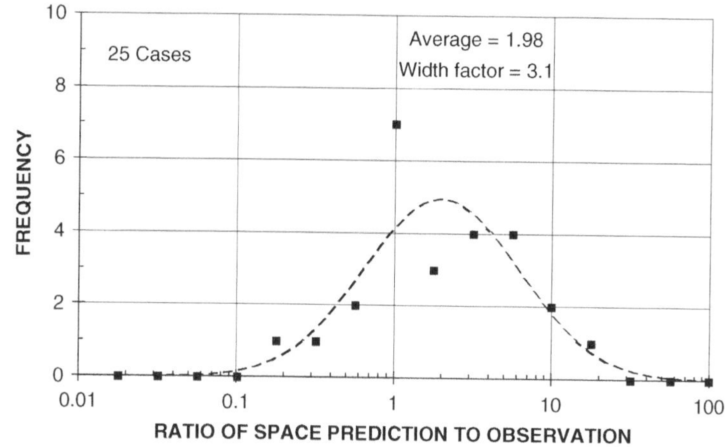

Figure 16-4 Quality of space prediction for heavy ion induced upsets for low Earth orbit satellites. These occur primarily near the north and south poles.

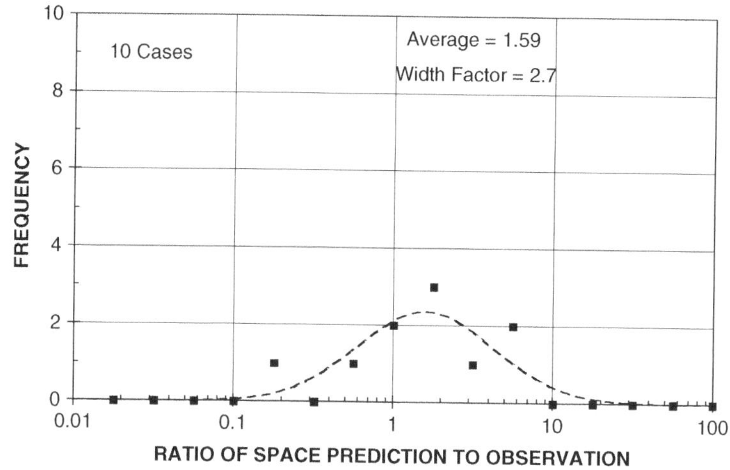

Figure 16-5 Quality of space predictions for heavy ion induced upsets for elliptic orbits. Some of the points with large ratios were obtained using nominal shielding, rather than sector analysis shielding. This pulls the curve right.

If we take a weighted average of the widths shown in Figures 16-4, 16-5, 16-6, and 16-7, we obtain a standard deviation of 0.47 for the 70 cases. The width factor corresponding to one standard deviation is then a factor of 3. The width factors for two and three standard deviations are 9 and 27, respectively. Therefore, predictions made for

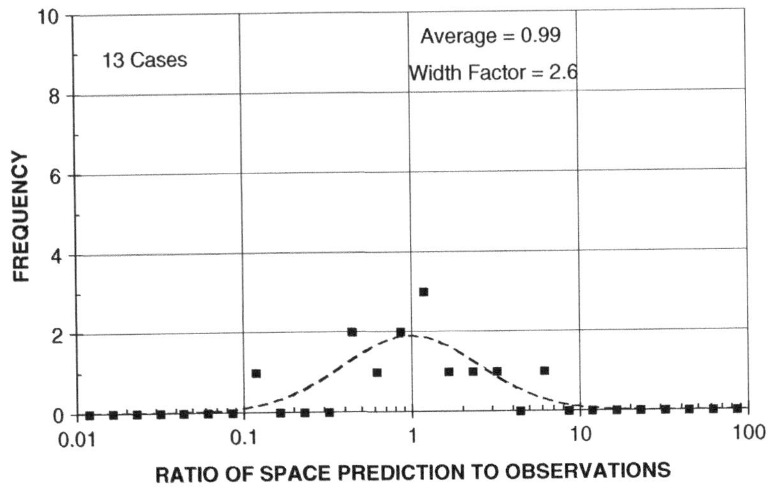

Figure 16-6 Quality of predictions for proton induced upsets observed in elliptical orbits. These upsets would occur in the heart of the proton radiation belt. The good agreement supports use of the basic AP8 proton environment.

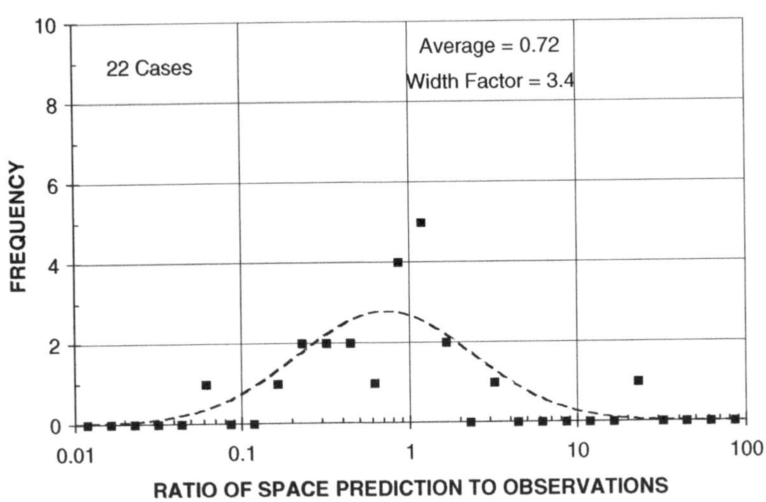

Figure 16-7 Quality of predictions for proton induced upsets for low Earth orbits. These upsets are produced in the South Atlantic anomaly. AP8 appears to underpredict the number of protons in this region.

generic parts have a spread of a factor of 3 about the average, and the average in turn may depend on the particular environment. The averages will be near one for cosmic rays at GEO and for protons in the heart of the belts. The average prediction is a factor of 2 high for heavy ions in LEO and a factor of 1.4 low for LEO protons. The heavy ion predictions at LEO may be high either due to environmental uncertainties or due to improper consideration of device shielding.

These results indicate that conservative designers, working with generic parts data, should in general allow for a factor of 10 under-predictions. This factor can be modified for the type of environment if desired.

16.2 ENVIRONMENTAL UNCERTAINTIES

Some authors in the past have defended poor predictions by saying that the environments were uncertain by factors of 2 to 10. It does not appear that this is a valid reason. The heavy ion predictions in general were made with the CREME environment, while the proton predictions were made with the AP8 trapped proton environments. In both cases the overall agreement of prediction and observation is equal to or better than a factor of 2. Both sets of predictions appear to be good for the basic environment, and slightly poorer for the LEO cases.

The radiation effects literature has several discussions of the environment used in the upset rate predictions. There is general agreement that the CREME86 solar event spectra and frequency are too conservative. This has been addressed at Cal Tech by Feynman and others and the results are summarized for the radiation effects community by McKerracher and co-workers [McKerracher 1994] Chenette and co-workers also present examples of the problem [Chenette 1994]. Majewski and co-workers presented an updated model [Majewski 1995]. CREME96 has addressed these problems, using additional data from GOES and IMP-8 and is probably the environment that should be used for solar particle predictions [Tylka 1996].

The basic CREME86 cosmic ray environment [Adams 1986] needed slight revisions. The problem with the time dependence of the variation from solar maximum to solar minimum has been addressed by Adams and others [Letaw 1986, Tylka 1997, O'Neill 1994]. The basic model of the interplanetary heavy ion environment also needed updating. The updated environments of the Moscow State group were discussed by Petersen and Adams [Petersen 1992b]. These indicate that the iron spectrum between 10 and 1000 MeV/nucleon was slightly overestimated by CREME86 for the CRRES orbit and time. The results of Chenette and co-workers indicate similar minor changes in the iron LET spectrum [Chenette 1994]. The CREME86 model has now been updated as CREME96 [Tylka 1996]. This is based on a very large amount of space data. The basic change in the galactic cosmic ray environment relative to single events is a very slight increase in the intensity of the iron peak during solar minimum. CREME-MC has an environment updated to 2009, but preliminary investigation indicates virtually no difference in predicted upset rates.

Dyer and co-workers examined the LET spectra and found the CREME86 predictions to be high by an average factor of 1.7 for high LET events in LEO orbit on UoSAT-3 [Dyer 1991]. There are some differences in the lower part of the LET spectrum ($< 10^3$ MeV/g/cm^2), thought to be due to effects of the spacecraft shielding. The environment in this region will have little impact on single event rates.

Underwood and co-workers made later comparisons of measured LET spectra with CREME86 predictions [Underwood 1994]. This paper gave results for Posat and Kitsat, and revised values for UoSAT-3. The ratios of CREME predictions to measured results were 1.3, 1.4, and 1.6. The results depend on LET and on the magnetic rigidity examined in the experiment. The authors also found underestimates for low LET events in the detector. The CREME spectra in both the Dyer and Underwood papers were calculated using a single shielding thickness and not with a shield distribution obtained using sector analysis.

Dyer and co-workers have updated the environmental LET measurements with results from STRV and APEX [Dyer 1996]. Both satellites were in elliptical orbits. The analysis was performed with shielding obtained using sector analysis, but CREME86 still overestimated the observed spectra by 30% to 50%. Tylka and co-workers have now analyzed some of this data with CREME96 showing basically good agreement, with a few minor unexplained discrepancies [Tylka 1997].

Figures 16-4 and 16-5 indicated overprediction of heavy ion upset rates below geosynchronous orbit. One factor that confuses the issue is that much of this data was obtained using predictions that did not include proper sector analysis of the shielding. As Smith pointed out, that can lead to overpredictions by factors as large as 2 [Smith 1994]. As an example, the data in Figure 16-5 is based on 4 points from Adoplphsen obtained with nominal shielding that have an average ratio of 3.47; combined with 6 CRRES points obtained using sector analysis that have an average ratio of 0.93. Fourteen of the data points in Figure 16-4 were based on either nominal or single valued shielding. Although we have assumed that systematic errors will cancel over a number of different experiments, this is not necessarily true when one experiment adds a number of points. If we remove Scott's 7 points from Figure 16-4, the mean will change to 1.6. For these reasons we cannot conclude that the factor of 2 overprediction is due to problems with the environment. The results shown in these figures contribute more to answering the question about how bad predictions can be rather than how good predictions can be.

The calculated heavy ion upset rates for low Earth orbits depend on the geomagnetic transmission of the interplanetary particles. CREME86 used the Shea and Smart 1980 grid of cutoff rigidities. CREME96 uses an updated version of these results and also an updated treatment of geomagnetic access during magnetic storms. CREME96 offers another option for 51.6° and 28.5° at 450 km. These are based on particle tracings through a complete field model. The two approaches give the same upset rates within 5% for the 51.6° orbits. The specific calculation gives results that are approximately 20% less for the 28.5° orbit. This indicates that one should obtain ray tracing transmission factors if one wants results better than 20% for low inclination LEOs.

Another factor that will enter is the east–west effect. This refers to the fact that particles arrive with greater abundance from the west due to the effects of the Earth's magnetic field [Barth 1997]. The effects are noticeable from 500 to 1200 km and may be a factor of 2 at the lower altitudes [Ginet 2007]. This may be important for spacecraft that maintain an east–west orientation. The author is unaware of any upset rate analysis that has considered this effect.

The uncertainty factor for protons in AP-8 is quoted as 2. It represents the uncertainty from combining the data sets from many instruments and satellites and is not meant to measure the variation in the actual environment. The magnitude of the variation of the environment depends on the region of space.

Takagi and co-workers presented proton results at the equator indicating that the radiation belt is more intense at lower altitudes than modeled by AP8 [Takagi 1993]. This may indicate higher proton fluxes in some orbits, although the impact on low Earth orbit is not apparent. Underwood and co-workers pointed out that the same parts flown in two different orbits had different ratios of predicted to observed upset rates [Underwood 1993]. Their data are shown in Table 16-1 for the S80/T and UoSAT-5.

Daly and co-workers summarized the problems with the models of the radiation belts [Daly 1996]. They reported shuttle measurements of particle fluxes that have a ratio of prediction to observation of 0.645 and LDEF measurements that indicated a ratio of 0.434 and 0.588. The present ratio of proton upsets has a broad one standard deviation range of 0.203 to 2.48. Therefore, these results are consistent.

The AP-8 model is a static model and it is well known that the proton radiation belts are dynamic. One aspect of this is the slight shift of the SAA from the position described by AP-8 [Barth 1997]. An extreme example of the dynamic nature is the production of new radiation belts by the flare on March 24, 1991 [Blake 1992, Gussenhoven 1992, Shea 1992]. The effects on SEU rates have been described by Campbell and co-workers [Campbell 1992, Campbell 1994]. Daly and co-workers show that the effects of the 1991 flare were still being seen through 1992 [Daly 1994]. This may partially explain the Underwood results.

The east–west effect may be very important for some spacecraft in the SAA. At low altitudes the directional difference may be a factor of 2 to 7. It is important to take the anisotropy into account if a spacecraft maintains a fixed axis throughout the mission [Barth 1997].

It appears from the results of these papers and from the near agreement of upset rate predictions and observations that there is nothing drastically wrong with the commonly available space environments. The proton environment may occasionally experience fluctuations that are not described by the static models, but the long-term averages appear to be well described.

The largest uncertainties in the environments arise because of the dynamic nature of the proton radiation belt(s). Barth describes several attempts to update the AP-8 model and to allow for the dynamic nature of the belts [Barth 1997]. None of the papers that we are discussing used these recent models. Lauenstein and Barth have compared several of these approaches for several standard orbits [Lauenstein 2005]. Xapsos and co-workers have given a complete updated analysis of the situation [Xapsos 2006a].

It would certainly be valuable to have continued detailed measurements on the population and energy spectra of the proton belts. The measurements should cover a range of B and L so that information is available about the SAA as well as about the central regions of both proton belts. B and L are the McIlwain space coordinates, where B represents rings of constant magnetic field strength and L represents the dipole shell parameter. The L parameter is most simply described as the value that marks the particle drift shells by their magnetic equatorial distance from the center of the Earth [Barth 1997]. The results of these measurements should be made available in a form so that they can be used for upset rate calculations.

16.3 EXAMINATION OF OUTLIERS

There are a number of data points that are far from the mean. However, the wide distributions that are observed do not lead to any of the results being excluded as true outliers. Here we use the standard data rejection criteria of outliers being points that are more than three standard deviations from the mean.

In some cases, there is a large deviation for either heavy ion or proton environments, but not for the other environment. This suggests that these were not caused by problems with SEU space instrumentation or analysis of space data. Harboe-Sorensen and co-workers report cases involving three part types [Harboe-Sorensen 1990, Harboe-Sorensen 1993]. These papers are good examples of how the problems should be approached and reported, and the only obvious problem is the use of nominal shielding rather than sector analysis shielding. Two of the three parts have heavy ion cross-section curves with discontinuities and the authors integrate over the cross-section curves. However, even though they may integrate over a curve going through the data points, and not a curve derived from a Weibull fit of the normal incidence points, this could only lead to a prediction that is too small, not the overly large one reported. Some of the devices are modern high density devices; however, the worst agreements are for 16 kbit parts, so we do not believe that the disagreements can be ascribed to rate prediction techniques breaking down at high densities.

The bipolar 93422 results of Campbell are particularly perplexing. All of the measurements of this device relative to the AMD 93L422 indicate that it is less sensitive to both proton and heavy ion upsets. The heavy ion results are consistent with this. However, the proton flight results indicate significantly more upsets for the 93422 than the 93L422.

There is a small chance that this could be explained by upsets due to direct ionization by protons. It may be that the 93422 devices are upsetting to low energy stopping protons and the 93L422 devices are not. The CRRES flight carried GaAs devices that are more sensitive than either of the bipolar devices. The results given by Campbell and co-workers show that they have their peak upset rates at a significantly higher altitude than the bipolar devices [Campbell 1992]. This is consistent with the increased direct ionization that will occur due to the more intense soft proton spectrum at higher altitudes. The GaAs devices do appear to be experiencing upset due to direct proton ionization by low energy protons. The 93422 has its peak upset rate at the same altitude as the 93L422, so the upsets in both of these cases must be due to proton reactions. The remaining possible explanation is a difference between the tested part and the flight part. This does not explain the relative proton and heavy ion behavior, however.

The D43100 device is reported by Underwood and co-workers to not have a well-defined proton threshold, and they say that the resulting uncertainty in the Bendel parameters may explain the poor results with that device [Underwood 1993]. They also state that the part tested was a nonflight spare part.

16.4 POSSIBLE REASONS FOR POOR UPSET RATE PREDICTIONS

Both Harboe-Sorensen and Underwood emphasize the importance of testing parts that are nearly identical with flight parts. They both observe large variations in test results when they use a variety of parts. This appears to be the most likely explanation for many of the large deviations observed in the predictions.

Adolphsen and co-workers have discussed the problem with generic parts in some detail [Adolphsen 1996a]. There can be large variations from lot to lot, and large variations within a lot. They present the basic questions as follows:

1. Are the parts really from the same mask set?
2. If the parts are from the same mask set, how broad is the upset distribution of those parts?
3. If the distribution is broad, from where in the distribution were the parts taken?

A second possible cause, and one that can be confused with part to part variation, is the effect of part location [Smith 1994]. Smith has shown that there can be large variations due to unequal shielding of the parts.

A third probable cause for heavy ion prediction deviations is a poor choice of the device depth. This is not an important factor for large sensitive devices, but as the devices become either smaller or less sensitive, it becomes more important. The uncertainties in depth for modern small devices may be cleared up by analyzing the discontinuities in experimental data in terms of the geometrical features of the device [Petersen 1993, Connell 1995a, Petersen 1997b]. Another promising approach uses proton upset simulation codes in combination with proton test results [Reed 1994b].

Highly integrated devices may have geometries that are not adequately approximated by the rectangular parallelepiped. This is a potential source of error in upset rate predictions, but there is no evidence of serious effects in the presently available data.

A fourth possible cause of deviations is the use of insufficient proton test data to thoroughly describe the part. Both the one-parameter and the two-parameter Bendel prescriptions are known to not completely describe the measurements of proton cross section for some devices. It is possible to use the more general Weibull function to describe proton data [Tylka 1996]. The Weibull or lognormal descriptions are the preferred approaches if adequate data are available in the threshold region and the high energy region [Petersen 1997b].

Another effect is multiple bit upsets. This is somewhat more difficult to consider, and not all of the authors clearly indicate that they have considered this factor. There is a minor problem of definition that can arise in this situation. If you are considering the number of upsets, then upsets in adjacent bits should be counted. However, the upset rate codes really calculate the number of single event events. Therefore, simultaneous upsets in adjacent bits should count as one event. This is the definition that should be used when calculating upset rates based on experimental results, either in the laboratory or in space.

There are other effects that can have an impact on the results. One is the inclusion of data during solar particle events in the heavy ion or proton data. Most authors clearly state that this is removed. None of the authors mentioned the possible influence of the east–west effects. Some parts are susceptible to the "stuck bits" phenomenon [Koga 1991]. This could also distort data. The experimenters also have to be aware of the possibility of sensitivity to the programmed state. This is most likely to be a difference between 0→1 and 1→0 transitions. It is also possible for the upset sensitivity to depend on the contents of adjacent cells.

16.5 CONSTITUENTS OF A GOOD RATE COMPARISON PAPER

There are several aspects to a good paper discussing space upset rates. The paper should present enough information to convince the reader that the work is definitive. The paper should include enough information so that someone else could repeat all calculations using the author's approach or using another approach.

Another valuable attribute of a paper is to include adequate information about experimental laboratory results so that others can use the results for predictions with other satellites.

The paper should address a number of small but important points that are unique for this combination of laboratory and space experiment. Much more is involved in making measurements and predictions for space experiments than is involved in making failure estimates for space systems. The rough worst case estimates that may be appropriate for failure predictions are not appropriate here.

The following lists show the various items that should be included in the report of ground results, predictions, and space results. All of these are present in one or more of the published reports, but none of the reports include all of them. The lists undoubtedly ignore necessary items, but they should help focus the experimenters to think of possible problems. The background for many of these observations is given above and in [Petersen 1997b]. Remember that the foundation of the entire program rests on the laboratory measurements and the proper analysis of the test results.

16.5.1 Reports on Laboratory and Space Measurements

1. Can the experiment observe, identify, and record all double and multiple upsets, including bit locations?
2. Can the experiment separate out the effects of bit line upset and other sources of spurious multiple-bit upsets?
3. Does the experiment report all upsets? Are appropriate dead-time corrections made?
4. Have the reported results been corrected for these effects?
5. What is the correspondence between the tested parts and the parts flown?
6. Is there a large spread from device to device? What is the distribution of upset sensitivity between parts?
7. What are the answers to Adolphsen's questions (Section 16.4)?
8. Are the 1→0 and 0→1 errors separated and identified? Is the bit pattern reported?
9. What tests were made for various types of pattern sensitivity? Is there evidence that the sensitivity depends on the contents of surrounding bits?
10. What do the experiments show about part SEU characteristics changing with total dose?
11. Is the data system fast enough so that there is not a false indication of multiple upsets and that there are not some bit reversals due to two hits on the same sensitive cell?
12. Are the results misleading due to combination of results taken under different test conditions, such as voltage or temperature?
13. Has the test software been thoroughly tested?
14. Have the data from multiple similar tests been combined to give the best single answer?
15. What are the statistical errors in the results?
16. What are the systematic errors in the results?
17. Have the experimental data been corrected for all possible factors?

16.5.2 Analysis of Ground Measurements

1. Are the proton cross-section measurements also presented with Bendel two-parameter characteristics? What are the parameters? Did the Bendel fits emphasize the threshold or the saturation region? Is the data better described using the Weibull function or the lognormal function?
2. Have the proton measurements been obtained over a wide enough range of proton energy so that the curve is adequately defined over the entire range that contributes to proton upsets?
3. Are the heavy ion upset characteristics presented in terms of the Weibull parameters? What are the parameters?
4. Does the heavy ion data have discontinuities?
5. Are the discontinuities used to determine depth and funnel parameters?
6. Does the heavy ion data really reach the saturated cross section when plotted on a linear plot?
7. Is the limiting cross section the value that would be expected from geometrical examination of the part?
8. Did the authors do any reverse engineering to examine cell area and depth?
9. Has the LET in the cross-section report been corrected for energy loss with depth?
10. Are the experimental statistics (accuracy and precision) adequate so that predictions can be made to a few percent?

16.5.3 Environment for Space Predictions

1. Does the environment include both heavy ions and protons from solar particle events?
2. Does the environment include the steady-state exposure of cosmic ray ions?
3. Does the environment include the trapped protons?
4. If the proton environment varies during the orbit and with time, is this quantified so that meaningful rate comparisons can be made?

5. Does the environment use the correct solar cycle variation?
6. Are the environments chosen as the most likely to be encountered rather than as being conservative?
7. Does the environment make the mistake of including the direct proton ionization contribution?
8. What is the satellite orientation? Will it be spinning; if so, in what orbital plane? Will it be stabilized; if so, what will be the orientation of the shielding?
9. Does the environment include the proper satellite shielding using sector analysis?
10. Do the authors include the shielding in the report so that others can repeat the calculation?
11. Is there a variation of shielding (and therefore upset rate) from part to part in the flight package?
12. If new environments are presented as improvements over the standard environments, do the authors present the standard environments and state what the differences will be in the predicted upset rates?

16.5.4 Upset Rate Calculations

1. What basic methods are used?
2. What device thickness was assumed for heavy ion calculations?
3. What method was used to determine the device thickness?
4. Do they separate or combine the results for protons and heavy ions as appropriate?
5. Do they assume one or two sensitive volumes per cell?
6. Are all parameters given so others can repeat or modify the calculations?
7. Do the parts allow any comments about device scaling?
8. Do the authors give some summary information about part technology?
9. Do the authors attempt to calculate double and multiple bit upset rates?

10. Are the results expressed in terms of upsets/bit-day averaged over many orbits?
11. Is it possible to make predictions for limited parts of the orbit, such as in proton belts, or outside proton belts, or in terms of L shell position or other description of position in orbit?
12. Will the flight data be susceptible to the same type of detailed analysis?
13. If the methods of prediction are presented as improvements over standard methods, do the authors give predicted rates under defined conditions for both types of calculation?
14. What are the likely variations with temperature?
15. What are the likely variations with total ionizing dose?
16. If the prediction parameters are varied to give a forced fit to observed flight data, what would the results be without a forced fit, or if standard methods were used?
17. Is the effect of Earth shadowing properly considered for low Earth orbits and elliptical orbits?
18. Are the effects of spacecraft orientation included in situations in which the east–west effect is present?

16.5.5 Characteristics of Space Experiment and Data

1. What are the details of the orbit?
2. What are the parts?
3. Are the results expressed in terms of upsets/bit-day averaged over many orbits?
4. Is it possible to extract data for limited parts of the orbit, such as in proton belts, or outside proton belts, or for given L shells or other description of the location in orbit?
5. In the latter case, is the form compatible with the form of the predictions?
6. Did the parts acquire total ionizing dose so that the upset rates changed with dose?
7. Was the shielding distribution really that used in the predictions? Is it appropriate for each part?

8. Are the results properly averaged over position and over time so that they can be compared with predictions that are averaged over the same times and positions? Do the results include orbit precession?
9. Are results during periods of solar activity separated from long-term results?
10. What is the part temperature and the possible range of variation of part temperatures?
11. Were the calculated rates accurately corrected for times during which data was not collected? These include satellite down time, periods in which data was specially filtered, and missing data tapes.

16.6 SUMMARY AND CONCLUSIONS

This chapter summarizes the results of 126 comparisons of predicted and observed rates from 23 satellites. The heavy ion predictions based on the CREME environments and the proton predictions based on the AP-8 environments are both successful. The results appear consistent with suggested slight modifications in these environments based on recent space measurements. Predictions may be wrong by a factor of 20 if based on generic parts data and nominal shielding. If flight parts data are used together with shielding obtained from sector analysis, it appears to be possible to make high quality space predictions, exceeding 20%, for all regions of space except the SAA. Recent improved models may now be making it possible to also achieve this accuracy in that region. Flight experiments that are attempting to test the methodology should have a goal of exceeding 10%.

There is no evidence of serious problems with the methods of upset rate prediction. The predictions do need data that has been carefully obtained.

Although it is possible to make good predictions, there are still a number of possible sources of error. The relatively large number of cases with a large difference between prediction and observation indicates that a great deal of care must be taken with the laboratory measurements, the predictions, and the space measurements.

The most common identified source of differences is due to part to part variation. The use of generic parts rather than flight parts can lead to large differences, especially for commercial parts. It is important to test parts that are nearly identical with flight parts. Commercial parts are likely to have a large variation from part to part, but there is some spread observed with many other parts. This problem appears to be the most likely explanation for many of the large deviations observed in the predictions.

The second probable cause of deviations is use of incorrect shielding distributions around the separate parts.

Another possible cause of deviations is the use of insufficient test data to thoroughly describe the part. A possible cause for heavy ion prediction deviations is a poor choice of the device depth.

It appears that there is nothing drastically wrong with the commonly available space environments. The heavy ion predictions for GEO orbits and the proton predictions for elliptical orbits through the heart of the belts both agree with measurements. The heavy ion predictions at LEO overestimate the upset rates by a factor of 2. This may be due to improper shielding estimates rather than environmental uncertainties. The proton environments lead to a factor of 1.4 underestimate for upset rates in the SAA. It appears that there are now improved models that eliminate this problem. The proton environment may occasionally experience fluctuations that are not described by the static models, but the long-term averages appear to be well described.

The largest uncertainties in the environments arise because of the dynamic nature of the proton radiation belt. It would be valuable to have continued detailed measurement on the population and energy spectra of the proton belts. The measurements should cover a range of B and L so that information is available about the SAA as well as about the central regions of both proton belts. The results of these measurements should be made available in a form so that they can be used for upset rate calculations.

Many of the authors do not give enough detail in their reports so that the results can be used by others either to check the author's approaches or to extend the results for predictions on other satellites. This chapter summarizes the helpful information that should be included in any comparison of ground data, predictions, and space data.

16.7 RECENT COMPARISONS

The above material was taken from [Petersen 1997a]. There have been a number of more recent papers that also show good agreement between predictions and flight data. Consider the following papers: [Inguimbert 2004, Kimoto 2005, Hansen 2007]. Figure 13.7 showed the effects of precession on predictions for the CRESS orbit. Petersen's analysis of the CRESS data did not have the data for this analysis [Petersen 1997a]. The help documentation for CREME-MC updates this situation.

16.8 COMPARISONS WITH EVENTS DURING SOLAR ACTIVITY

There have been a number of reports of data taken during solar activity [Normand 1990, Croley 1995, Tylka 1996, Campbell 2002, Hansen 2007]. Individual comparisons showed reasonable agreement with models, but there has not been a systematic examination of the data. There have also been developments in modeling of the solar environment [Xapsos 2000, Xapsos 2006b, Xapsos 2007]. The solar proton fluence has been refined and extended to high energies.

CHAPTER 17

Limitations of the IRPP Approach

The RPP and IRPP approaches were initially derived in the 1970s and early 1980s. At that time devices were large, flat, and separated. These models were developed using a large number of approximations. It is very reasonable to expect that later technology would test these approximations and introduce scenarios that would not fit in the IRPP approach. With the development of the technology, we have also had development of the physical modeling tools. These tools enable detailed modeling of modern small scale devices and let us test the approximations in the IRPP approach [Hubert 2009, Kobayashi 2009, Sierawski 2009, Weller 2009].

17.1 THE IRPP AND DEEP DEVICES

The IRPP approach was greatly strengthened when Petersen suggested and Connell implemented the geometrical corrections that extended the IRPP approach to deep devices [Petersen 1993, Connell 1995a, Connell 1995b]. Although the effective LET concept no longer holds in these situations, that does not weaken the IRPP approach. If someone claims that the IRPP approach is not working because their plot of cross section versus effective LET is discontinuous, they are not

Single Event Effects in Aerospace, First Edition. By Edward Petersen.
© 2011 the Institute of Electrical and Electronics Engineers, Inc.
Published 2011 by John Wiley & Sons, Inc.

taking advantage of the geometrical corrections. This was discussed in Section 5.3.4. The Monte Carlo calculations such as CREME-MC are also useful in this case. There are other possible problems with the concept of LET, and we will examine them below.

17.2 THE RPP WHEN TWO HITS ARE REQUIRED

It was recognized from the beginning that single event effects that required two hits in adjacent regions normally required an extended RPP modeling not available in the CREME type calculation. We have discussed this in Section 6.11. This modeling can be performed either geometrically or using the modern Monte Carlo codes. We also pointed out in Section 6.11 that if the regions are slightly separated, the upset rates can be calculated by using a CREME IRPP calculation for one volume, reduced by the included solid angle of the second volume. One can estimate the effect rate with this approach, then examine the geometrical approaches, and then use the Monte Carlo methods if necessary.

17.3 THE RPP APPROACHES NEGLECT TRACK SIZE

A possible fundamental problem was suggested by Xapsos in 1992. The RPP approach assumes that the energy deposition is entirely characterized by the particle LET. The problem is that the energy lost by the ion is not the same as the energy deposited in the sensitive volume. There are high energy delta rays that pass out of a small volume. Xapsos showed that this could become a major effect for very small sensitive volumes. We have discussed this in Section 6.9. It appears that as devices shrink—so that the expected volume shrinks—various internal electronic effects, such as parasitic bipolar effects, come into play, and larger fractions of the device become sensitive. The microscopic electrical modeling of the device becomes more challenging, but the gross macroscopic modeling of the IRPP is still valid.

17.4 THE IRPP CALCULATES NUMBER OF EVENTS, NOT TOTAL NUMBER OF UPSETS

As devices become smaller, there are many more multiple upsets. The experiments need to differentiate the number of events from the total observed upsets. They then combine the IRPP calculation of the number of events using the observed event cross-section curve, with the observed multiplicity. This was discussed in Section 6.11.

17.5 THE RPP APPROACHES NEGLECT EFFECTS THAT ARISE OUTSIDE THE SENSITIVE VOLUME

The IRPP model of single event effects considers only the ionization produced by passing ions inside the sensitive volume. On the other hand, the calculation of proton rates is based on the measurement of proton cross sections, so that the proton reactions outside the sensitive volume are also considered. It now appears that upsets can be produced by reaction products from heavy ion reactions with heavy metals outside the sensitive volume. The cross-section curve has two components: the upsets at high LET due to direct ionization and the upsets indicated at low LET from the reaction products. We showed in Section 10.2 that the total upset rate could be calculated by considering these two components separately. The low LET component is treated as a measured cross section, similar to the approach for proton upsets. The IRPP approach is easily modified to include ionization produced outside the sensitive volume.

17.6 THE IRPP APPROACHES ASSUME THAT THE EFFECT OF DIFFERENT PARTICLES WITH THE SAME LET IS EQUIVALENT

There are two separate problems here. We have discussed the questions about the ionization effects of high energy and low energy particles with the same LET in Section 6.8. It appears that there is no difference for ionization effects.

The second aspect of the problem is that nuclear reactions in adjacent heavy metals can also occur [Reed 2007]. This will lead to upset cross sections that are completely inconsistent with the ionization cross sections. The apparent cross sections will vary with energy, ion, and amount of tungsten. The cross-section curves for reaction products and for direct ionization need to be considered separately. See Section 10.2. In Chapter 10 we showed that there is a small window of susceptibility in very hard devices where the additional contributions of heavy ion reactions need to be considered in upset rate calculations. Although the reaction contributions need further research, it does not appear that the IRPP approaches are breaking down.

Cross sections measured for devices that require two hits for an upset also give results that are not consistent with effective LET. These results can be analyzed geometrically, but are probably best analyzed using the Monte Carlo calculations.

17.7 THE IRPP APPROACHES ASSUME THAT THE LET OF THE PARTICLE IS NOT CHANGING IN THE SENSITIVE VOLUME

Generally, the upset environment consists of very high energy particles that lose very little energy in the sensitive volume so the LET can be considered constant. The only situation where the particles are changing LET is for stopping particles at the end of their range. In this situation, the population is very small, so the particles make little contribution. As the particle energy approaches zero, the particle density approaches zero. The one possible exception is for stopping protons in some special environments. We discussed this in Section 8.6. There are situations where the constant LET approximation would overestimate the number of events. Edmonds has presented an approach for detecting and handling this situation [Edmonds 2008]. Although the IRPP approach breaks down in these rare cases, there is no need to resort to physical modeling.

17.8 THE IRPP APPROACH ASSUMES THAT THE CHARGE COLLECTION DOES NOT CHANGE WITH DEVICE ORIENTATION

The IRPP approach assumes that the cross-section curve is determined by the charge-collection variation across the device, measured at normal incidence, and that this curve is a specific property of the device. There may be exceptions. The device modeling of Pellish seems to show that their structure has different charge-collection mechanisms that depend on track orientation [Pellish 2007]. However, the critical charge depends only on the circuit, and not on the track orientation. Therefore, the critical LET does not change. The calculated upset rate varies little with minor changes in the shape of the cross-section curve, so these effects will have little impact on upset rate. This was discussed in Section 5.3.5. This is a situation where detailed physical modeling will augment the IRPP rate calculations.

17.9 THE STATUS OF SINGLE EVENT RATE ANALYSIS

The primary method of calculating single event rates caused by heavy ions is the integral rectangular parallelepiped (IRPP) method. This method works very well in virtually all situations. Some of the situations that appeared problematic in the early days have been shown not to be problems. There are a few rare cases showing up in modern small scale devices that need additional or alternate methods of analysis.

The secondary method of calculating heavy ion upset rates is the figure of merit (FOM) method. This has intrinsic small errors, but has the advantage of being quick and computationally simple.

The tertiary method of calculating heavy ion upset rates is the Monte Carlo method. This allows detailed physical modeling of the processes involved. It is computationally intense and generally more valuable for detailed physical analysis of rare processes.

The prime approach for predicting space proton upset rates is the integral cross-section approach integrating the device cross-section curve with the proton spectrum. This is a simple and straightforward approach.

The secondary approach to proton upset rates is the figure of merit approach. This has the advantage that the FOM can be determined from either heavy ion or proton cross-section measurements.

Proton upset rates have also been calculated using nuclear reaction analysis. This can be done either with Monte Carlo methods or with analytic methods based on heavy ion upset data. These methods have helped our understanding of the basic processes, but have not passed into general use for proton rate predictions.

Device modelers using mixed-mode simulations can determine the device limiting cross section, depth of the device's sensitive volume, critical charge, and charge efficacy distribution. This complete set of SEU characteristics completely defines the sensitivity of the device, so the modeler can predict SEU rates in any proton, neutron, or heavy ion environment.

In many situations, we are examining SEE in nonmemory parts such as controllers, microprocessors, or FPGAs. In these cases we obtain a cross section per device, which is some unknown number times the unknown cross section per bit. The IRPP approach for rate calculations cannot be used in these cases as there is no information about basic bit dimensions. In these cases we can calculate the FOM per device using the cross section per device. The SEE rate per device is then calculated using the orbit specific rate coefficient.

APPENDIX A

Useful Numbers

1 nautical mile = 1853 meters

1 barn = 10^{-24}cm^2

e = 1.6 × 10^{-19} coulomb

1 year = 365.256 days = 3.156 × 10^7 seconds

1 erg = 10^{-7} joule = 6.24 × 10^5 MeV

1 inch = 6.855 g/cm^2 Al

1 μm = 10^{-4}cm; 1 μm^2 = 10^{-8} cm^2

1 MeV/g = 1.602 e^{-10} Gy = 1.602 e^{-8} rd

1 electron = 1.60 e^{-19} coulomb

For silicon

Silicon density 2.33 g/cm^3

3.6 ev/electron–hole pair

1 pC/μm = 98 MeV/mg/cm^2 in silicon

22.5 MeV = 1 pC

1 MeV/mg/cm^2 = 0.233 MeV/μm = 0.0104 pC/μm = 0.044pC/mg/cm^2

$E_c = 22.5\ Q_c$ MeV/pC in silicon

Single Event Effects in Aerospace, First Edition. By Edward Petersen.
© 2011 the Institute of Electrical and Electronics Engineers, Inc.
Published 2011 by John Wiley & Sons, Inc.

For GaAs

5.316 g/cm^3

4.8 eV/pair

30.03 MeV = 1 pC

1 MeV/mg/cm^2 = 0.532 MeV/μm = 0.0179 pC/μm = 0.033 pC/mg/cm^2

$Q_c = 1.03e^{-2} L_{th} \times t$ pC

Plotting Steps

```
1.122 20 steps per decade
1.2115 12 steps per decade
```

Replot from Digitized Plots

```
Read coordinates from graphs
  read in with scanner
attempt to align axes so that vertical
  are vertical and horizontal are horizontal
Then coordinates translate directly
  to x and y values.
Transformation for linear scale
x = (c-c0)/(c1-c0) * (x1-x0) + x0
transformation for 1 or more decade log y1>y0
y = y0 * (y1/y0) (c0-c)/(c0-c1)
```

APPENDIX B

Reference Equations

B.1 WEIBULL FUNCTION

The use of the Weibull function was introduced in 1992 by Petersen [Petersen 1992b].

$$F(L) = 1 - \exp\{-[(L - L_0)/W]^s\} \qquad L > L_0$$

$$F(L) = 0 \qquad L < L_0$$

where L is the effective LET, L_0 is the device threshold, W is the width parameter, and s is the shape parameter. This function is then normalized to the limiting cross section of the device. The L values corresponding to 25% and 50% of the limiting cross sections are

$$L_{0.25} = L_0 + W \times 0.288^{1/s}$$

$$L_{0.5} = L_0 + W \times 0.693^{1/s}$$

B.2 LOGNORMAL DISTRIBUTION

The lognormal function is the normal distribution with the variable being log of the LET. Let

Single Event Effects in Aerospace, First Edition. By Edward Petersen.
© 2011 the Institute of Electrical and Electronics Engineers, Inc.
Published 2011 by John Wiley & Sons, Inc.

$$z = \frac{\ln(LET) - m}{\sqrt{2}\sigma}$$

where m = mean of the function in terms of ln(LET) = $\ln(L_c)$ and σ is the standard deviation of the function in terms of ln(LET), sometimes designated as s. The lognormal distribution is

$$f(z) = \frac{1}{\sqrt{2\pi}\sigma} e^{-z^2}$$

The cumulative lognormal distribution is

$$F(z) = 0.5\text{erfc}(-z) : z \leq 0$$

$$F(z) = 0.5 + 0.5\text{erf}(z) : z \geq 0$$

In Excel, it is calculated using LOGNORMDIST (LET, m, σ).

For the lognormal distribution $\ln(L_{0.25}) = m - 0.674\ \sigma$.

The cumulative Weibull function (four parameters) and the cumulative lognormal function (three parameters) can have very nearly the same shape and will only be distinguished by small details in the threshold region.

B.3 EXPONENTIAL DISTRIBUTION

An exponential distribution has been used to fit SEU data [Smith 1992].

$$\sigma = \sigma_0 e^{-K/(L-L_0)}$$

Edmonds [Edmonds 1996] suggested an exponential without the threshold term

$$\sigma = \sigma_0 e^{-k/L}$$

B.4 LEAST SQUARES FITTING

For least squares fitting, you calculate and minimize the quantity S:

$$S = \sum \Delta y^2$$

For linear least squares fitting Δy is the difference between the measured and calculated cross section while for logarithmic least squares fitting it is the differences of the logarithms of the measured and calculated cross sections. The Excel Solver routine is useful for calculating the minimum by varying the parameters in the fitting equation.

B.5 WEIGHTED LEAST SQUARES FITTING

For weighted least squares fitting, you calculate and minimize the quantity S_W [Beers 1958, p. 37]. This places more weight on the points with small statistical errors and less weight on the points with large statistical errors.

$$S_W = \sum \left(\frac{\Delta y}{s_y{}^2} \right)^2$$

Δy is defined above. s_y is the fractional standard deviation for the data point.

$$s_y = 1/\sqrt{N}$$

where N is the observed number of upsets.

A discussion of the propagation of errors through least squares analysis into the figure of merit has been given by Mavis [Mavis 2007].

B.6 CALCULATION OF COEFFICIENT OF DETERMINATION R^2

We use the *coefficient of determination* R^2 to evaluate the explained sum of the squares. It is derived by examining the ratios of least squares deviation from the fitting line to the least squares value of the data points about their average. The fit is very good if R^2 approaches one. An R^2 of 0.80 says that we are describing 80% of the data variation from its average value, while an R^2 of 0.25 says that we are explaining only 25% of the variation.

For nonlinear regression analysis R^2 is defined as [Hamilton 1992, p. 170]

$$R^2 = 1 - \frac{S}{\sum (Y_i - \bar{Y})^2}$$

S in the unweighted sum of the squares defined above and Y_i represents the individual data points that have an average of $\overline{Y}$. R^2 will be different for linear and logarithmic fitting, so the type of fitting should be specified.

B.7 THE BENDEL–PETERSEN EQUATION FOR THE PROTON CROSS SECTION

Initially, the Bendel–Petersen two-parameter semiempirical equation was used to describe the cross-section curve [Bendel 1983, Shimano 1989, Stapor 1990]. The initial 1983 paper used a single threshold (A) and a $(B/A)^{14}$ term instead of the limiting cross section, but that was replaced after the Shimano paper.

$$\sigma = \sigma_{PL}[1 - \exp(-0.18Y^{0.5})]^4$$

$$y = (18/A)^{0.5}(E - A)$$

where A is the threshold, σ_{PL} is the limiting cross section, and E and A are in MeV. More recent good low energy data [Duzellier 1997, Harboe-Sorensen 1997, Poivey 1998] show that the threshold region is better described by the integral lognormal distribution [Petersen 1998]. CREME96 and CREME-MC use the Weibull function rather than the lognormal distribution. See Section 4.3 for more details.

B.8 THE FIGURE OF MERIT AND UPSET RATE CALCULATIONS

The figure of merit is defined as

$$FOM = \frac{\sigma_{HL}}{L_{0.25}^2}\left[\frac{(\mathrm{MeV/mg/cm^2})^2}{\mathrm{cm^2}}\right] \tag{B.1}$$

where σ_{HL} is the limiting heavy ion cross section per bit at large LET and $L_{0.25}$ is the LET at 25% of the limiting cross section. It is common to describe the heavy ion cross-section curve using the cumulative Weibull distribution, with threshold L_0, width w, and shape parameter s. In this case the $L_{0.25}$ term can be easily calculated:

$$L_{0.25} = L_0 + w \times 0.288^{(1/s)} \tag{B.2}$$

The figure of merit can also be directly related to the limiting proton cross section:

$$FOM = 4.5 \times 10^4 \times \sigma_{PL}[1/\text{cm}^2] \tag{B.3}$$

The upset rate R in a particular orbit is calculated by introducing orbit-specific rate coefficients C_p and C_{hi} with units of upsets/bit-day:

$$R = (C_p + C_{hi}) \times FOM \tag{B.4}$$

In many situations, we are examining SEEs in nonmemory parts such as controllers, microprocessors, or FPGAs. In these cases we obtain a cross section per device, which is some unknown number times the cross section per bit. In these cases we can calculate the FOM per device using the cross section per device. The SEE rate per device is calculated using the same orbit-specific rate coefficients.

Sometimes there is a classified mission, so that manufacturers and suppliers need to meet upset criteria but are not allowed to know the orbit! The figure of merit approach is ideal for this situation. Someone who knows the orbit can calculate the upset rates for a few standard devices, and then determine the orbit-specific rate coefficients. The part specifications can then give the desired rate coefficients, without any hint of the orbit.

B.9 RPP DIFFERENTIAL AND INTEGRAL PATH LENGTH DISTRIBUTIONS

Upset rate calculations utilizing the differential path length distribution were introduced by Pickel and Blandford, using a path length distribution derived by Petroff [Pickel 1980]. Bendel derived a simplified expression that is now commonly used [Bendel 1984a, Adams 1986]. The distribution is a function of the path length s and the x,y,z dimensions of the RPP. The values of x,y,z are cycled through the six possible combination of width W, length L, and height H. A series of constant values are calculated for each permutation. We will write the equations in a spreadsheet format and retain Bendel's symbols in the equations.

```
k=     SQRT(X^2+Y^2)
max=r= SQRT(X^2+Y^2+Z^2)
T=     SQRT(X^2+Z^2)
V=     12*X*Y*Z^2
B1=     -(3*X*Y/(r*t))^2
B2=     (3*Y/k)^2+b1
B3=     -(3*X*Z/(k*r))^2
B4=     v*ATAN(Y/X)-(Y*Z^2/k)^2
B5=     v^2*Z^2*(Z^2/k^2-3)+v*ATAN(Y/X)
B6=     3*y^2*Z^2/k^2+6*X*Y* ATAN(Y/X)
norm=   3*3.1415926*(X*Y+X*Z+Y*Z)
```

These values are then used for the calculation as a function of the chord length s for each x,y,z permutation. There are three regions: $0<Z$, $Z<t$, and $t<r$.

```
p=    IF(s>Z,SQRT(s^2-Z^2),0)
Q=    IF(s^2>X 2+Z^2,SQRT(s^2-X 2-Z^2),0)
N1E= IF(AND(s>=0,s<Z),b1*s+4*Z,0)
N2=   IF(AND(s>=Z,s<t),b2*s+b4/s 3-X*(p/s)
   *(8+4*Z^2/s^2),0)
N3E= IF(AND(s>=t,s<=r),b3*s+b5/s 3+Y*(q/s)*
   (4+2*t^2/s^2)-(v/s^3)*ACOS(X/p),0)
F=    (N1E+N2+N3E)/NORM
```

The values for the complete chord length distribution are now a sum of the six F terms for each chord length. Figure B-1 shows a representative result.

Bendel then integrated his simplified path length distribution to determine the integral path length distribution [Bendel 1984a].

First produce the indefinite integrals over regions 1, 2, and 3.

```
I1=    IF(AND(s>=0,s<Z),b1*s 2/2 + 4*Z*s,0)
I2=    IF(AND(s>=Z,s<t),b2*s 2/2-b4/(2*s 2)
  +X*p*(2*z 2/s 2-8) + 6*X*Z * ACOS(Z/s),0)
I3E=   IF(AND(s>=t,s<=r),b3*s 2/2-b5/(2*s 2)+ Y*
  q*(4 - t 2/s 2) -6*X*Y*(p/s) 2 * ACOS (X/p),0)
```

We are now interested in the definite integrals from 0 to s:

```
J1=     IF(AND(s>=0,s<Z),I1(s),0)
J2=     IF(AND(s>=Z,s<t),I2(s) + b6,0)
J3E=    IF(AND(s>=t,s<=r),I3E(s) + b6 -3*X 2
  + 1.5*3.1415926 * X * Z,0)
sumJ = sum over 3 J terms and six
  permutation of W,L,H
```

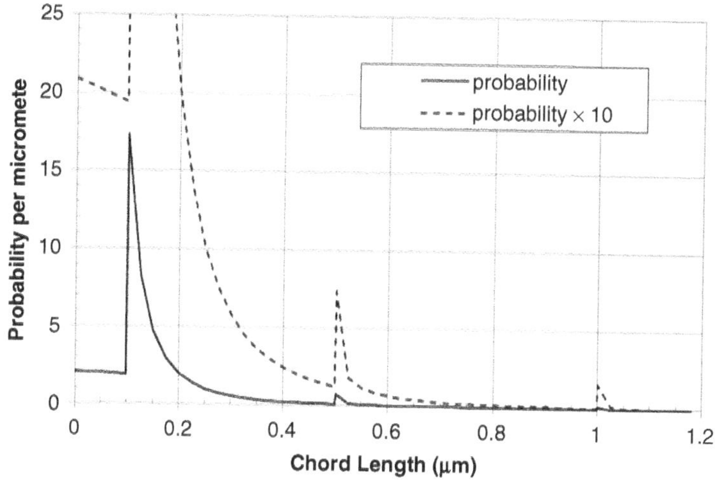

Figure B-1 Chord length distribution in a rectangular volume 1 μm × 0.5 μm × 0.1 μm.

The fraction of chords longer than s is $C(s)$

```
C(s) = 1 - sumJ/norm
```

It is sometimes useful to have simplified analytic expressions for the integral path distribution. Bradford developed an approximation for the case where the minimum dimension H is not more than a third of L or W [Bradford 1979]:

$$C = 1-0.25\ s/H \text{ for } s \le H$$
$$C = 0.75\ (H/s)^{2.2} \text{ for } s > H$$

Bendel developed an approximation for deep devices where the long path lengths are important [Bendel 1984a]. Define a coefficient:

$$A = 1.27/(1.80 + R/H)$$

The two part formula is

$$C = 1-As/H \text{ for } s \le H$$
$$C = (1-A)\ [(R^{3.8}-s^{3.8})/(R^{3.8}-H^{3.8})]^{3}\ (H/s)^{2} \text{ for } s>H$$

Figure B-2 shows the integral chord length distribution $C(s)$ for a device 1.0 μm × 0.5 μm × 0.1 μm with the Bradford and Bendel approximations. Sometimes there is a question about the number of chords nearly in the plane of the device. In this case, if we define

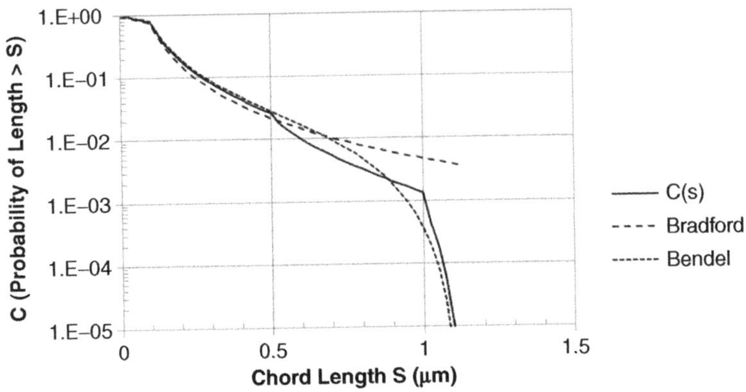

Figure B-2 Integral chord length distribution $C(s)$ for the device in Figure B-1. The curves labeled Bradford and Bendel are approximations for thin and thick devices.

those chords as having a length greater than 0.5 μm, it is 2.6% of the total chords, while if we define those chords as having length greater than 0.6 μm, it is 1% of the total chords.

APPENDIX C

Quick Estimates of Upset Rates Using the Figure of Merit

We have presented a number of plots of upset rate versus orbit and versus figure of merit. These can be used for quick estimates of upset rates in various orbits and for various parts. Consider the following figures. These figures were prepared at different times for different purposes, using different types of data fits, but give a good overall picture of the situation. When the numbers are critical, you should do your own calculations using Weibull fits to the proton cross-section data.

Figure 7.22 Geosynchronous upset rates as a function of the figure of merit.

Figure 7.23 Coefficient for generalized figure of merit equation in various orbits.

Figure 7.26 The variation of the rate coefficient with shielding at geosynchronous orbit.

Figure 7.27 The variation of the rate coefficient with shielding for low Earth orbit.

Figure 9.6 The variation of single event rates with altitude at ground level (after Ziegler) [Ziegler 1996b]. The data points indicate IBM results at sea level (New York), Denver, and Tombstone.

Single Event Effects in Aerospace, First Edition. By Edward Petersen.
© 2011 the Institute of Electrical and Electronics Engineers, Inc.
Published 2011 by John Wiley & Sons, Inc.

Figure 11.14 Scatter plot of space upset rates versus the device figure of merit. There is a much better correlation than for the device threshold. The upset rates are calculated in CREME96 for the solar minimum geosynchronous environment with 100 mils of shielding.

Figure 12.1 Proton upset rate for low polar orbits: 500, 750, 1000, 1500, 3000 km.

Figure 12.2 Proton upset rates for low equatorial orbits: 750, 1000, 1500, 3000 km.

Figure 12.3 Proton upset rates in high equatorial orbits.3000, 6000, 8000, 10,000 km

Figure 13.1 Upset rates for 93L422 due to cosmic rays at various inclinations and altitudes near earth. The figure of merit for this device is 1.4 e−5.

Figure 13.2 Upset rates for 93L422 due to protons at various altitudes and inclinations near earth. Figure of merit is 1.4e−5

Figure 13.3 Upset rate for 93L422 as a function of orbit in the proton radiation belt for the combined environments. Figure of merit is 1.4e−5. You can scale for other devices. They will have much lower figurers of merit and many more bits.

Figure 13-4 FOM calculation of both heavy ion and proton upset rates for a high altitude low Earth orbit. This is an 1111 km, 63° orbit.

Figure 13.5 FOM calculation of proton and heavy ion upset rates for a low Earth orbit. The proton data has been displaced upward by two decades. This is a 500 km, 51.6° orbit.

Figure 13.6 FOM calculation of both heavy ion and proton upset rates for an elliptical orbit. This is an 18.2°, 33,500 km × 348 km orbit.

Figure 13.8 The proton rate coefficient for FOM calculations as a function of altitude for several inclination circular orbits. These curves show the geographical dependence of proton upset rates.

Figure 13.9 The proton rate coefficient as a function of inclination for circular orbits at several altitudes. The proton upset rate in any circular orbit can be determined from these plots.

Figure 13.10 The heavy ion rate coefficient as a function of altitude for circular orbits at several inclinations. These curves show the geographical dependence of heavy ion upset rates.

Figure 13.11 Heavy ion rate coefficients as a function of inclination for circular orbits at several altitudes. The heavy ion upset rate in any circular orbit can be determined from these plots.

Figure 13.13 The altitude variation of single event effects at 45° from ground to outer space. For satellites it is 45° inclination and approaching ground it is 45° latitude. There is slight variation from 30° to 60°, and a larger but still small variation from 0° to 90° so this curve can be used for all common circular orbits. At high altitudes, there is no variation with inclination, so this chart can also be used for geosynchronous orbits.

Figure 14.5 The upset rate as a function of the device figure of merit for the worst day spectrum for a 63°, 1111 km spectrum. The rate coefficient is 1193 for this case, compared with the average rate coefficient of 500 for this orbit as shown in Figure 13-4.

APPENDIX D

Part Characteristics

Table D-1 gives the characteristics of a variety of parts used as examples.

Single Event Effects in Aerospace, First Edition. By Edward Petersen.
© 2011 the Institute of Electrical and Electronics Engineers, Inc.
Published 2011 by John Wiley & Sons, Inc.

Table D-1
Part Characteristics [a]

Device	Type	L_0	w (MeV/mg/cm^2)	s	σ_{HL} (μm^2)	$x = y$ (μm)	z (μm)	A (MeV)	σ_{PL} (cm^2)	FOM
93L422	Bipolar	0.6	4.4	0.7	2600	51	2	6.5	1.418×10^{-10}	1.441×10^{-5}
93422	Bipolar	0.58	5.5	0.8	3700	60.8	2	NA[b]	NA	1.222×10^{-5}
93L422AM	Bipolar	0.547	5.49	0.66	1500	38.7298	2	6.5	3.136×10^{-10}	7.88×10^{-6}
82S212	Bipolar	1	6	0.8	870	29.5	1	18		1.695×10^{-6}
93L422fsc	Bipolar	0.98	6.45	1.2	1400	37.4166	2	NA	NA	1.313×10^{-6}
2901B	Bipolar uP	4.2	10	1.5	3750	61.2372		48	1.52×10^{-11}	5.117×10^{-7}
92L44	Bipolar	0.8	25	2.5	10300	100	5	20.3		4.026×10^{-7}
54ALS374	Bipolar	0.3	19.5	1.9	4375	66.1438	1	NA	NA	4.024×10^{-7}
21L47	Bipolar	0.8	27	2.7	10600	100	5	20.48		3.335×10^{-7}
2164	MOS	0.487	4.95	1.42	170	13	2	3.5	7.2×10^{-12}	2.62×10^{-7}
54as374	Bipolar	3.04	16.5	2.1	3750	61.2372	1	NA	NA	2.536×10^{-7}
54F374	Bipolar	2	12	2	875	29.5804	1	NA	NA	1.228×10^{-7}
HM6116	16K SRAM	4.2	7.9	2.5	400	20		40	2.116×10^{-12}	4.937×10^{-8}
D424100v	4M DRAM	0.8	10	1.1	35.8	5.96	1	11	3.983×10^{-13}	2.21×10^{-8}
62256R	SRAM	1.6	20	1.65	240	15.4919		9	5.292×10^{-13}	1.981×10^{-8}
HYB51410	4M DRAM	0.86	14	1.1	50	7.07	1	12	3.357×10^{-13}	1.731×10^{-8}
6516	Bulk CMOS	4	12.4	2.7	220	14.8	4	NA	NA	1.575×10^{-8}

(continued overleaf)

Table D-1
Part Characteristics [a]

Device	Type	L_0	w	s	σ_{HL} (μm^2)	$x = y$ (μm)	z (μm)	A (MeV)	σ_{PL} (cm^2)	FOM
		(MeV/mg/cm^2)								
MB814100	4M DRAM	1.15	15	1.35	76.3	8.73	1	8	1.542×10^{-13}	1.507×10^{-8}
xcdm62256	RMOS	1.49	16	2	153	12.3693	1	NA	NA	1.507×10^{-8}
HM6516	16K SRAM	5	14	1.9	180	13.4164		42	4.545×10^{-13}	1.195×10^{-8}
OW-62256	256K SRAM	2.4	16.5	2.25	160	12.6491		15	3.443×10^{-13}	1.132×10^{-8}
smj44100	4M DRAM	1.39	15	1.21	47.7	6.9	1	14	1.633×10^{-13}	1.046×10^{-8}
TC514100	4M DRAM	0.86	18	1.15	50	7.08	1	22	2.455×10^{-13}	1.033×10^{-8}
IDT71256	RMOS	1.99	18	1.2	68.7	8.28855	1	NA	NA	9.808×10^{-9}
HM-65656	256K SRAM	1.5	12	1.75	40	6.32456		15	1.086×10^{-13}	7.32×10^{-9}
1280a-s	CMOS FPGA	2.8	100	1.2	1000	31.6228	1	NA	NA	6.839×10^{-9}
IDT71681b	RMOS	2.35	10	2	40	6.32456	1	NA	NA	6.718×10^{-9}
R4-25	CMOS	23.7	70	1.1	1200	34.6	1	NA	NA	5.604×10^{-9}
mt4c1004	4M DRAM	1.54	14.5	1.45	31	5.57	1	11	9.534×10^{-14}	5.249×10^{-9}
IDT6167	RMOS	2.9	60	1.2	275	16.5831	1	NA	NA	4.71×10^{-9}
MT4C4001	4M DRAM	1.49	20	1.2	30	5.47723		9	6.454×10^{-14}	4.077×10^{-9}
km41c4000	4M DRAM	1.52	18	1.45	31	5.57	1	10.5	8.435×10^{-14}	3.704×10^{-9}
IDT6116v	RMOS	4.9	70	1.25	336	18.3303	1	NA	NA	3.551×10^{-9}
IDT7164	RMOS	2.98	90	1	259	16.0935	1	NA	NA	3.101×10^{-9}
6283H	256K SRAM	3.4	20	1.5	40	6.32456		10	1.028×10^{-13}	2.722×10^{-9}
ID771681	RMOS	2.35	14.7	3.2	40	6.32456	1	NA	NA	2.638×10^{-9}

6508RH	EPI	38.8[c]	57	0.75	420	20.5	2	NA	NA	1.704×10^{-9}
Samsung	16M 3.3 V DRAM	0.60	16.39	1.85	9.87	3.14	0.1	1.17	3.5×10^{-14}	1.23×10^{-9}
HM6508	EPI	48[c]	50	1.15	490	22.1	2	NA	NA	1.162×10^{-9}
1280a-c	CMOS FPGA	28	125	1.3	500	22.3607	1	NA	NA	8.661×10^{-10}
6504RH	EPI	30.75[c]	40	1.4	170	13	2	NA	NA	7.634×10^{-10}
Hitachi	16M 3.3 V DRAM	0.5[c]	7.9	4.11	2.27	1.5	0.64	1.4	1.6×10^{-14}	5.66×10^{-10}
R50-25	Hard CMOS	72[c]	166	1.6	1200	34.6	1	NA	NA	5.46×10^{-10}
Micron	16M 3.3 V DRAM	0[c]	8.98	5.37	1.92	1.38	0.81	2.21	8.0×10^{-15}	3.79×10^{-10}
4042	SOS	16.2[c]	25	1.4	12	3.5	0.5	NA	NA	1.712×10^{-10}
R160-25	Hard CMOS	136.8[c]	350	3	1200	34.6	1	NA	NA	8.864×10^{-11}
IBM E	16M 3.3 V DRAM	−0.4[c]	7.89	5.39	0.26	0.51	0.22	2.64	1.73×10^{-15}	7.54×10^{-11}
LUNA-c	IBM 16M DRAM	3.2[c]	14	3	0.894	0.94	1	11	1.19×10^{-15}	5.772×10^{-11}
01G9274	IBM 4M DRAM	1.6[c]	28	3.25	2.3	1.52	1	10	9.877×10^{-16}	5.373×10^{-11}
IBM 16M	IBM DRAM	1.7[c]	20	3	0.77	0.88	1	5	1.24×10^{-15}	3.465×10^{-11}
TCS130	SOS	40[c]	40	1.4	5	2.24	0.4	NA	NA	1.57×10^{-11}

[a] L_0, w, s, and σ_{HL} are the Weibull parameters for the part; x, y, and z are its dimensions. A and σ_{PL} are Bendel parameters.
[b] NA, not available.
[c] Hardened part.

APPENDIX E

Sources of Device Data

E.1 NUCLEAR AND SPACE RADIATION EFFECTS CONFERENCE (NSREC)

There are a variety of sources for device data for the various types of single event effect. The majority of the data is presented at the annual IEEE Nuclear and Space Radiation Effects Conference (NSREC). The data that has some accompanying technical discussion is presented in technical discussions, such as "Single-Event Effects: Devices and Integrated Circuits." These papers are then normally published in the December issue of the *IEEE Transactions on Nuclear Science*. The transactions in turn are also archived on the IEEE/Xplore® website. All previous issues are available.

Other radiation test data is presented at the IEEE NSREC, "Radiation Effects Data Workshop," which is a part of NSREC. This has a separate Workshop Record Publication. These papers may be a compendium of results from active test groups [George 2008, Miyahira 2008, O'Bryan 2008], or they may be test results (usually positive) from commercial manufacturers [Lawrence 2008]. Besides a cumulative index, there is also a guide that helps the search for data on a particular part type or radiation effect [Hiemstra 2006, Hiemstra 2007, Hiemstra 2008]. Tables are available on the NSREC website,

Single Event Effects in Aerospace, First Edition. By Edward Petersen.
© 2011 the Institute of Electrical and Electronics Engineers, Inc.
Published 2011 by John Wiley & Sons, Inc.

http://www.nsrec.com/redw/. Individual copies of the papers are available online at IEEE Xplore®.

E.2 DATA SETS

Several groups keep websites with compilations of data. One of the largest is that of the Radiation Effects and Analysis Group (REAG) at NASA Goddard: http://radhome.gsfc.nasa.gov/. A parallel site is that of the NASA Electronic Parts and Packaging (NEPP) Program: http://nepp.nasa.gov/. These groups give a very large number of references to other active groups and interesting locations. They will also include up-to-date links to the major SEE test facilities. The NASA JPL group has data at http://radcentral.jpl.nasa.gov/. The principal data site in Europe is the ESA site at https://escies.org/ReadArticle?docId = 227. A site that discusses ground level environments and test facilities is www.seutest.com. Note that links on the Internet are subject to change, or even removal, so are not as stable as the published literature. We believe that these groups are relatively stable.

References

ABDUL-KADER 1987. W. G. ABDUL-KADER, P. J. MCNULTY, S. El-TELEATY, J. E. LYNCH, and A. N. KHONDKER, Estimating the Dimensions of the SEU-Sensitive Volume, *IEEE Trans. Nucl. Sci.*, **NS-34**, 1300 (1987).

ADAMS 1981. J. H. ADAMS, Jr., R. SILBERBERBERG, and C. H. TSAO, Cosmic Ray Effects on Microelectronics, Part I: The Near-Earth Particle Environment, NRL Memorandum Report 4506 (1981).

ADAMS 1983. J. H. ADAMS, Jr., The Variability of Single Event Upset Rates in the Natural Environment, *IEEE Trans. Nucl. Sci.*, **NS-30**, 4475 (1983).

ADAMS 1984. J. H. ADAMS, Jr. and A. GELMAN, The Effects of Solar Flares on Single Event Upset Rates, *IEEE Trans. Nucl. Sci.*, **NS-31**, 1212 (1984).

ADAMS 1986. J. H. ADAMS, Jr., Cosmic Ray Effects on Microelectronics, Part IV, NRL Memorandum Report 5901 (1986).

ADAMS 1989. L. ADAMS, R. HARBOE-SORENSEN, E. DALY, and J. WARD, Proton Induced Upsets in Low Altitude Polar Orbit, *IEEE Trans. Nucl. Sci.*, **NS-36**, 2339 (1989).

ADAMS 1991. L. ADAMS, E. J. DALY, R. HARBOE-SORENSEN, A. G. HOLMES-SIEDLE, A. K, WARD, and R. A. BULL, Measurements of SEU and Total Dose in Geostationary Orbit Under Normal and Solar Flare Conditions, *IEEE Trans. Nucl. Sci.*, **NS-38**. 1686 (1991).

ADOLPHSEN 1995. J. ADOLPHSEN, J. L. BARTH, E. G. STASSINOPOULOS, T. GRUNER, M. WENNERSTEN, K. A. LABEL, and C. M. SEIDLECK, Single Event Upset Rates on 1 Mbit and 256 Kbit Memories: CRUX Experiment on APEX, *IEEE Trans. Nucl. Sci.*, **NS-42**, 1964 (1995).

ADOLPHSEN 1996a. J. ADOLPHSEN, J. L. BARTH, E. G. STASSINOPOULOS, T. GRUNER, M. WENNERSTEN, K. A. LABEL, and C. M. SEIDLECK, SEP Data from the APEX Cosmic Ray Upset Experiment: Predicting the Performance of Commercial Devices in Space, *IEEE Trans. Nucl. Sci.*, **NS-43**, 1069 (June 1996).

ADOLPHSEN 1996b. J. W. ADOLPHSEN, J. L. BARTH, and G B. GEE, First Observations of Proton Induced Power MOSFET Burnout in Space, *IEEE Trans. Nucl. Sci.*, **NS-43**, 2924 (1996).

AEROSPACE 2003. http://www.aero.org/publications/crosslink/summer2003/backpage.html.

Single Event Effects in Aerospace, First Edition. By Edward Petersen.
© 2011 the Institute of Electrical and Electronics Engineers, Inc.
Published 2011 by John Wiley & Sons, Inc.

AKKERMAN 2002a. A. AKKERMAN, J. BARAK, and Y. LIFSHITZ, Nuclear Models for Proton Induced Upsets: A Critical Comparison, *IEEE Trans. Nucl. Sci.*, **NS-49**, 1539 (June 2002).

AKKERMAN 2002b. A. AKKERMAN and J. BARAK, Ion Track Structure and Its Effects in Small Size Volumes in Silicon, *IEEE Trans. Nucl. Sci.*, **NS-49**, 3022–3031 (2002).

ALEXANDER 1996. Design Issues for Radiation Tolerant Microcircuits for Space, IEEE NSREC Short Course Notes, July 1996.

AWST 1992. Superbird Team Revising Procedures to Prevent Future Spacecraft Loss, *Aviation Week & Space Technology*, December 7, 1992.

BAGGIO 2004. J. BAGGIO, V. FERLET-CAVROIS, H. DUARTE, and O. FLAMENT, Analysis of Proton/Neutron SEU Sensitivity of commercial SRAMs—Application to the Terrestrial Environment Test Method, *IEEE Trans. Nucl. Sci.*, **NS-51**, 3420–3425 (2004).

BARAK 1996. J. BARAK, J. V LEVINSON, M. VICTORIA, and W. HAJDAS, Direct Processes in the Energy Deposition of Protons in Silicon, *IEEE Trans. Nucl. Sci.*, **NS-43**, 2820 (1996).

BARAK 1999a. J. BARAK, J. LEVINSON, A. AKKERMAN, E. ADLER, A. ZENTER, D. DAVID, Y. LIFSHITZ, M. HASS, B. FISCHER, M. SCHLOGL, M. VICTORIA, and W. HAJDAS, Scaling of SEU Mapping and Cross Section, and Proton Induced SEU at Reduced Supply Voltage, *IEEE Trans. Nucl. Sci.*, **NS-46**, 1342–1353 (1999).

BARAK 1999b. J. BARAK, R. A. REED, and K. A. LABEL, On the Figure of Merit Model and SEU Rate Calculations, *IEEE Trans. Nucl. Sci.* **NS-46**, 1504 (1999).

BARAK 2001. J. BARAK, Analytical Microdosimetry Model for Proton-Induced SEU in Modern Devices, *IEEE Trans. Nucl. Sci.*, **NS-048**, 1937 (2001).

BARAK 2004. J. BARAK, A. HARAN, E. ADLER, Azoulay, J. LEVINSON, A. ZENTNER, D. DAVID, B. E. FISCHER, M. HEISS, and D. BETEL, Use of Light-Ion-Induced SEU in Devices Under Reduced Bias to Evaluate Their SEU Cross Section, *IEEE Trans. Nucl. Sci.*, **NS-51**, 3486 (2004).

BARAK 2005. J. BARAK and A. AKKERMAN, Straggling and Extreme Cases in the Energy Deposition by Ions in Submicron Silicon Volumes, *IEEE Trans. Nucl. Sci.*, **NS-52**, 2175 (2005).

BARAK 2006. J. BARAK, Simple Calculations of Proton SEU Cross Sections from Heavy Ion Cross Sections, *IEEE Trans. Nucl. Sci.*, **NS-53**, 3336 (2006).

BARTH 1997. J. BARTH, Modeling Space Radiation Environment, 1997 NSREC Short Course, p. I-1 (1997).

BAUMAN 2005. R. BAUMANN, Single-Event Effects in Advanced CMOS Technology, 2005 IEEE NSREC Short Course Notebook (July 2005).

BEDINGFIELD 1996. K. L. BEDINGFIELD, R. D. LEACH, and M. B. ALEXANDER, Spacecraft System Failures and Anomalies Attributed to the Natural Space Environment, NASARP -1390 (1996).

BEERS 1958. Y. BEERS, *Introduction to the Theory of Error*, 2nd ed. Reading, MA: Addison-Wesley, 1958.

BENDEL 1983. W. L. BENDEL and E. L. PETERSEN, Proton Upsets in Orbit, *IEEE Trans. Nucl. Sci.*, **NS-30**, 4481 (1983).

BENDEL 1984a. W. L. BENDEL, Length Distribution of Chords Through a Rectangular Volume, NRL Memorandum Report 5369 (July 1984). Available at: http://torpedo.nrl.navy.mil/tu/ps/doc.html?dsn=352511&hi=1&p=1

BENDEL 1984b. W. L. BENDEL and E. L. PETERSEN, Predicting Single Event Upsets in the Earths Proton Belts, *IEEE Trans. Nucl. Sci.*, **NS-31**, 1201 (1984).

BENEDETTO 2005. J. M. BENEDETTO, P H. EATON, D. G. MAVIS, M. GADLAGE, and T. TURFLINGER, Variation of Digital SET Pulse Widths and the Implications for Single Event Hardening of Advanced CMOS Processes, *IEEE Trans. Nucl. Sci.*, **NS-52**, 2114–2119 (2005).

BENEDETTO 2008. J. BENEDETTO, Soft Error Case Study: Single Event Functional Interrupsts (SEFIs) in COTS SDRAMS, 2008 IEEE NSREC Short Course Notebook (July 2008).

BERG 2007. M. BERG, C. POIVEY, D. PETRICK, K. LABEL, M. FRIENDLICH, S. STANSBERRY, and H. KIM, Risk Reduction for Use of Complex Devices in Space Projects, *IEEE Trans. Nucl. Sci.*, **NS-54**, 2337–2140 (2007).

BINDER 1975. D. BINDER, E. C. SMITH, and A. B. HOLMAN, Satellite Anomalies from Galactic Cosmic Rays, *IEEE Trans. Nucl. Sci.*, **NS-22**, 2675 (1975).

BINDER 1988. D. BINDER, Analytic SEU Rate Calculations Compared to Space Data, *IEEE Trans. Nucl. Sci.*, **NS-35**, 1570 (1988).

BION 1989. T. BION and J. BOURRIEAU, A Model for Proton-Induced SEU, *IEEE Trans. Nucl. Sci.*, **NS-36**, 2281 (1989).

BISGROVE 1986. J. M. BISGROVE, J. E. LYNCH, P. J. MCNULTY, W. G. ABDEL-KADER, V. KLETNICKS, and W. A. KOLASINSKI, Comparison of Soft Errors Induced by Heavy Ions and Protons, *IEEE Trans. Nucl. Sci.*, **NS-33**, 1571 (1986).

BLACK 2005. J. D. BLACK, A. L. STERNBERG, M. L. ALLES, A. F. WITULSKI, B. L. BHUVA, L. W. MASSENGILL, J. M. BENEDETTO, M. P. BLAZE, J. W. WERT, and M. G. HUBERT, HBD Layout Isolation Techniques for Multiple Node Charge Collection Mitigation, *IEEE Trans. Nucl. Sci.*, **NS-52**, 2536 (2005).

BLACK 2006. J. BLACK and T. HOLMAN, Circuit Modeling of Single Event Effects, 2006 IEEE NSREC Short Course (2006).

BLAKE 1986. J. B. BLAKE and R. NANDEL, On-Orbit Observations of Single Event Upset in Harris HM-6508 1K RAMS, *IEEE Trans. Nucl. Sci.*, **NS-33**, 1616 (1986).

BLAKE 1992. J. B. BLAKE, M. S. GUSSENHOVEN, E. G. MULLEN, and R. W. FILLIUS, Identification of an Unexpected Space Radiation Hazard, *IEEE Trans. Nucl. Sci.*, **NS-39**, 1761 (1992).

BLOEMBERGEN 1987. N. BLOEMBERGEN et al., Report to the American Physical Society of the Study Group on Science and Technology of Directed Energy Weapons, *Rev. Mod. Phys.*, **59**, (July 1987).

BOURDARIE 2008. S. BOURDARIE and M. XAPSOS, The New Near-Earth Radiation Environment, *IEEE Trans. Nucl. Sci.*, **NS-55**, 1810–1853 (August 2008).

BRADFORD 1979. J. N. BRADFORD, A Distribution Function for Ion Track Lengths in Rectangular Volumes, *J. Appl. Phys.*, **50**(16), 3799 (June 1979).

BRADFORD 1980a. J. N. BRADFORD, Cosmic Ray Effects in Very Large Scale Integration, in Space Systems and Their Interactions with the Earth's Space Environment, H. B. GARRETT and C. P. PIKE, Eds., p. 349. Cambridge: MIT Press, 1980.

BRADFORD 1980b. J. N. BRADFORD, Geometrical Analysis of Soft Errors and Oxide Damage Produced by Heavy Cosmic Rays and Alpha Particles, *IEEE Trans. Nucl. Sci.*, **NS-27**, 942 (Feb. 1980).

BRADFORD 1982. J. N. BRADFORD Micro-volume Energy Deposition from High Energy Proton–Silicon Reactions, *IEEE Trans. Nucl. Sci.*, **NS-29**, 2085 (1982).

BROTHERS 1997. C. BROTHERS, R. PUGH, P. DUGGAN, J. CHAVEZ, D. SCHEPIS, D. YEE, and S. WU, Total-Dose and SEU Characterization of 0.25 Micron CMOS/SOI Integrated Circuit Memory Technologies, *IEEE Trans. Nucl. Sci.*, **NS-44**, 2134 (1997).

BUCHNER 1990. S. BUCHNER, K. KANG, W. J. STAPOR, A. B. CAMPBELL, A. R. KNUDSON, P. MCDONALD, and S. RIVIT, Pulsed Laser-Induced SEU in Integrated Circuits: A Practical Method for Hardness Assurance Testing, *IEEE Trans. Nucl. Sci.*, **NS-37**, 1825 (1990).

BUCHNER 1992. S. BUCHNER, K. KANG W. J. STAPOR, and S. RIVIT, Spatial and Temporal Dependence of SEU in a 64K SRAM, *IEEE Trans. Nucl. Sci.*, **NS-39**, 1630 (1992).

BUCHNER 1994. S. BUCHNER, J. B. LANGWORTHY, W. J. STAPOR, A. B. CAMPBELL, and S. RIVIT, Implications of the Spatial Dependence of the Single-Event-Upset Threshold in SRAMs Measured with a Pulsed Laser, *IEEE Trans. Nucl. Sci.*, **NS-41**, 2195 (1994).

BUCHNER 1996. S. BUCHNER, A. B. CAMPBELL, D. MCMORROW, J. MELINGER, M. MASTI, and Y. J. CHEN, Modification of Single Event Upset Cross Section of an SRAM at High Frequencies, in *Proceeding of RADECS 95*, pp. 326–332 (1996).

BUCHNER 2000. S. BUCHNER, A. B. CAMPBELL, T. MEEHAN, K. A. CLARK, D. MCMORROW, C. DYER, C. SANDERSON, C. COMBER, and S. KUBOYAMA, Investigation of Single-Ion Multiple Bit Upsets in Memories on Board a Space Experiment, *IEEE Trans. Nucl. Sci.*, **NS-47**, 705–711 (June 2000).

BUCHNER 2001. S. BUCHNER and M. P BAZE, Single Event Transients in Fast Electronic Systems, IEEE NSREC Short Course Notes (July 2001).

BUCHNER 2004. S. BUCHNER, A. CAMPBELL, R. REED, B. FONDNESS, and S. KUBOYAMA, Angular Dependence of Multiple-Bit Upset Induced by Protons in a 16 Mbit DRAM, *IEEE Trans. Nucl. Sci.*, **NS-51**, 3270 (2004).

BUCHNER 2005. S. BUCHNER and D. MCMORROW, Single-Event Transients in Linear Integrated Circuits, 2005 IEEE NSREC Short Course Notebook (July 2005).

BURKE 1981. E, A. BURKE, Technical Memorandum RADC-TM-8 1-ES-03 (1981).

CALIN 1996. T. CALIN, M. NICOLAIDIS, and R. VELAZCO, Upset Hardened Memory Design for Submicron CMOS Technology, *IEEE Trans. Nucl. Sci.*, **NS-43**, 2874 (1996).

CALVEL 1996. P. CALVEL, C. BARILLOT, P. LAMOTHE, R. ECOFFET, S. DUZELLIER, and D. FALGUERE, An Empirical Model for Predicting Proton Induced Upset, *IEEE Trans. Nucl. Sci.*, **NS-43**, 2827 (1996).

CAMPBELL 1985. A. B. CAMPBELL, W. J. STAPOR, R. KOGA, and W. A. KOLASINSKI, Correlated Proton and Heavy Ion Upset Measurements on IDT Static RAMs, *IEEE Trans. Nucl. Sci.*, **NS-32**, 4150 (1985).

CAMPBELL 1991. A. B. CAMPBELL, SEU Flight Data from the CRRES MEP, *IEEE Trans. Nucl. Sci.*, **NS-38**, 1647 (1991).

CAMPBELL 1992. A. CAMPBELL, P. MCDONALD, and K. RAY, Single Event Upset Rates in Space, *IEEE Trans. Nucl. Sci.*, **NS-39**, 1828 (1992).

CAMPBELL 1994. A. CAMPBELL, P. MCDONALD, R. GONYEA, and M. REEVES, Results from the CRRES MEP Experiment, *IEEE Trans. Nucl. Sci.*, **NS-41**, 432 (June 1994).

CAMPBELL 2002. A. CAMPBELL, S. BUCHNER, E. B. BLAKE, J. MAZUR, and C. DYER, SEU Measurements and Predictions for MPTB for a Large Energetic Particle Event, *IEEE Trans. Nucl. Sci.*, **NS-40**, 1340 (2002).

CHADWICK 1999. M. B. CHADWICK and E. NORMAND, Use of New ENDF/B-VI Proton and Neutron Cross Sections for Single Event Upset Calculations, *IEEE Trans. Nucl. Sci.*, **NS-46**, 1386–1394 (1999).

CHENETTE 1994. D. L. CHENETTE, J. CHEN, E. CLAYTON, T. G. GUZIK, J. P. WEFEL, M. GARCIA-MUNOZ, C. LOPATE, K. R. PYLE, K. P. HARDY, E. G. MULLEN, and D. A. HARDY, The CRRES SPACERAD Heavy Ion Model of the Environment (CHIME) for Cosmic Ray and Solar Particle Effects on Electronic and Biological Systems in Space, *IEEE Trans. Nucl. Sci.*, **NS-41**, 2332 (1994).

CHERN 1986. J. S. CHERN, P. YANG, P. PATTNICK, and J. A. SEITCHIK, Alpha-Particle Induced Charge Transport Between Closely Spaced Memory Cells, *IEEE Trans. Electron Devices*, **ED-33**, 822 (1986).

CHLOUBER 1990. D. CHLOUBER, P. O'NEILL, and J. POLLOCK, General Upper Bound on Single Event Upset Rates, *IEEE Trans. Nucl. Sci.*, **NS-37**, 1065 (April 1990).

CONNELL 1995a. L. W. CONNELL, P. J. MCDANIEL, A. K. PRINJA, and F. W. SEXTON, Modeling the Heavy Ion Upset Cross Section, *IEEE Trans. Nucl. Sci.*, **NS-42**, 73–82 (April 1995).

CONNELL 1995b. L. W. CONNELL, F. W. SEXTON, and A. K. PRINJA, Further Development of Heavy Ion Cross Section of Single Event Upset: Model (HICUP), *IEEE Trans. Nucl. Sci.*, **NS-42**, 2026–2034 (1995).

CONNELL 1996. L. W. CONNELL, F. W. SEXTON, P. J. MCDANIEL, and A. K. PRINCE, Modeling of Heavy Ion Cross Section for Single Event Upset with Track Structure Effects: The HIC-UP-TS Model, *IEEE Trans. Nucl. Sci.*, **NS-43**, 2814 (1996).

CORREAS 2007. V. CORREAS, F. SAIGNE, B. SAGNES, J. BOCH, G. GASIOT, D. GIOT, and P. ROCHE, Innovative Simulations of Heavy Ion Cross Sections in 130 nm CMOS SRAM, *IEEE Trans. Nucl. Sci.*, **NS-54**, 2413 (2007).

CORREAS 2008. V. CORREAS, F. SAIGNE, B. SAGNES, J. BOCH, G. GASSIOT, D. GIOT, and P. ROCHE, Simulation Tool for the Prediction of Heavy Ion Cross Sections of Innovative 130nm SRAMs, *IEEE Trans. Nucl. Sci.*, **NS-55**, 2036–2041 (August 2008).

CRAIN 2001. S. H. CRAIN, J. E. MAZUR, R. B. KATZ, R. KOGA, M. D. LOOPER, and R. LORENTZEN, Analog and Digital Single-Event Effects Experiments in Space, *IEEE Trans. Nucl. Sci.*, **NS-48**, 1841–1848 (2001).

CRESSLER 2003. J. CRESSLER, Radiation Effects in SiGe HBT BiCMOS Technology, IEEE NSREC Short Course Notes (July 2003).

CRISWELL 1984. T. L CRISWELL, P. R. MEASEL, and K. L. WAHLIN, Single Event Upset Testing with Relativistic Heavy Ions, *IEEE Trans. Nucl. Sci.*, **NS-31**, 1559–1563 (1984).

CRISWELL 1987. T. L. CRISWELL, D. L. OBERG, J. L. WERT, P. R. MEASEL, and W. E. WILSON, Measurement of SEU Thresholds and Cross Sections at Fixed Incidence Angles, *IEEE Trans. Nucl. Sci.*, **NS-34**, 1316–1321 (1987).

CROLEY 1995. D. R. CROLEY, H. B. GARRETT, G. B. MURPHY, and T. L. GARRARD, Solar Particle Induced Upsets in the TDRS- 1 Attitude Control System RAM During the October 1989 Solar Particle Events, *IEEE Trans. Nucl. Sci.*, **NS-42**, 1489 (1995).

CUTCHIN 1993. J. H. CUTCHIN, P. W. MARSHALL, T. R. WEATHERFORD, J. LANGWORTHY, E. L. PETERSEN, A. B. CAMPBELL, S HANKA, and A. PECZALSKI, Heavy Ion and Proton Analysis of a GaAs C-Higfet SRAM, *IEEE Trans. Nucl. Sci.*, **NS-40**, 1660 (1993).

DALY 1994. E. J. DALY, F. VAN LEEUWEN, H. D. R. EVANS, and M. A. C. PERRYMAN, Radiation-Belt and Transient Solar-Magnetospheric Effects on Hipparcos Radiation Background, *IEEE Trans. Nucl. Sci.*, **NS-41**, 2376 (1994).

DALY 1996. E. J. DALY, J. LEMAIRE, D. HEYNDERICKX, and D. J. ROGERS, Problems with Models of the Radiation Belts, *IEEE Trans. Nucl. Sci.*, **NS-43**, 403 (April 1996).

DARRACQ 2002. F. DARRACQ, H. LAPUYADE, N. BUARD, F. MOUNSI, B. FOUCHER, P. FOUILLAT, M. C. CALVET, and R. DUFAYEL, Backside SEU Laser Testing for Commercial-Off-The-Shelf SRAM, *IEEE Trans. Nucl. Sci*, **NS-49**(6), 2977–2983 (2002).

DAVIS 1985. G. E. DAVIS, L. R. HITE, T. G. W. BLAKE, C. E. CHEN, H. W. LAM, and R. DEMOYER, Jr., Transient Radiation Effects in SOI Memories, *IEEE Trans. Nucl. Sci.*, **NS-32**, 4432 (1985).

DETCHEVERRY 1997. C. DETCHEVERRY, C. DACHS, E. LORFEVRE, C. SUDRE, G. BRUGUIER, J. M. PALAU, J. GASIOT, and R. ECOFFET, SEU Critical Charge and Sensitiver Area in a Submicron CMOS Technology, *IEEE Trans. Nucl. Sci.*, **NS-44**, 2266 (1997).

DIEHL 1982. S. E. DIEHL, A. OCHOA, Jr., P. V. DRESSENDORFER, R. KOGA, and W. A. KOLASINSKI, Error Analysis and Prevention of Cosmic Ion-Induced Soft Errors in Static CMOS RAMs, *IEEE Trans. Nucl. Sci.*, **NS-29**, 2032 (1982).

DIEHL-NAGLE 1984. S. E. DIEHL-NAGLE, J. E. VINSON, and E. L. PETERSEN, Single Event Upset Rate Predictions for Complex Logic Systems, *IEEE Trans. Nucl. Sci.*, **NS-31**, 1132–1138 (1984).

DODD 1994. P. E. DODD, F. W. SEXTON, and P. S. WINOKUR, Three Dimensional Simulation of Charge Collection and Multiple-bit Upset in Si Devices, *IEEE Trans. Nucl. Sci.*, **NS-41**, 2005 (1994).

DODD 1995. P. E. DODD and F. W. SEXTON, Critical Charge Concepts for High-Density SRAMS, *IEEE Trans. Nucl. Sci.*, **NS-42**, 1764 (1995).

DODD 1996a. P. E. DODD, Device Simulation of Charge Collection and Single-Event Upset, *IEEE Trans. Nucl. Sci.*, **NS-43**, 561 (February 1996).

DODD 1996b. P. E. DODD, F. W. SEXTON, G. L. HASH, M. R. SHANEYFELT, B. L. DRAPER, A. J. FARINO, and R. S. FLORES, Impact of Technology Trends on SEU in CMOS SRAMs, *IEEE Trans. Nucl. Sci.*, **NS-43**, 2797 (1996).

DODD 1997. P. E. DODD, M. R. SHANEYFELT, and F. W. SEXTON, Charge Collection and SEU from Angled Ion Strikes, *IEEE Trans. Nucl. Sci.*, **NS-44**, 2256 (1997).

DODD 1998. Impact of Ion Energy on Single-Event Upset, *IEEE Trans. Nucl. Sci.*, **NS-45**, 2483 (1998).

DODD 1999. Basic Mechanisms in Single-Event Effects, IEEE NSREC Short Course Notebook (July 1999).

DODD 2001. P. E. DODD, M. R. SHANEYFELT, K. M. HORN, D. S. WALSH, G. L. HASH, T. A. HILL, B. L. DRAPER, J. R. SCHWANK, F. W. SEXTON, and P. S. WINOKUR, SEU-Sensitive- Volumes in Bulk and SOI SRAMs from First-Principles Calculations and Experiments, *IEEE Trans. Nucl. Sci.*, **NS-48**, 1893–1903 (December 2001).

DODD 2003. P. E. DODD and L. W. MASSENGILL, Basic Mechanisms and Modeling of Single-Event Upset in Digital Microelectronics, *IEEE Trans. Nucl. Sci.*, **NS-50**, 583–602 (June 2003).

DODD 2007a. P. E. DODD, J. R. SCHWANK, M. R. SHANEYFELT, J. A. FELIX, P. PAILLET, V. FERLET-CAVROIS, J. BAGGIO, R. A. REED, K. M. WARREN, R. A. WELLER, R. D. SCHRIMPF, G. L. HASH, S. M. DALTON, K. HIROSE, and H. SAITO, Impact of Heavy Ion Energy and Nuclear Interactions on Single-Event Upset and Latchup in Integrated Circuits, *IEEE Trans. Nucl. Sci.*, **NS-54**, 2303 (2007).

DODD 2007b. P. E. DODD, J. R. SCHWANK, M. R. SHANEYFELT, V. FERLET-CAVROIS, P. PAILLET, J. BAGGION, G. L. HASH, J. A. FELIX, K HIROSE, and H. SAITO, Heavy Ion Energy Effects in CMOS SRAMs, *IEEE Trans. Nucl. Sci.*, **NS-54**, 889 (August 2007).

DOUCIN 1995. B. DOUCIN, T. CARRIERE, C. POIVEY, P. GARNIER, J. BEAUCOUR, and Y. PATIN, Model of Single Event Upsets by Space Protons in Electronic Devices, in *Proceedings of RADECS 95*, p. 402 (1995).

DRAPER 1981. N. DRAPER and H. SMITH, *Applied Regression Analysis*, 2nd ed. Hoboken, NJ: John Wiley & Sons, 1981.

DRESSENDORFER 1998. P. V. DRESSENDORFER, Basic Mechanisms for the New Millennium, IEEE NSREC Short Course Notes (July 1998).

DREUTE 1994. J. DREUTE, H. HORCHER, W. HEINRICH, R. HARGOVE-SORENSEN, and L. ADAMS, Study of SEUs Generated by High Energy Ions, *IEEE Trans. Nucl. Sci.*, **NS-41**, 601 (June 1994).

DUSSAULT 1995. H DUSSAULT, J. W. HOWARD, Jr., R. C. BLOCK, W. J. STAPOR, A. R. KNUDSON, P. T. MCDONALD, and M. R. PINTO, High-Energy Heavy-Ion-Induced Transport Across Multiple Junctions, *IEEE Trans. Nucl. Sci.*, **NS-42**, 1780 (1995).

DUZELLIER 1995. S. DUZELLIER, D. FALGUERE, L. MOULIERE, R. ECOFFET, and J. BUISSON, SEE Results Using High Energy Ions, *IEEE Trans. Nucl. Sci.*, **NS-42**, 1797 (1995).

DUZELLIER 1997. S. DUZELLIER, R. ECOFFET, D. FALGUERE, T. NUNS, L. GUIBERT, W. HAJDAS, and M. C. CALVET, Low Energy Proton Induced SEE in Memories, *IEEE Trans. Nucl. Sci.*, **NS-44**, 2306 (1997).

DYER 1991. C. S. DYER, A. J. SIMS, J. FARREN, J. STEPHEN, and C. UNDERWOOD, Radiation Environment Measurements and Single Event Upset Observations in Sun-Synchronous Orbit, *IEEE Trans. Nucl. Sci.*, **NS-38**, 1700–1707 (1991).

DYER 1996. C. S. DYER, C. J. WILSON, C. L. PEERLESS, A. J. SIMS, and J. BARTH, Measurements of the Radiation Environment from CREDO-II on STRV and APEX, *IEEE Trans. Nucl. Sci.*, **NS-43**, 2751 (1996).

DYER 1997. C. S. DYER, private communication.

DYER 2004. C. S. DYER, S. N. CLUCAS, C. SANDERSON, A. D. FRYLAND, and R. T. GREEN, An Experimental Study of Single Event Effects Induced in Commercial SRAMs by Neutrons and Protons from Thermal Energies to 500 MeV, *IEEE Trans. Nucl. Sci.*, **NS-51**, 2817–2824 (October 2004).

DYER 2009. C. DYER, A. HANDS, F. LEI, P. TRUSCOTT, K. A. RYDEN, P. MORRIS, I. GETLEY, L. BENNETT, B. BENNETT, and B. LEWIS, Advance in Measuring and Modeling the Atmospheric Radiation Environment, *IEEE Trans. Nucl. Sci.*, **NS-56**, 3415–3422 (2009).

ECOFFET 1992. R. ECOFFET, M. ALBRUNEE, S. DUZELLIER, and D. FALGURERE, Heavy Ion Test Results on Memories, 1992 IEEE Radiation Effects Data Workshop, p. 27 (1992).

ECOFFET 1995. R. ECOFFET, M. PRIEUR, M. F. DEL CASTILLO, S. DUZELLIER, and D. FALGUERE, Influence of Solar Cycle on SPOT -1, -2, -3 Upset Rates, *IEEE Trans. Nucl. Sci.*, **NS-42**, 1983 (1995).

ECOFFET 1997. R. ECOFFET, S. DUZELLIER, D. FALGUERE, L. GUIBERT, and C. INGUIMBERT, Low LET Cross-Section Measurements Using High Energy Carbon Beam, *IEEE Trans. Nucl. Sci.*, **NS-44**, 2230 (1997).

EDMONDS 1989. L. D. EDMONDS, A Distribution Function for Double-Bit Upsets, *IEEE Trans. Nucl. Sci.*, **NS-36**, 1344 (April 1989).

EDMONDS 1991. L. D. EDMONDS, A Simple Estimate of Funneling-Assisted Charge Collection, *IEEE Trans. Nucl. Sci.*, **NS-38**, 828 (February 1991).

EDMONDS 1996. L. D. EDMONDS, SEU Cross Sections Derived for a Diffusion Analysis, *IEEE Trans. Nucl. Sci.*, **NS-43**, 3207 (1996).

EDMONDS 1997. L. D. EDMONDS, Charge Collection from Ion Tracks in Simple EPI Diodes, *IEEE Trans. Nucl. Sci.*, **NS-44**, 1448 (June 1997).

EDMONDS 1998. L. D. EDMONDS, Electric Currents Through Ion Tracks in Silicon Devices, *IEEE Trans. Nucl. Sci.*, **NE-45**, 3153 (1998).

EDMONDS 2000. L. D. EDMONDS, Proton SEU Cross-sections Derived from Heavy Ion Test Data, *IEEE Trans. Nucl. Sci.*, **NS-47**, 1713 (October 2000).

EDMONDS 2001. L. D. EDMONDS, A Time-Dependent Charge Collection-Efficiency for Diffusion, *IEEE Trans. Nucl. Sci.*, **NS-48**, 1609 (October 2001).

EDMONDS 2005. L. D. EDMONDS, Recommendations Regarding the Use of CREME96 for Heavy-Ion SEU Rate Calculations, JPL Unpublished (July 29, 2005).

EDMONDS 2008. L. D. EDMONDS and K. J. EDMONDS, A Method for Estimating SEU Rates from Protons by Direct Ionization, *IEEE Trans. Nucl. Sci.*, **NS-55**, 2666–2678 (October 2008).

EDWARDS 2004. R. EDWARDS, C. DYER, and E. NORMAND, Technical Standard for Atmospheric Radiation Single Event Effects (SEE) on Avionics Electronics, 2004 IEEE Radiation Effects Data Workshop, pp. 1–5 (2004).

EL TELEATY 1983. S. EL TELEATY, G. E. FARRELL, and P. J. MCNULTY, Charge Deposition Spectra in Thin Slabs of Silicon Induced by Energetic Protons, *IEEE Trans. Nucl. Sci.*, **NS-30**, 4394 (1983).

EVANS 1955. R. D. EVANS, The Atomic Nucleus. New York: McGraw-Hill, 1955.

FALGUERE 1994. D. FALGUERE, S. DUZELLIER, and R. ECOFFET, SEE In-flight Measurement on the MIR Orbital Station, *IEEE Trans. Nucl. Sci.*, **NS-41**, 2346 (1994).

FARRELL 1982. G. E. FARRELL and P. J. MCNULTY, Microdosimetric Aspects of Proton-Induced Nuclear Reaction in Thin Layers of Silicon, *IEEE Trans. Nucl. Sci.*, **NS-29**, 2012 (1982).

FARRELL 1984. G. E. FARRELL, P. J. MCNULTY, and W. ABDEL-KADER, Microdosimetric Analysis of Proton Induced Reaction in Silicon and Gallium Arsenide, *IEEE Trans. Nucl. Sci.*, **NS-31**, 1073 (1984).

FERLET-CAVROIS 2006. V. FERLET-CAVROIS, P. PAILLET, M. GAILLARDIN, D. LAMBERT, J. Z. GAGGIO, J. R. SCHWANK, G. VIZKELETHY, M. R. SHANEYFELT, K. HIROSE, E. W. BLACKMORE, O. FAYNOT, C. JAHAN, and L. TOSTI, Statistical Analysis of the Charge Collected in SOI and Bulk Devices under Heavy Ion and Proton Irradiation-Implications for Digital SETs, *IEEE Trans. Nucl. Sci.*, **NS-53**, 3242 (2006).

FLEETWOOD 2004. M. FLEETWOOD and R. D. SCHRIMPF, *Radiation Effects and Soft Errors in Integrated Circuits and Electronic Devices* (*Selected Topics in Electronics and Systems*). Hackensack, NJ: World Scientific Publishing Company, (2004).

FOSTER 1996. C. C. FOSTER, S. L. CASEY, A. L. JOHNSTON, P. MIESLE, N. SIFRI, A. H. SKEES, and K. M. MURRAY, Opportunities for Single Event and Other Radiation Effects Testing and Research at the Indiana University Cyclotron Facility, 1996 IEEE Radiation Effects Data Workshop, p. 84 (1996).

FREEMAN 1996. L. B. FREEMAN, Critical Charge Calculations for a Bipolar SRAM Array, *IBM J. Res. Dev.*, **40**(1), 119 (January 1996).

FU 1985. J. S. FU, C. L. AXNESS, and H. T. WEAVER, Memory SEU Simulations Using 2-D Transport Calculations, *IEEE Electron Device Lett.*, **6**, 422 (August 1985).

FULKERSON 2005. D. E. FULKERSON and E. E. VOGT, Prediction of SOI Single-Event Effects Using a Simple Physics-Based SPICE Model, *IEEE Trans. Nucl. Sci.*, **NS-52**, 2168 (2005).

FULKERSON 2006. D. E. FULKERSON, D. K. NELSON, and R. M. CARLSON, Boxes: An Engineering Methodology for Calculating Soft Error Rates in Integrated Circuits, *IEEE Trans. Nucl. Sci.*, **NS-53**, 3329 (2006).

FULKERSON 2007. D. E. FULKERSON, D. K. NELSON, R. M. CARLSON, and E. E. VOGT, Modeling Ion-Induced Pulses in Radiation Hard SOI Integrated Circuits, *IEEE Trans. Nucl. Sci.*, **NS-54**, 1406 (August 2007).

GADLAGE 2004. M. J. GADLAGE, R. D. SCHRIMPF, J. M. BENEDETTO, P. H. EATON, D. G. MAVIS, M. SIBLEY, K. AVERY, and T. L. TURFLINGER, Single Event Transient Pulsewidths in Digital Microcircuits, *IEEE Trans. Nucl. Sci.*, **NS-51**, 3285 (2004).

GALLOWAY 1996. K. F. GALLOWAY and G. H. JOHNSON, Catastrophic Single-Event Effects in the Natural Space Environment, IEEE NSREC Short Course Notes (July 1996).

GASIOT 2002. G. GASIOT, V. FERLET-CAVROIS, J. BAGGIO, and P. FLATRESSE, SEU Sensitivity of Bulk and SOI Technologies to 14 MeV Neutrons, *IEEE Trans. Nucl. Sci.*, **NS-49**, 3032 (2002).

GEORGE 2008. J. S. GEORGE, R. KOGA, and M. A. MCMAHAN, Neutron Soft Errors in Xilinx FPGAs at Lawrence Berkeley National Laboratory, 2008 IEEE Radiation Effects Data Workshop, pp. 118–123 (July 2008).

GILBERT 1985. R. M. GILBERT, G. K. OVREBO, and J. SCHIFANO, Plasma Screening of Funnel Fields, *IEEE Trans. Nucl. Sci.*, **NS-32**, 4098 (1985).

GINET 2007. G. P. GINET, B. K. DICHTER, D. H. BRAUTIGAM, and D. MADDEN, Proton Flux Anisotropy in Low Earth Orbit, *IEEE Trans. Nucl. Sci.*, **NS-54**, 1975 (2007).

GIOT 2007. D. GIOT, P. ROCHE, G. GASIOT, and R. HARBOE-SORENSEN, Multiple-Bit Upset Analysis in 90nm SRAMs: Heavy Ions Testing and 3D Simulations, *IEEE Trans. Nucl. Sci.*, **NS-54**, 904–911 (August 2007).

GOKA 1991. T. GOKA, S. KUBOYAMA, Y. SHIMANO, and T. KAWANISHI, The On-Orbit Measurements of Single Event Phenomena by ETS-V Spacecraft, *IEEE Trans. Nucl. Sci.*, **NS-38**, 1693 (1991).

GOLKE 1993. K. W. GOLKE, Determination of Funnel Length from Cross Section Versus LET Measurements, *IEEE Trans. Nucl. Sci.*, **NS-40**, 1910–1917 (1993).

GORDON 2004. M. S. GORDON, P. GOLDHAGEN, K. P. RODBELL, T. H. ZABEL, H. H. K. TANG, J. M. CLEM, and P. BAILEY, Measurement of the Flux and Energy Spectrum of Cosmic-Ray Induced Neutrons on the Ground, *IEEE Trans. Nucl. Sci.*, **NS-51**, 3427–3434 (2004).

GUENZER 1979. C. S. GUENZER, E. A. WOLICKI, and R. G. ALLAS, Single Event Upset of Dynamic RAMs by Neutrons and Protons, *IEEE Trans. Nucl. Sci.*, **NS-26**, 5048 (1979).

GUENZER 1981. C. S. GUENZER, A. B. CAMPBELL, and P. SHAPIRO, Single Event Upsets in NMOS Microprocessors, *IEEE Trans. Nucl. Sci.*, **NS-28**, 3955 (1981).

GUSSENHOVEN 1992. M. S. GUSSENHOVEN, E. G. MULLEN, M. SPERRY, K. J. KERNS, and J. B. BLAKE, The Effect of the March 1991 Storm on Accumulated Dose for Selected Satellite Orbits: CRRES Dose Models, *IEEE Trans. Nucl. Sci.*, **NS-39**, 1765 (1992).

HADDAD 2006. N. F. HADDAD, T. BACH, T. CONWAY, D. LAWSON, J. ROSS, J. RODGERS, A. TIPTON, D. BALL, K. WARREN, and R. SCHRIMPF, Eliminating Low LET Sensitivities in Deep Sub-Micrometer SRAM Through Non-intrusive Technology Features, 2006 SEE Symposium, Paper A2, (2006).

HAFER 2006a. C. HAFER J. MABRA, D. SLOCUM, T. FARRIS, and A. JORDAN, SEE and TID Results for a RADHard-by-Design 16 Mbit SRAM with Embedded EDAC, 2006 IEEE Radiation Effects Data Workshop Record, pp. 131–135 (2006).

HAFER 2006b. C. HAFER, J. MABRA, D. SLOCUM, and T. S. KALKUR, Next Generation Radiation-Hardened SRAM for Space Applications, IEEE Aerospace Conference, March 2006, Digital Object Identifier, 10.1109/AERO. 2006.1655956.

HAFER 2007. C. HAFER, M. LAHEY, H. GARDNER, D. HARRIS, A. JORDAN, T. FARRIS, and M. JOHNSON, Radiation Hardness Characterization of a 130 nm Technology, 2007 IEEE Radiation Effects Data Workshop Record, pp. 123–130 (2007).

HAFER 2008. G. HAFER, Ground-Based Testing and Evaluation of Soft Errors, IEEE NSREC Short Course Notes (July 2008).

HAMILTON 1992. L. C. HAMILTON, *Regression with Graphics, A Second Course in Applied Statistics*. Belmont, CA: Duxbury Press, 1992.

HAMM 1981. R. N. HAMM, M. L. RUSTGI, H. A. WRIGHT, and J. E. TURNER, Energy Spectra of Heavy Fragments from the Interaction of Protons with Communications Material, *IEEE Trans. Nucl. Sci.*, **NS-28**, 4004 (1981).

HANDS 2009. A. HANDS, C. S. DYER, and F. LEI, SEU Rates in Atmospheric Environments: Variations Due to Cross-Section Fits and Environmental Models, *IEEE Trans. Nucl. Sci.*, **NS-56**, 2026–2034 (August 2009).

HANSEN 2007. D. L. HANSEN, K. JOBE, J. WHITTINGTON, M. SHOGA, and D. A. SUNDERLAND, Correlation of Prediction to On-Orbit SEU Performance for a Commercial 0.25-μm CMOS SRAM, *IEEE Trans. Nucl. Sci.*, **NS-54**, 2525–2533 (2007).

HARBOE-SORENSEN 1986. R. HARBOE-SORENSEN et al., The SEU Risk Assessment of Z80A 8086 and 80C86 Microprocessors Intended for Use in a Low Altitude Polar Orbit, *IEEE Trans. Nucl. Sci.*, **NS-33**, 1626 (1986).

HARBOE-SORENSEN 1990. R. HARBOE-SORENSEN, E. J. DALY, C. L. UNDERWOOD, J. WARD, and L. ADAMS, The Behaviour of Measured SEU at Low Altitude During Periods of High Solar Activity, *IEEE Trans. Nucl. Sci.*, **NS-37**, 1938 (1990).

HARBOE-SORENSEN 1993. R. HARBOE-SORENSEN, E. J. DALY, L. ADAMS, C. I. UNDERWOOD, and R. MULLER, Observation and Prediction of SEU in Hitachi SRAMs in Low Altitude Polar Orbits, *IEEE Trans. Nucl. Sci.*, **NS-40**, 1498 (1993).

HARBOE-SORENSEN 1997. R. HARBOE-SORENSEN, Proton SEE Results—A Summary of ESA's Ground Test Data, 1997 RADECS Conference Data Workshop, p. 89 (1997).

HARBOE-SORENSEN 1998. R. HARBOE-SORENSEN, M. BRUGGEMANN, R. MULLER, and F. J. ROMBECK, Radiation Evaluation of 3.3 Volt 16M-bit DRAMs for Solid State Mass Memory Space Application, 1998 NSREC Data Workshop, p. 74 (1998).

HARBOE-SORENSEN 2008. R. HARBOE-SORENSEN, C. POIVEY, R. S. GUERRE, A. ROSENG, F. LOCHON, G. BERGER, W. HAJDAS, A. VIRTANEN, H. KETTUNEN, and S. DUZELLIER, From the Reference SEU Monitor to the Technology Demonstration Module On-Board BROBA-II, *IEEE Trans. Nucl. Sci.*, **NS-55**, 3082–3087 (2008).

HARTWELL 2004. M. HARTWELL, C. HAFER, P. MILLIKEN, and T. FARRIS, Megarad Total Ionizing Dose and Single Event Effects Test Results of a RadHard-By-Design 0.25Micron ASIC, 2008 IEEE Radiation Effects Data Workshop Record, pp. 104–109 (2004).

HEIDERGOTT 2001. W. F. HEIDERGOTTT, R. LADBURY, P. W. MARSHALL, S. BUCHNER, A. B. CAMPBELL, R. A. REED, J. HOCKMUTH, N. KHA, C. HAMMOND, C. SEIDLECK, and A. ASSAD, Complex SEU Signatures in High-Speed Analog-to-Digital Conversion, *IEEE Trans. Nucl. Sci.*, **NS-48**, 1828–1832 (2001).

HEINRICH 1977. W. HEINRICH, Calculation of LET-Spectra of Heavy Cosmic Ray Nuclei at Various Absorber Depths, *Radiat. Eff.*, **34**, 143–148 (1977).

HELENE 1984. O. HELENE, Errors in Experiments with Small Numbers of Events, *Nucl. Instrum. Methods Phys. Res.*, **228**, 120–128 (1984).

HIEMSTRA 2003. D. M. HIEMSTRA and E. W. BLACKMORE, LET Spectra of Proton Energy Levels from 50 to 500 MeV and Their Effectiveness for Single Event Effects Characterization of Microelectronics, *IEEE Trans. Nucl. Sci.*, **NS-50**, 2245 (2003).

HIEMSTRA 2006. D. HIEMSTRA, Guide to the IEEE Radiation Effects Data Workshop Record, IEEE Radiation Effects Data Workshop, pp. 1–5 (2006).

HIEMSTRA 2007. D. HIEMSTRA, Guide to the 2006 IEEE Radiation Effects Data Workshop Record, IEEE Radiation Effects Data Workshop, pp. 211–214 (2007).

HIEMSTRA 2008. D. HIEMSTRA, Guide to the 2007 IEEE Radiation Effects Data Workshop Record, IEEE Radiation Effects Data Workshop, pp. 1–4 (2008).

HIROSE 2002. K. HIROSE, H. SAITO, Y. KURODA, S. ISHII, Y. FUKUOKA, and D. TAKAHASHI, SEU Resistance in Advanced SOI-SRAMs Fabricated by Commercial Technology Using a Rad-Hard Circuit Design, *IEEE Trans. Nucl. Sci.*, **NS-49**, 2965 (2002).

HOFFMANN 2000. L. HOFFMANN and R. C. DIBARI, Radiation Testing and Characterization of Programmable Logic Devices (PLDs), IEEE NSREC Short Course Notes (July 2000).

HORN 1992. K. M. HORN, B. L. DOYLE, and F. W. SEXTON, Nuclear Microprobe Imaging of Single-Event Upsets, *IEEE Trans. Nucl. Sci.*, **NS-39**, 7 (1992).

HOWE 2005. C. L. HOWE, R. A. WELLER, R. A. REED, M. H. MENDENHALL, R. D. SCHRIMPF, K. M. WARREN, D. R. BALL, L. W. MASSENGILL, K. A. LABEL, J. W. HOWARD, Jr., and N. R. HADDAD, Role of Heavy-Ion Nuclear Reactions in Determining On-Orbit Single Event Error Rates, *IEEE Trans. Nucl. Sci.*, **NS-52**, 2182 (2005).

HSIEH 1981a. C. M. HSIEH, P. C. MURLEY, and R. R. O'BRIEN, A Field-Funneling Effect on the Collection of Alpha-Particle Generated Carriers in Silicon Devices, *IEEE Electron Device Lett.*, **EDL-2**, 102 (April 1981).

HSIEH 1981b. C. M. HSIEH, P. C. MURLEY, and R. R. O'BRIEN, Dynamics of Charge Collection from Alpha-Particle Tracks in Integrated Circuits, Proceedings of the IEEE International Reliability Physics Symposium p. 38 (1981).

HSIEH 1983. C. M. HSIEH, P. C. MURLEY, and R. R. O'BRIEN, Collection of Charge from Alpha-Particle Tracks in Silicon Devices, *IEEE Trans. Electron Devices*, **30**, 686 (1983).

HUBERT 2001. G. HUBERT, J. M. PALAU, K. CASTELLANI-COULI, M. D. CALVET, and S. FOURTINE, Detailed Analysis of Secondary Ions' Effect for the Calculation of Neutron-Induced SER in SRAMs, *IEEE Trans. Nucl. Sci.*, **NS-48**, 1953 (2001).

HUBERT 2009. G. HUBERT, S. DUZELLIER, C. INGUIMBERT, C. BOATELLA-POLO, F. BEZERRA, and R. ECOFFET, Operational SER Calculations on the SAC-C Orbit Using the Multi-Scales Single Event Phenomena Predictive Platform (MUSCA SEP^3), *IEEE Trans. Nucl. Sci.*, **NS-56**, 3032–3042 (2009).

HUGHLOCK 1990. B. W. HUGHLOCK G. S. LARUE, and A. H. JOHNSTON, Single-Event Upset in GaAs E/D MESFET Logic, *IEEE Trans. Nucl. Sci.*, **NS-37**, 1894 (1990).

IKEDA 2005. N. IKEDA, S. KUBOYAMA, S. MATSUDE, and T. HANDA, Analysis of Angular Dependence of Proton-Induced Multiple Bit Upsets in a Synchronous SRAM, *IEEE Trans. Nucl. Sci.*, **NS-52**, 2200 (2005).

INGUIMBERT 2004. C. INGUIMBERT and S. DUZELLIER, SEU Rate Calculation with GEANT4: Comparison with CREME86, *IEEE Trans. Nucl. Sci.*, **NS-51**, 2805 (October 2004).

IROM 2005. F. IROM and T. F. MIYAHIRA, Test Results of Single-Event Effects Conducted by the Jet Propulsion Laboratory, 2005 IEEE Radiation Effects Data Workshop Record, p. 36 (2005).

JOHLANDER 1996. R. JOHLANDER, R. HARBOE-SORENSEN, G. OHLSSON, and L. BYLANDER, Ground Verification of In-Orbit Anomalies in the Double Probe Electric Field Experiment on Freja, *IEEE Trans. Nucl. Sci.*, **NS-43**, 2767 (1996).

JOHNSON 1985. R. L. JOHNSON, Jr., S. E. DIEHL-NAGLE, and J. R. HAUSER, Simulation Approach for Modeling Single Event Upsets on Advanced CMOS SRAMs, *IEEE Trans. Nucl. Sci.*, **NS-32**, 4122 (1985).

JOHNSTON 1998. A. H. JOHNSTON, G. M. SWIFT, T. MIYAHIRA, S. GUERTIN, and L. D. EDMONDS, Single-Event Upset Effects in Optocouplers, *IEEE Trans. Nucl. Sci.*, **NS-45**, 2867–2875 (1998).

JOHNSTON 1999. A. H. JOHNSTON, T. MIYAHIRA, G. M. SWIFT, S. M. GUERTIN, and L. D. EDMONDS, Angular and Energy Dependence of Proton Upset in Optocouplers, *IEEE Trans. Nucl. Sci.*, **NS-46**, 1335–1341 (1999).

JOHNSTON 2000. A. H. JOHNSTON, G. M. SWIFT, T. F. MIYAHIRA, and L. D. EDMONDS, A Model for Single-Event Transients in Comparators, *IEEE Trans. Nucl. Sci.*, **NS-47**, 2624–2623 (2000).

JONES 2000. R. JONES, A. M. CHUGG, C. M. S. JONES, P. H. DUNCAN, C. S. DYER, and C. SANDERSON, Comparison Between SRAM SEE Cross-Sections from Ion Beam Testing with Those Obtained Using a New Picosecond Pulsed Laser Facility, *IEEE Trans. Nucl. Sci.*, **47**(3), 539–544 (2000).

KASTENSMIDT 2007. F. L. KASTENSMIDT, SEE Mitigation Strategies for Digital Circuit Design Applicable to ASICs and FPGAs, 2007 IEEE NSREC Short Course Notebook (July 2007).

KERNS 1989. S. E. KERNS, L. W. MASSENGILL, D. V. KERNS, Jr., M. L. ALLES, T. W. HOUSTON, H. LU, and L. R. HITE, Model for CMOS/SOI Single-Event Vulnerability, *IEEE Trans. Nucl. Sci.*, **NS-36**, 2305 (1989).

KIMOTO 2005. Y. KIMOTO, N. NEMOTO, H. MATSUMOTO, K. UENO, T. GOKA, and T. OMODAKA, Space Radiation Environment and Its Effects on Satellites: Analysis of the First Data from TEDA on Board ADEOS-II, *IEEE Trans. Nucl. Sci.*, **NS-52**, 1474 (October 2005).

KIRKPATRICK 1979. S. KIRKPATRICK, Modeling Diffusion and Collection of Charge from Ionizing Radiation in Silicon Devices, *IEEE Trans. Electron Devices*, **ED-26**, 1742 (November 1979).

KNUDSON 1984. A. R. KNUDSON, A. B. CAMPBELL, P. SHAPIRO, W. J. STAPOR, E. A. WOLICKI, E. L. PETERSEN, S. E. DIEHL-NAGLE, J. HAUSER, and P. V. DRESSENDORFER, Charge Collection in Multilayer Structures, *IEEE Trans. Nucl. Sci.*, **NS-31**, 1149 (1984).

KOBAYASHI 2004. A. S. KOBAYASHI, A. L. STERNBERG, L. W. MASSENGILL, R. D. SCHRIMPF, and R. A. WELLER, Spatial and Temporal Characteristics of Energy Deposition by Protons and Alpha Particles in Silicon, *IEEE Trans. Nucl. Sci.*, **NS-51**, 3312–3317 (2004).

KOBAYASHI 2009. D. KOBAYASHI, K. HIROSE, V. FERLET-CAVROIS, D. MCMORROW, T. MAKINO, H. IKEDA, Y. ARAI, and M. OHNO, Device-Physics-Based Analytical Model for Single-Event Transients in SOI CMOS Logic, *IEEE Trans. Nucl. Sci.*, **NS-56**, 3043–3049 (2009).

KOGA 1984. R. KOGA and W. A. KOLASINSKI, Heavy Ion-Induced Single Event Upsets of Microcircuits, A Summary of the Aerospace Corporation Test Data, *IEEE Trans. Nucl. Sci.*, **NS-31**, 1190 (1984).

KOGA 1988. R. KOGA, W. A. KOLASINSKI, J. V. OSBORN, J. H. ELDER, and R. CHITTY, SEU Test Techniques for 256K Static RAMs and Comparisons of Upsets Induced by Heavy Ions and Protons, *IEEE Trans. Nucl. Sci.*, **NS-35**, 1638 (1988).

KOGA 1990. R. KOGA, N. KATZ, S. D PINKERTON, W. A. KOLASINSKI, and D. L. OBERG, Bevelac Ion Beam Characterizations for Single Event Phenomena, *IEEE Trans. Nucl. Sci.*, **NS-37**, 1923–1928 (1990).

KOGA 1991. R. KOGA, W. R. CRAIN, K. B. CRAWFORD, D. D. LAU, S. D. PINKERTON, B. K. YI, and R. CHITTY, On the Suitability of Non-hardened High Density SRAMs for Space Applications, *IEEE Trans. Nucl. Sci.*, **NS-38**, 1507 (1991).

KOGA 1993a. R. KOGA, S. D. PINKERTON, T. J. LIE, and K. B. CRAWFORD, Single-Word, Multiple-Bit Upsets in Static Random Access Devices, *IEEE Trans. Nucl. Sci.*, **NS-40**, 1941 (1993).

KOGA 1993b. R. KOGA, K. B. CRAWFORD, P. B. GRANT, W. A. KOLASINSKI, D. L. LEUNG, T. J. LIE, D. C. MAYER, S. D. PINKERTON, and T. K. TSUBOTA, Single Ion Induced Multiple-Bit Upset in IDT 256K SRAMs, RADECS-93, p. 485 (1993).

KOGA 1996. R. KOGA, Single-Event Effect Ground Test Issues, *IEEE Trans. Nucl. Sci.*, **NS-43**, 661 (1996).

KOHLER 1989. R. A. KOHLER and R. KOGA, SEU Characterization of Hardened CMOS SRAMs Using Statistical Analysis of Feedback Delay in Memory Cells, *IEEE Trans. Nucl. Sci.*, **NS-36**, 2318 (1989).

KOLASINSKI 1979. W. A. KOLASINSKI, J. B. BLAKE, J. K. ANTHONY, W. E. PRICE, and E. C. SMITH, Simulation of Cosmic-Ray Induced Soft Errors and Latchup in Integrated Circuit Computor Memories, *IEEE Trans. Nucl. Sci.*, **NS-26**, 5087 (1979).

KOLASINSKI 1981. W. A. KOLASINSKI, R. KOGA, J. B. BLAKE, and S. E. DIEHL, Soft Error Susceptibility of CMOS RAMs: Dependence upon Power Supply Voltage, *IEEE Trans. Nucl. Sci.*, **NS-28**, 4013–4016 (1981).

KRESKOVSKY 1985. J. P. KRESKOVSKY and H. L. GRUBIN, Simulation of Charge Collection in a Multilayer Device, *IEEE Trans. Nucl. Sci.*, **NS-32**, 4140–4144 (1985).

LABEL 1993. K. A. LABEL, C. M. CRABTREE, J. HENGEMIHLE, and M. M. GATES, Solid State Tape Recorders: Spaceflight SEU Data for SAMPEX and TOMS/Meteor-3, 1993 IEEE Radiation Effects Data Workshop, p. 77 (1993).

LABEL 2008. K. A. LABEL, R. L. LADBURY, L. M. COHN, and T. R OLDHAM, Radiation Test Challenges for Scaled Commercial Memories, *IEEE Trans. Nucl. Sci.*, **NS-55**, 2174–2180 (August 2008).

LACOE 2008. R. LACOE, Improving Integrated Circuit Performance Through the Application of Hardness-by-Design Methodology, *IEEE Trans. Nucl. Sci.*, **NS-55**, 1903–1925 (August, 2008).

LADBURY 2004. R. LADBURY, R. A. REED, P. MARSHALL, K. A. LABEL, A. ANANTRAMAN, R. FOX, D. P. SANGERSON, A. STOLZ, J. YURKON, A. F. ZELLER, and J. W. STETSON, Performance of the High-Energy Single-Event Effects Test Facility (SEE) at Michigan State University's National Superconducting Cyclotron Laboratory (NSCL), *IEEE Trans. Nucl. Sci.*, **NS-51**, 3664 (2004).

LADBURY 2007a. R. LADBURY, Radiation Hardening at the System Level, 2007 IEEE NSREC Short Course Notebook (July 2007).

LADBURY 2007b. R. LADBURY, Statistical Properties of SEE Rate Calculation in the Limits of Large and Small Event Counts, *IEEE Trans. Nucl. Sci.*, **NS-54**, 2113 (2007).

LADBURY 2009. R. L. LADBURY, J. BENEDETTO, D. MCMORROW, S. P. BUCHNER, K. A. LABEL, M. D. BERG, H. S. KIM, A. B. SANDERS, M. R FRIENDLICH, and A. PHAN, TPA Laser and Heavy-Ion Testing: Complementary Techniques for SDRAM Single-Event Evaluation, *IEEE Trans. Nucl. Sci.*, **NS-56**, 3334–3340 (2009).

LANGWORTHY 1982. J. B. LANGWORTHY, Recoil Distributions in Some Proton Reactions, NRL Memorandum Report 4731 (1982).

LANGWORTHY 1989a. J. B. LANGWORTHY, Depletion Region Geometry Analysis Applied to Single Event Sensitivity, *IEEE Trans. Nucl. Sci.*, **NS-36**, 2427 (1989).

LANGWORTHY 1989b. J. B. LANGWORTHY, A General Approach to Chord Length Distributions Applied to a Hemisphere, *Radiat. Res.*, **118**, 21 (1989).

LANGWORTHY 1991. J. B, LANGWORTHY, Device Upset Predictions for the Chemical Release and Radiation Effects Satellite (CRRES): A Summary, NRL Memorandum Report 6913 (November 1991).

LAUENSTEIN 2005. J.-M. LAUENSTEIN and J. L. BARTH, Radiation Belt Modeling for Spacecraft Design: Model Comparison for Common Orbits, 2005 IEEE Radiation Effects Data Workshop Record, pp. 102–109 (2005).

LAW 2006. M. LAW, Device Modeling of Single Event Effects, 2006 IEEE NSREC Short Course Notebook (July 2006).

LAWRENCE 2008. R. K. LAWRENCE, Single Event Gate Rupture Testing on 22A Gate Oxide Structures from a 90nm Commercial CMOS Process, 2008 IEEE Radiation Effects Data Workshop Record, pp. 76–81 (July 2008).

LEI 2006. F. LEI, A. HANDS, S. CLUCAS, C. DYER, and P. TRUSCOTT, Improvement to and Validations of the QinetiQ Atmospheric Radiation Model(QARM), *IEEE Trans. Nucl. Sci.*, **NS-53**, 1851–1858 (August 2006).

LERAY 2007. J. L. LERAY, Rayonnements ionisants et Compatibillte electronique et Nucleaire Effets, Les Conferences CEM Mars 2007. Available at http://www.rfhyper.com/commun/CEM/10h40J_L_LERAY_Rayonnements_ionisants.pdf. http://www.afcem.org/documents/RF%20HYPER%202007/LERAY_Rayonnements_ionisants.pdf.

LETAW 1985. J. R. LETAW, R. SILBERBERG, C. H. TSAO, and J. H. ADAMS, Jr., Geometric Considerations in Single Event Upset Estimation, *IEEE Trans. Nucl. Sci.*, **NS-32**, 4212 (1985).

LETAW 1986. J. R. LETAW and J. H. ADAMS, Jr., Comparison of CREME Model LET Spectra with Spaceflight Dosimetry Data, *IEEE Trans. Nucl. Sci.*, **NS-33**, 1620–1625 (1986).

LETAW 1991. J. R. LETAW and E. NORMAND, Guidelines for Prediction Single Event Upsets in Neutron Environments, *IEEE Trans. Nucl. Sci.*, **NS-38**, 1500 (1991).

LEWIS 2001. D. LEWIS, V. POUGET, F. BEAUDOIN, P. PERDU, H. LAPUYADE, P. FOUILLAT, and A. TOUBOUL, Backside Laser Testing of ICs for SET Sensitivity Evaluation, *IEEE Trans. Nucl. Sci.*, **48**(6), 2193–2201 (2001).

LIU 2004. S. T. LIU, H. Y. LIU, D. ANTHONY, W. HEIKKILA, H. HUGHES, A. CAMPBELL, E. L. PETERSEN, and P. J. MCMARR, Proton-Induced Upset in SOI CMOS SRAMs, *IEEE Trans. Nucl. Sci.*, **NS- 512**, 3475–3479 (2004).

LIU 2006a. M. S. LIU, H. Y. LIU, N. BREWSTER, D. NELSON, K. W. GOLKE, G. KIRCHNER, H. L. HUGHES, A. CAMPBELL, and J. F ZIEGLER, Limiting Upset Cross Sections of SEU Hardened SOI SRAMs, *IEEE Trans. Nucl. Sci.*, **NS-53**, 3487–3493 (2006).

LIU 2006b. H. Y. LIU, M. S. LIU, and H. L. HUGHES, Proton Induced Single Event Upset in 6T SOI SRAMs, *IEEE Trans. Nucl. Sci.*, **NS-53**, 3502–3505 (2006).

LUKE 1988. K. L. LUKE and M. G. BUEHLER, An Exact, Closed Form Expression of the Integral Chord-Length Distribution for the Calculation of Single-Event Upsets Induced by Cosmic Rays, *J. Appl. Phys.*, **64**(10), 5132 (1988).

LUM 2004. G. K. LUM, Hardness Assurance for Space Systems, IEEE 2004 NSREC Short Course Notebook, p. I-1 (2004).

MAJEWSKI 1995. P. P. MAJEWSKI, E. NORMAND, and D. L. OBERG, A New Solar Flare Heavy Ion Model and Its Implementation Through MACREE, An Improved Modeling Tool to Calculate Single Event Effect Rates in Space, *IEEE Trans. Nucl. Sci.*, **NS-42**, 2043 (1995).

MAKIHARA 2000. A. MAKIHARA, H. SHINDOU, N. NEMOTO, S. KUBOYAMA, S. MATSUDA, T. OSHIMA, T. HIRAO, H. ITOH, S. BUCHNER, and A. B. CAMPBELL, Analysis of Single-Ion Multiple Bit Uset in High-Density DRAMs, *IEEE Trans. Nucl. Sci.*, **NS-47**, 2400 (2000).

MARSHALL 1995. P. W. MARSHALL, C. J. DALE, M. A. CARTS, and K. A. LABEL, Particle-Induced Bit Errors in High Performance Fiber Optic Data Links for Satellite Data Management, *IEEE Trans. Nucl. Sci.*, **NS-41**, 1958 (1995).

MARSHALL 2004. P. MARSHALL, M. CARTS, A. CAMPBELL, R. LADBURY, R. REED, C. MARSHALL, S. CURRIE, D. MCMORROW, S. BUCHNER, C. SEIDLECK, P. RIGGS, K. FRITZ, B. RANDALL, and H. GILBERT, A Comparative Study of Heavy-Ion and Proton-Induced Bit-Error Sensitivity and Complex Burst-Error Modes in Commercially Available High-Speed SiGe and BiCMOS, *IEEE Trans. Nucl. Sci.*, **NS-51**, 3457–3463 (2004).

MARSHALL 2005. P. MARSHALL, M. CARTS, S. CURRIE, R. REED, B RANDALL, K. FRITZ, K. KENNEDY, M. BERG, R. KRITHIVASAN, C. SIEDLECK, R. LADBURY, C. MARSHALL, J. CRESSSLER, G. NIU, K. LABEL, and B. GILBERT, Autonomous Bit Error Rate Testing at Multi-Gbit/s Rates Implemented in a 5AM SiGe Circuit for

Radiation Effects Self Test (CREST), *IEEE Trans. Nucl. Sci.*, **NS-52**, 2446–2454 (2005).

MARTIN 1987. R. C. MARTIN, N. M. GHONIEM, Y. SONG, and J. S. CABLE, The Size Effect of Ion Charge Tracks on Single Event Multiple-Bit Upset, *IEEE Trans. Nucl. Sci.*, **NS-34**, 1305 (1987).

MASSENGILL 1990. L. W. MASSENGILL, D. V. KERNS, S. E. KERNS, and M. L. ALLES, Single-Event Charge Enhancement in SOI Devices, *IEEE Electron Device Lett.*, **11**, 98 (February 1990).

MASSENGILL 1993a. L. W. MASSENGILL, M. L. ALLES, S. E. KERNS, and L. L JONES, Effects of Process Parameter Distributions and Ion Strike Locations on SEU Cross-Section Data, *IEEE Trans. Nucl. Sci.*, **NS-40**, 1804–1811 (1993).

MASSENGILL 1993b. L. W. MASSENGILL, M. L. ALLES, and S. E. KERNS, SEU Error Rates in Advanced Digital CMOS, RADECS-93, p. 546 (1993).

MASSENGILL 1993c. L. W. MASSENGILL, SEU Modeling and Prediction Techniques, Chapter III in IEEE Nuclear and Space Radiation Effects Conference Short Course Notes (1993).

MASSENGILL 1996. L. W. MASSENGILL, Cosmic and Terrestrial Single-Event Radiation Effects in Dynamic Random Access Memories, *IEEE Trans. Nucl. Sci.*, **NS-43**, 576 (April 1996).

MAVIS 2007. D. MAVIS, P. EATON, M. SIBLEY, E. SMITH, and K. AVERY, Novel Test Methods and Analysis Approaches for Ultra-Deep Submicron Technologies, 2007 SEE Symposium Invited Talk (2007).

MAY 1979. T. C. MAY and M. H. WOODS, Alpha-Particle-Induced Soft Errors in Dynamic Memories, *IEEE Trans. Electron Devices*, **ED-26**, 2 (1979).

MAY 1984. T. C. MAY, G. L. SCOTT, E. S. MEICAN, P. WINER, and V. R. RAO, Dynamic Fault Imaging of VLSI Random Logic Devices, 1984 International Reliability Physics Symposium, Las Vega, Nevada, April 2–5 (1984).

MCDONALD 1989. P. T. MCDONALD, W. J. STAPOR, A. B. CAMPBELL, and L. W. MASSENGILL, Non-random Single Event Upset Trends, *IEEE Trans. Nucl. Sci.*, **NS-36**, 2324 (1989).

MCKERRACHER 1994. P. L. MCKERRACHER, J. D. KINNISON, and R. H. MAURER, Applying New Solar Particle Event Models to Interplanetary Satellite Programs, *IEEE Trans. Nucl. Sci.*, **NS-41**, 2368 (1994).

MCLEAN 1982. F. B. MCLEAN and T. R. OLDHAM, Charge Funneling in N- and P-Type Si Substrates, *IEEE Trans. Nucl. Sci.*, **NS-29**, 2018 (1982).

MCMAHON 2008. M. A. MCMAHON, E. BLACKMORE, E. W. CASCIO, C CASTANEDA, B. VON PRZEWSKI, and H. EISEN, Standard Practice for Dosimetry of Proton Beams for Use in Radiation Effects Testing of Electronics, 2008 IEEE Radiation Effects Data Workshop, IEEE NSREC, pp. 135–141 (July 2008).

McMorrow 1994. D. McMorrow, J. S. Melinger, N. Thantu, A. B. Campbell, T. R. Weatherford, A. R. Knudson, L. H. Tran, and A. Peczalski, Charge-Collection Mechanisms of Heterostructure FETs, *IEEE Trans. Nucl. Sci.*, **NS-41**, 2055–2062 (1994).

McMorrow 2000. D. McMorrow, J. S. Melinger, S. Buchner, T. Scott, R. D. Brown, and N. F. Haddad, Application of a Pulsed Laser for Evaluation and Optimization of SEU-Hard Designs, *IEEE Trans. Nucl. Sci.*, **47**(3), 559–565 (2000).

McMorrow 2005. D. McMorrow, W. T. Lotshaw, J. S. Melinger, S. Buchner, J. D. Davis, R. K. Lawrence, D. Haddad, J. H. Bowman, R. D. Brown, D. Carlton, J. Pena, J. Vasquez, K. M. Warren, and L. W. Massengill, Single-Event Upset in Flip-Chip SRAM Induced by Through-Wafer, Two-Photon Absorption, *IEEE Trans. Nucl. Sci.*, **NS-52**, 2421 (2005).

McNulty 1980. P. J. McNulty, G. E. Farrell, R. C. Wyatt, P. L. Rothwell, R. C. Filz, and N. N. Bradford, Upset Phenomena Induced by Energetic Protons and Electrons, *IEEE Trans. Nucl. Sci.*, **NS-27**, 1516 (1980).

McNulty 1981. P. J. McNulty, G. E. Farrell, and W. P, Tucker, Proton-Induced Nuclear Reactions in Silicon, *IEEE Trans. Nucl. Sci.*, **NS-28**, 4007 (1981).

McNulty 1985. P. J. McNulty, W. G. Abdel-Kader, and J. M. Bisgrove, Methods for Calculating SEU Rates for Bipolar and NMOS Circuits, *IEEE Trans. Nucl. Sci.*, **NS-32**, 4180 (1985).

McNulty 1990. P. J. McNulty, Predicting Single Event Phenomena in Natural Space Environments, 1990 International Nuclear and Space Radiation Effects Conference Short Course.

McNulty 1991a. P. J. McNulty, W. J. Beauvais, W. G. Abdel-Kader, S. El-Teleaty, E. G. Mullen, and K. P. Ray, Test of SEU Algorithms Against Preliminary CRRES Satellite Data, *IEEE Trans. Nucl. Sci.*, **NS-38**, 1642 (1991).

McNulty 1991b. P. J. McNulty, W. J, Beauvais, and D. R. Roth, Determination of SEU Parameters of NMOS and CMOS SRAMs, *IEEE Trans. Nucl. Sci.*, **NS-38**, 1463 (1991).

Melinger 1998. J. S. Melinger, D. McMorrow, A. B. Campbell, S. Buchner, L. H. Tran, A. R. Knudson, and W. R. Curtice, Pulsed Laser Induced Single Event Upset and Charge Collection Measurements as a Function of Optical Penetration Depth, *J. Appl. Phys.*, **84**, 690 (1998).

Merelle 2005. T. Merelle, H. Chabane, J. M. Palau, K. Castellani-Coulic, F. Wrobel, F. Saigne, B. Sagnes, J. Boch, R. R. Vaille, G. Gasiot, P. Roche, M. C. Palau, and T. Carriere, Criterion for SEU Occurences in SRAM Deduced from Circuit and Devices Simulation in Case of Neutron-Induced SER, *IEEE Trans. Nucl. Sci.*, **NS-52**, 1148 (August 2005).

Messenger 1982. G. C. Messenger, Collection of Charge on Junction Nodes from Ion Tracks, *IEEE Trans. Nucl. Sci.*, **NS-29**, 2024 (1982).

Metzger 1994. S. Metzger, J. Dreute, W. Heinrich, H. Rocher, B. E. Fischer, R Harboe-Sorensen, and L. Adams, Heavy Ion Microscopy of Single Event Upsets in CMOS SRAMs, RADECS-93, p. 499 (1993). Also, *IEEE Trans. Nucl. Sci.*, **NS-41**, 589 (1994).

MILLER 2006. F. MILLER, N. BUARD, G. HUBERT, S. ALESTRA, G. BAUDRILLARD, T. CARRIERE, R. GAILLARD, J. M. PAULAU, F. SAIGNE, and P. FOUILLAT, Laser Mapping of SRAM Sensitive Cells: A Way to Obtain Input Parameters for DASIE Calculation Code, *IEEE Trans. Nucl. Sci.*, **NS-53**, 1863 (August 2006).

MIROSHKIN 1994. V. V. MIROSHKIN and M. G, TVERSKOY, Two Parameter Model for Predicting SEU Rates, *IEEE Trans. Nucl. Sci.*, **NS-41**, 2085 (1994).

MIROSHKIN 1995. V. V. MIROSHKIN and M. G. TVERSKOY, Some Aspects of Application of the Two Parameter SEU Model, *IEEE Trans. Nucl. Sci.*, **NS-42**, 1809 (1995).

MIYAHIRA 2008. T. F. MIYAHIRA and F. IROM, Results of Single-Event Latchup Measurements Conducted by the Jet Propulsion Laboratory, 2008 IEEE Radiation Effects Data Workshop, pp. 53–57 (July 2008).

MONTES 2008. E. J. MONTES, R. A. REED, J. A. PELLISH, M. ALLES, R. D. SCHRIMPF, R. A. WELLER, M. VARADHARAJAPERUMAL, G. NIU, A. K. SUTTON, R. DIESTELHORST, J. P. COMEAU, J. D. CRESSLER, P. W. MARSHALL, and G. VIZKELETHY, Single Event Upset Mechanisms for Low-Energy Deposition Events in SiGe HBTs, *IEEE Trans. Nucl. Sci.*, **NS-55**, 1581–1586 (June 2008).

MORRIS 2006. Wesley Morris, private communication, Silicon Space Technology Corporation.

MORRIS 2007. Wesley Morris, Silicon Space Technology, 804 Las Cimas Parkway, Suite 140, Austin TX, 78746-5195.

MOSS 1995. S. C. MOSS, S. D. LALUMONDIERE, J. R. SCARPULLA, K. P. MACWILLIAMS, W. R. CRAIN, and R. KOGA, Correlation of Picosecond Laser-Induced Latchup in CMOS Test Structures, *IEEE Trans. Nucl. Sci.*, **NS-45**, 1948 (1995).

MOSTELLER 1977. F. MOSTELLER and J. W. TUKEY, *Data Analysis and Regression, a Second course in Statistics*. Reading MA: Addison-Wesley, 1977.

MULLEN 1995. E. G. MULLEN, K. P. RAY, R. KOGA, E. G. HOLEMAN, and D. E. DELOREY, SEU Results from the Advanced Photovoltaic and Electronics Experiments (APEX) Satellite, *IEEE Trans. Nucl. Sci.*, **NS-42**, 1988 (1995).

MUNTEANU 2008. D. MUNTEANU and J. L. AUTRAN, Modeling and Simulation of Single-Event Effects in Digital Devices and ICs, *IEEE Trans. Nucl. Sci.*, **NS-55**, 1854–1878 (August 2008).

MURLEY 1996. P. C. MURLEY and G. R. SRINIVASAN, Soft Error Monte Carlo Modeling Program, SEMM, *IBM J. Res. Dev.*, **40**(1), 109 (January 1996).

MURRAY 1989. K. M. MURRAY, W. J. STAPOR, and C. CASTENEDA, Calibrated Charged Particle Radiation System with Precision Dosimetric Measurement and Control, *Nucl. Instrum. Methods A*, 616 (1989).

MUSSEAU 1996a. O. MUSSEAU, Single Event Effect in SOI Technologies and Devices, *IEEE Trans. Nucl. Sci.*, **NS-43**, 603 (February 1996).

MUSSEAU 1996b. O. MUSSEAU, F. GARDIC, P. ROCHE, T. CORBIERE, R. A. REED, S. BUCHNER, P. MCDONALD, J. MELINGER, L. TRAN, and A. B. CAMPBELL, Analysis of Multiple Bit Upsets (MBU) in a CMOS RAM, *IEEE Trans. Nucl. Sci.*, **NS-43**, 2879 (1996).

MUSSEAU 1997. O. MUSSEAU, V. FERLET-CAVROIS, A. B. CAMPBELL, A. R. KNUDSON, W. J. STAPOR, P. T. MCDONALD, J. L. PELLOIE, and C. RAYNAUD, Charge Collection in Submicron CMOS/SOI Technology, *IEEE Trans. Nucl. Sci.*, **NS-44**, 2124 (1997).

MUSSEAU 2000. O. MUSSEAU, V. FERLET-CAVROIS, J. L. PELLOIE, S. BUCHNER, D. MCMORROW, and A. B. CAMPBELL, Laser Probing of Bipolar Amplification in 0.25 μm MOS/SOI Transistors, *IEEE Trans. Nucl. Sci.*, **NS-47**, 2196 (2000).

NATRELLA 1963. M. G. NATRELLA, *Experimental Statistics*, National Bureau of Standards Handbook 91. Reprinted 1966 with corrections.

NEWBERRY 1986. D. M. NEWBERRY, J. D. TOWNEY, D. H. KAYE, and S. P. SEXTON, Single Event Induced Errors in VLSI Logic Circuits, *J. Radiat Effects Res. Eng.*, **5**(1), 43 (1986).

NEWBERRY 1988. D. M. NEWBERRY, SEU Hardening Approaches for VLSI Logic Devices, *J. Radiat Effects Res. Eng.*, **6**(2), 146 (1988).

NEWBERRY 1992. D. M. NEWBERRY, Single Event Upset Error Propagation Between Interconnecting VLSI Devices, *IEEE Trans. Nucl. Sci.*, **NS-39**, 446 (June 1992).

NICHOLS 1988. D. K. NICHOLS, M. A. HUEBNER, W. E. PRICE, L. S. SMITH, and J. R. COSS, Heavy Ion Induced Single Event Phenomena Data for Semiconductor Devices from Engineering Testing, JPL Publication 88–17 (1988).

NICHOLS 1995. D. K. NICHOLS, J. R. COSS, K. P. MCCARTY, H. R. SCHWARTZ, G. M. SWIFT, R. K. WATSON, R. KOGA, W. R. CRAIN, K. B. CRAWFORD, and S. J. HANSEL, Trends in SEE Susceptibility from Heavy Ions, 1995 IEEE Radiation Effects Data Workshop, p. 1 (1995).

NICHOLS 1997. D. K. NICHOLS, J. R. COSS, T. F MIYAHIRA, H. R. SCHWARTZ, G. M. SWIFT, R. KOGA, W. R. CRAIN, K. B. CRAWFORD, and S. H. PENZIN, Device SEE Susceptibility from Heavy Ions (1995–1996), 1997 IEEE Radiation Effects Data Workshop Record, p. 1 (1997).

NORMAND 1990. E. NORMAND and W. J. STAPOR, Variation in Proton-Induced Upsets Rates from Large Solar Flares Using an Improved SEU Model, *IEEE Trans. Nucl. Sci.*, **NS-37**, 1947 (1990).

NORMAND 1993. E. NORMAND and T. J. BAKER, Altitude and Latitude Variations in Avionics SEU and Atmospheric Neutron Flux, *IEEE Trans. Nucl. Sci.*, **NS-40**, 1484 (1993).

NORMAND 1994a. E. NORMAND, Single Event Effects in Systems Using Commercial Electronics in Harsh Environments, IEEE NSREC Short Course Notes (July 1994).

NORMAND 1994b. E. NORMAND, D. L. OBERG, J. L. WERT, J. D. NESS, P. P. MAJEWSKI, S. WENDER, and A. GAVRON, Single Event Upset and Charge Collection Measurements Using High Energy Protons and Neutrons, *IEEE Trans. Nucl. Sci.*, **NS-41**, 2203 (1994).

NORMAND 1996. E. NORMAND, Single Event Effects in Avionics, *IEEE Trans. Nucl. Sci*., **NS-43**, 461–474 (April 1996).

NORMAND 1998. E. NORMAND, Extensions of the Burst Generation Rate Method for Wider Application to Proton/Neutron-Induced Single Event Effects, *IEEE Trans. Nucl. Sci*., **NS-45**, 2904 (1998).

NORMAND 2001. E. NORMAND, Correlation of In-flight Neutron Dosimeter and SEU Measurements with Atmospheric Neutron Model, *IEEE Trans. Nucl. Sci*., **NS-48**, 1996–2003 (2001).

NORMAND 2004. E. NORMAND, Extensions of the FOM Method—Proton SEL and Atmospheric Neutron SEU, *IEEE Trans. Nucl. Sci*., **NS-51**, 3494–3504 (2004).

NORMAND 2006. E. NORMAND, K. VRANISH, A. SHEETS, M. STITT, and R. KIM, Quantifying the Double-Sided Neutron SEU Threat, from Low Energy (Thermal) and High Energy (>10 MeV) Neutrons, *IEEE Trans. Nucl. Sci*., **NS-53**, 3587–3595 (2006).

OBERG 1996. D. L. OBERG, J. L. WERT, P. P. MAJEWSKI, and S. A. WENDER, First Observations of Power MOSFET Burnout with High Energy Neutrons, *IEEE Trans. Nucl. Sci*., **NS-43**, 2913 (1996).

O'BRYAN 2005. M. V. O'BRYAN, K. A. LABEL, S. D. KNIFFEN, C. POIVEY, J. W. HOWARD, Jr., R. L. LADBURY, S. P. BUCHNER, T. R. OLDHAM, P. W. MARSHALL, A. B. SAUNDERS, D. K. HAWKINS, J. D. FORNEY, T. IRWIN, C. M. SEIDLECK, S. R. COX, C. PALOE, D. PERTICK, W. POWELL, and B. L. WILLITS, Recent Single Event Effects Results for Candidate Spacecraft Electronics for NASA, IEEE 2005 IEEE Radiation Effects Data Workshop Record, p. 26 (2005).

O'BRYAN 2008. M. V. O'BRYAN, K. A. LABEL, S. P. BUCHNER, R. L. LADBURY, C. F. POIVEY, T. R. OLDHAM, M. J. CAMPOLA, M. A. CARTS, M. D. BERG, A. B. SANDERS, and S. R. MACKEY, Compendium of Recent Single Event Effects Results for Candidate Spacecraft Electronics for NASA, 2008 IEEE Radiation Effects Data Workshop Record, pp. 11–20 (July 2008).

O'GORMAN 1996. T. J. O'GORMAN et al., Field Testing for Cosmic Ray Soft Errors in Semiconductor Memories, *IBM J. Res. Dev*., **40**(1), 41 (January 1996).

O'NEILL 1994. P. M. O'NEILL and G. D. BADHWAR, Single Event Upsets for Space Shuttle Flights of New General Purpose Computer Memory Devices, *IEEE Trans. Nucl. Sci*., **NS-41**, 1755 (October 1994).

O'NEILL 1998. P. M. O'NEILL, G. D. BADHWAR, and W. X. CULPEPPER, Internuclear Cascade–Evaporation Model for LET Spectra of 200 MeV Protons Used for Parts Testing, *IEEE Trans. Nucl. Sci*., **NS-45**, 2467 (1998).

OLDHAM 2003. T. OLDHAM, Scaling and Single Event Effects (SEE) Sensitivity, IEEE NSREC Short Course Notes (July 2003).

OLSEN 2005. B. D. OLSON, D. R. BALL, K. M. WARREN, L. W. MASENGILL, N. F. HADDAD, and D. MCMCMORROW, Simultaneous SE Charge Sharing and Parasitic Bipolar Conduction in a Highly-Scaled SRAM Design, *IEEE Trans. Nucl. Sci*., **NS-52**, 2132 (2005).

PALAU 2002. J. M. PALAU, F. WROBEL, K. CASTELLANI-COULE, M. C. CALVET, P. E. DODD, and F. W. SEXTON, Monte Carlo Exploration of Neutron-Induced SEU-Sensitive Volumes in SRAMs, *IEEE Trans. Nucl. Sci.*, **NS-49**, 3075 (2002).

PEASE 2008. R. L. PEASE, Modeling Single Event Transients in Bipolar Linear Circuits, *IEEE Trans. Nucl. Sci.*, **NS-55**, 1879–1890 (August 2008).

PELLISH 2007. J. A. PELLISH, R. A. REED, A. K. SUTTON, R. A. WELLER, M. A. CARTS, P. W. MARSHALL, C. J. MARSHALL, R. KRITHIVASAN, J. D. CRESSLER, M. H. MENDENHALL, R. D. SCHRIMPF, K. M. WARREN, B. D. SIERAWSKI, and G. F. NIU, A Generalized SiGe HBT Single-Event Effects Model for On-Orbit Event Rate Calculations, *IEEE Trans. Nucl. Sci.*, **NS-54**, 2322–2329 (2007).

PETERSEN 1967. E. L. PETERSEN, I. SLAUS, J. W. VERBA, R. F. CARLSON, and J. R. RICHARDSON, Inelastic Scattering of 45 MeV Protons from Carbon, *Nucl. Phys.*, **102**, 145–160 (1967).

PETERSEN 1980. E. L. PETERSEN, Nuclear Reactions in Semiconductors, *IEEE Trans. Nucl. Sci.*, **NS-27**, 1494 (1980).

PETERSEN 1981. E. L. PETERSEN, Soft Errors Due to Protons in the Radiation Belt, *IEEE Trans. Nucl. Sci.*, **NS-28**, 3981 (1981).

PETERSEN 1982. E. L. PETERSEN, P. SHAPIRO, J. H. ADAMS, Jr., and E. A. BURKE, Calculation of Cosmic-Ray Induced Soft Upsets and Scaling in VLSI Devices, *IEEE Trans. Nucl. Sci.*, **NS-29**, 2055 (1982).

PETERSEN 1983a. E. L. PETERSEN, J. B. LANGWORTHY, and S. E. DIEHL, Suggested Single Event Upset Figure of Merit, *IEEE Trans. Nucl. Sci.*, **NS-30**, 4533–4539 (1983).

PETERSEN 1983b. E. L. PETERSEN, Radiation Induced Soft Fails in Space Electronics, *IEEE Trans. Nucl. Sci.*, **NS-30**, 1638–1641 (April 1983).

PETERSEN 1983c. E. L. PETERSEN, Single Event Upsets in Space: Basic Concepts, in IEEE Nuclear and Space Radiation Effects Conference Short Course Text (1983).

PETERSEN 1988. E. L. PETERSEN and P. W. MARSHALL, Single Event Phenomena in the Space and SDI Arenas, *J. Radiat Effects Res. Eng.*, **6**, 1 (1988).

PETERSEN 1992a. E. L. PETERSEN and J. H. ADAMS, Jr., Comparison of SEU Rate Prediction Techniques, *IEEE Trans. Nucl. Sci.*, **NS-39**, 1836 (1992).

PETERSEN 1992b. E. L. PETERSEN, J. C. PICKEL, J. H. ADAMS, Jr., and E. C. SMITH, Rate Prediction for Single Event Effects—A Critique, *IEEE Trans. Nucl. Sci.*, **NS-39**, 1577–1599 (1992).

PETERSEN 1992c. E. L. PETERSEN, The Relationship of Proton and Heavy Ion Upset Thresholds, *IEEE Trans. Nucl. Sci.*, **NS-39**, 1600 (1992).

PETERSEN 1993. E. L. PETERSEN, J. C. PICKEL, E. C. SMITH, P. J. RUDEK, and J. R. LETAW, Geometrical Factors in SEE Rate Predictions, *IEEE Trans. Nucl. Sci.*, **NS-40**, 1888–1909 (December 1993).

PETERSEN 1995. E. L. PETERSEN, SEE Rate Calculations Using the Effective Flux Approach and a Generalized Figure of Merit Approximation, *IEEE Trans. Nucl. Sci.*, **NS-42**, 1995 (1995).

PETERSEN 1996a. E. L. PETERSEN, Approaches to Proton Single-Event Rate Calculations, *IEEE Trans. Nucl. Sci.*, **NS-43**, 496 (April 1996).

PETERSEN 1996b. E. L. PETERSEN, Interpretation of Heavy Ion Cross Section Measurements, *IEEE Trans. Nucl. Sci.*, **NS-43**, 952 (June 1996).

PETERSEN 1996c. E. L. PETERSEN, Cross Section Measurements and Upset Rate Calculations, *IEEE Trans. Nucl. Sci.*, **NS-43**, 2805–2813 (December 1996).

PETERSEN 1997a. E. L. PETERSEN, Predictions and Observations of SEU Rates in Space, *IEEE Trans. Nucl. Sci.*, **NS-44**, 2174 (1997).

PETERSEN 1997b. E. L. PETERSEN, Single Event Analysis and Prediction, 1997 IEEE NSREC Short Course Notes (July 1997).

PETERSEN 1998. E. L. PETERSEN, The SEU Figure of Merit and Proton Upset Rate Calculations, *IEEE Trans. Nucl. Sci.*, **NS-45**, 2550–2562 (1998).

PETERSEN 2005. E. L. PETERSEN, V. POUGET, L. W. MASSENGILL, S. P. BUCHNER, and D. MCMORROW, Rate Predictions for Single-Event Effects—Critique II, *IEEE Trans. Nucl. Sci.*, **NS-52**, 2158–2167 (2005).

PETERSEN 2007. E. L. PETERSEN, Parametric and Threshold Studies of Single Event Sensitivity, *IEEE Trans. Nucl. Sci.*, **NS-54**, 1392–1405 (August 2007)

PETERSEN 2008a. E. L. PETERSEN, Single Event Data Analysis, *IEEE Trans. Nucl. Sci.*, **NS-55**, 2819–2841 (2008).

PETERSEN 2008b. E. L. PETERSEN, Soft Error Results Analysis and Error Rate Prediction, 2008 IEEE NSREC Short Course Notebook (July 2008).

PETERSEN 2010. E. L. PETERSEN, Dosimetry Comparisons Between Various Single Event Test Facilities, *IEEE Trans. Nucl. Sci*, **57**, 3477–3482 (2010).

PICKEL 1978. J. C. PICKEL and J. T. BLANDFORD, Jr., Cosmic Ray Induced Errors in MOS Memory Cells, *IEEE Trans. Nucl. Sci.*, **NS-25**, 1166 (1978).

PICKEL 1980. J. C. PICKEL and J. T. BLANDFORD, Cosmic-Ray Induced Errors in MOS Devices, *IEEE Trans. Nucl. Sci.*, **NS-27**, 1006–1014 (April 1980). Erratum, *ibid.*, **NS-29**, 2122 (December 1982).

PICKEL 1982. J. C. PICKEL, Effect of CMOS Miniaturization on Cosmic-Ray-Induced Error Rate, *IEEE Trans. Nucl. Sci.*, **NS-29**, 2049 (1982).

PICKEL 1983. J. C. PICKEL, Single Event Upset Mechanisms and Predictions, in IEEE Nuclear and Space Radiation Effects Conference Short Course Text (1983).

PICKEL 1984. J. C. PICKEL and J. T. BLANDFORD, Modeling for Single Event Error Rate Prediction, DNA-TR-84- 317 (June 1984).

PICKEL 1989. J. C. PICKEL, B. LAWTON, A. L. FRIEDMAN, and P. J. MCNULTY, Proton-Induced SEU in CMOS/SOS, *J. Radiat Effects Res. Eng.*, **7**, 67 (1989).

PICKEL 1996. J. C. PICKEL, Single-Event Effects Rate Predictions, *IEEE Trans. Nucl. Sci.*, **NS-43**, 483 (April 1996).

POIVEY 1998. C. POIVEY, B. DOUCIN, M. BRUGGEMANN, and R. HARGOE-SORENSEN, Radiation Characterisation of Commercially Available 1 Mbit/4 Mbit SRAMs for Space Applications, 1998 NSREC Data Workshop, p. 98 (1998).

POUGET 2000. V. POUGET, P. FOUILLAT, D. LEWIS, H. LAPUYADE, F. DARRACQ, A. TOUBOUL, Laser Cross-Section Measurement for the Evaluation of Single-Event Effects in Integrated Circuits, *Microelectronics Reliability*, **40**, 1371–1375 (2000).

POUGET 2004. V. POUGET, D. LEWIS, and P. FOUILLAT, Time-Resolved Scanning of Integrated Circuits with a Pulsed Laser: Application to Transient Fault Injection in an ADC, *IEEE Trans. Instrum. Meas.*, **53**(4), 1227–1231 (2004).

QUINN 2005. H. QUINN, P. GRAHAM, J. KRONE, M. GAFFREY, and S REZGUI, Radiation-Induced Multi-bit Upsets in SRAM-Based FPGAs, *IEEE Trans. Nucl. Sci.*, **NS-52**, 2455 (2005).

QUINN 2007. H. QUINN, K. MORGAN, P. GRAHAM, J. KRONE, M. CAFFREY, and K. LUNDGREEN, Domain Crossing Errors: Limitations on Single Device Triple-Modular Redundancy Circuits in Xilinx FPGAs, *IEEE Trans. Nucl. Sci.*, **NS-54**, 2037 (2007).

REED 1994a. R. A. REED, P. J. MCNULTY, and W. G. ABDEL-KADER, Implications of Angle of Incidence in SEU Testing of Modern Circuits, *IEEE Trans. Nucl. Sci.*, **NS-41**, 2049 (1994).

REED 1994b. R. A. REED, P. J. MCNULTY, W. J. BEAUVAIS, W. G. ABDEL-KADER, E. G. STASSINOPOULOS, and J. C. L BARTH, A Simple Algorithm for Predicting Proton SEU Rates in Space Compared to the Rates Measured on the CRESS Satellite, *IEEE Trans. Nucl. Sci.*, **NS-41**, 2389 (1994).

REED 1995. R. A. REED and P. J. MCNULTY, Effects of Geometry on the Proton SEU Dependence on the Angle of Incidence, *IEEE Trans. Nucl. Sci.*, **NS-42**, 1803 (1995).

REED 1996. R. A. REED, M. A. CARTS, P. W. MARSHALL, S. BUCHNER, M. LAMACCHIA, B. MATHER, and D. MCMORROW, Single Event Upset Cross Section at Various Data Rates, *IEEE Trans. Nucl. Sci.*, **NS-43**, 2862–2867 (1996).

REED 1997. R. A. REED, M. A. CARTS, P. W. MARSHALL, C. J. MARSHALL, O. MUSSEAU, P. J. MCNULTY, D. R. ROTH, S. BUCHENER, J. MELINGER, and T. CORBIERE, Heavy Ion and Proton-Induced Single Event Multiple Upset, *IEEE Trans. Nucl. Sci.*, **NS-44**, 2224 (1997).

REED 2006. R. A. REED, R. A. WELLER, R. D. SCHRIMPF, M. H. MENDENHALL, K. M. WARREN, and L. W. MASSENGILL, Implications of Nuclear Reactions for Single Event Effects Test Methods and Analysis, *IEEE Trans. Nucl. Sci.*, **NS-53**, 3356 (2006).

REED 2007. R. A. REED, R. A. WELLER, M. H. MENDENHALL, J. M. LAUENSTEIN, K. M. WARREN, J. A. PELLISH, R. D. SCHRIMPF, B. D. SIERAWSKI, L. W. MASSENGILL, P. E. DODD, M. R. SHANEYFELT, J. A. FELIX, J. R. SCHWANK, N. F. HADDAD, R. K. LAWARENCE, J. H. BOWMAN, and R. CONDE, Impact of Ion Energy and Species on Single Event Effects Analysis, *IEEE Trans. Nucl. Sci.*, **NS-54**, 2312 (2007).

REED 2008. R. REED, Fundamental Mechanisms for Single Particle-Induced Soft Error, 2008 IEEE NSREC Short Course Notebook (July 2008).

Remke 1989. R. L. Remke, S. B. Witmer, S. D. F. Jones, F. E. Barber, L. D. Flesner, and M. E. O'Brien, SEU Measurements on HFETs and HFET RAMs, *IEEE Trans. Nucl. Sci.*, **NS-36**, 2362 (1989).

Rezgui 2008. S. Rezgui, J. J. Wang, Y. Sun, B. Cronquist, and J. McCollum, Configuration and Routing Effects on the SET Propagation in Flash-Based FPGAs, *IEEE Trans. Nucl. Sci.*, **NS-55**, 3328–3335 (2008).

Ritter 1996. J. C. Ritter, Spacecraft Anomalies and Future Trends, in IEEE Nuclear and Space Radiation Effects Conference Short Course Text (1996).

Rodbell 2007. K. P. Rodbell, D. F. Heidel, H. H. K. Tang, M. S. Gordon, P. Oldiges, and C. E. Murray, Low-Energy Proton-Induced Single-Event-Upsets in 65 nm Node, Silicon-on-Insulator Latches and Memory Cells, *IEEE Trans. Nucl. Sci.*, **NS-54**, 2474 (2007).

Rollins 1987. J. G. Rollins, J. Choma, Jr., and W. A. Kolasinski, Single Event Effects in SOS Integrated Circuits, *IEEE Trans. Nucl. Sci.*, **NS-34**, 1713 (1987).

Rollins 1990. J. G. Rollins, Estimation of Proton Upset Rates from Heavy Ion Test Data, *IEEE Trans. Nucl. Sci.*, **NS-37**, 1961–1965 (1990).

Roth 1993. D. R. Roth, P. J. McNulty, W. G. Abdel-Kader, L. Strauss, and E. G. Stassinopoulos, Monitoring SEU Parameters at Reduced Bias, *IEEE Trans. Nucl. Sci.*, **NS-40**, 1721–1724 (1993).

Rudie 1996. N. J. Rudie, unpublished (1996).

Santin 2006. G. Santin, D. Wright, and M. Asai, Space Radiation Transport Models, 2006 IEEE NSREC Short Course Notebook (July 2006).

Schrimpf 2008. R. D. Schrimpf, K. M. Warren, D. R. Ball, R. A. Weller, R. A. Reed, D. M. Fleetwood, L. W. Massengill, M. H. Mendhall, S. N. Rashkeev, S. T. Pantelides, and M. A. Alles, Multi-Scale Simulation of Radiation Effects in Electronic Devices, *IEEE Trans. Nucl. Sci.*, **NS-55**, 1891–1902 (August 2008).

Schwank 2003. J. R. Schwank, V. Ferlot-Cavrois, M. R. Sheneyfelt, P. Paillet, and P. E. Dodd, Radiation Effects in SOI Technologies, *IEEE Trans. Nucl. Sci.*, **NS-50**, 522–538 (June 2003).

Schwank 2006. J. R. Schwank, M. R. Shaneyfelt, J. A. Felix, P. E. Dodd, J. Baggio, V. Ferlet-Cavrois, P. Paillet, G. L. Hash, R. S. Flores, L. W. Massengill, and E. Blackmore, Effects of Total Dose Irradiation on Single-Event Upset Hardness, *IEEE Trans. Nucl. Sci.*, **NS-53**, 1772–1778 (August 2006).

Scott 1989a. T. Scott, Cosmic Ray Upset Experiment, IBM Report 89-PN6-023 (November 1989).

Scott 1989b. T. M. Scott, A Single Event Rate Calculation Technique, IBM Report 89-PN6-004 (February 1989).

Segre 1964. E. Segre, Nuclei and Particles. New York: W. A. Benjamin, 1964.

Sexton 1989. F. W. Sexton, J. S. Fu, R. A. Kohler, and R. Koga, SEU Characterization of a Hardened CMOS 64K and 256K SRAM, *IEEE Trans. Nucl. Sci.*, **NS-36**, 2311–2317 (1989).

SEXTON 1991. F. W. SEXTON, W. T. CORBETT, R. K. TREECE, K. J. HASS, K. L. HUGHES, C. L. AXNESS, G. L HASH, M. R. SHANEYFELT, and T. F. WUNSCH, SEU Simulation and Testing of Resistor-Hardened D-Latches in the SA3300 Microprocessor, *IEEE Trans. Nucl. Sci.*, **NS-38**, 1521–1528 (1991).

SEXTON 1992. F. W. SEXTON, Measurement of Single Event Phenomena in Devices and ICs, in IEEE Nuclear and Space Radiation Effects Conference Short Course Text (1992).

SEXTON 1993. F. W. SEXTON, K. M. HORN, B. L. DOYLE, J. S. LAIRD, M CHOLEWA, A. SAINT, and G. J. F. LEGGE, Relationship Between IBICC Imaging and SEU in CMOS ICs, *IEEE Trans. Nucl. Sci.*, **NS-40**, 1787 (1993).

SHAPIRO 1982. P. SHAPIRO, E. L. PETERSEN, and J. H. ADAMS, Jr., NRL Memorandum Report 4864 (1982).

SHAPIRO 1983. P. SHAPIRO, Calculation of Cosmic Ray Induced Single Event Upsets: Program CRUP, Cosmic Ray Upset Program, NRL Memorandum Report 5171 (September 1983).

SHEA 1992. M. A. SHEA, D. F. SMART, J. H. ALLEN, and D. C. WILKINSON, Spacecraft Problems in Association with Episodes of Intense Solar Activity and Related Terrestrial Phenomena During March 1991, *IEEE Trans. Nucl. Sci.*, **NS-39**, 1754 (1992).

SHIMANO 1989. Y. SHIMANO, T. GOKA, S. KUBOYAMA, K. KAWACHI, T. KANAI, and Y. TAKAMI, The Measurement and Prediction of Proton Upset, *IEEE Trans. Nucl. Sci.*, **NS-36**, 2344 (1989).

SHINN 1995. J. L. SHINN, F. A. CUCINOTTA, J. W. WILSON, G. D. BADHWAR, P. M. O'NEILL, and F. F. BADAVI, Effects of Target Fragmentation on Evaluation of LET Spectra from Space Radiation in Low-Earth Orbit (LEO) Environment: Impact on SEU Predictions, *IEEE Trans. Nucl. Sci.*, **NS-42**, 2017 (1995).

SHIONA 1986. N. SHIONA, Y. SAKAGAW, M. SEKIGUCHI, K. SATE, I. SUGAI, T. HATTORI, and Y. HIRAO, Single Event Effects in High Density CMOS RAMS, *IEEE Trans. Nucl. Sci.*, **NS-33**, 1632 (1986).

SHOGA 1987. M. SHOGA, P. ADAMS, D. L. CHENETTE, R. KOGA, and E. C. SMITH, Verification of Single Event Upset Rate Estimations Methods with On-Orbit Observations, *IEEE Trans. Nucl. Sci.*, **NS-34**, 1256 (1987).

SIERAWSKI 2009. B. D. SIERAWSKI, J. A. PELLISH, R. A. REED, R. D. SCHRIMPF, K. M. WARREN, R. A. WELLER, M. H. MENDENHALL, J. D. BLACK, A. D. TIPTON, M. A. XAPSOS, R. C. BAUMAN, X. DENG, M. J. CAMPOLA, M. R. FRIENDLICH, H. S. KIM, A. M. PHAN, and C. M. SEIDLECK, Impact of Low-Energy Proton Induced Upsets on Test Methods and Rate Predictions, *IEEE Trans. Nucl. Sci.*, **NS-56**, 3085–3092 (2009).

SILBERBERG 1984. R. SILBERBERG, C. H. TSAO, and J. R. LETAW, Neutron Generated Single-Event Upsets in the Atmosphere, *IEEE Trans. Nucl. Sci.*, **NS-31**, 1183–1185 (1984).

SILBURT 2008. A. SILBURT, A. EVANS, A. BURGHELEA, S. WEN, D. WARD, R. NORRISH, and D. HOGLE, Specifications and Verification of Soft Error Performance in

Reliable Internet Core Routers, *IEEE Trans. Nucl. Sci.*, **NS-55**, 2389–2398 (August 2008, Part II).

SMITH 1992. E. C. SMITH and M. SHOGA, Double Upsets from Glancing Collisions: A Simple Model Verified with Flight Data, *IEEE Trans. Nucl. Sci.*, **NS-39**, 1859 (1992).

SMITH 1994. E. C. SMITH, Effects of Realistic Satellite Shielding on SEE Rates, *IEEE Trans. Nucl. Sci.*, **NS-41**, 2396 (1994).

SMITH 1995. E. C. SMITH, E. G. STASSINOPOULOS, K. LABEL, G. BRUCKER, and C. M. SEIDLICK, Applications of a Diffusion Model to SEE Cross Section of Modern Devices, *IEEE Trans. Nucl. Sci.*, **NS-42**, 1772 (1995).

SOKOL 1987. J. H. SOKOL, W. A. KOLASINSKI, M. WONG, R. KOGA, R. SUHRKE, and T. H. FREY, Advantage of Advanced CMOS Over Advanced TTL in a Cosmic Ray Environment, *IEEE Trans. Nucl. Sci.*, **NS-34**, 1338–1340 (1987).

SOLIN 2005. J. R. SOLIN, A. J. TYLKA, M. A. SHEA, and D. F. SMART, The Cosmic Ray Environment of Tactical ABMs, *IEEE Trans. Nucl. Sci.*, **NS-52**, 546 (April 2005).

SONG 1988. Y. SONG, K. N. VU, J. S. CABLE, A. A. WITTELES, W. A. KOLASINSKI, R. KOGA, J. H. ELDER, J. V. OSBORN, R. C. MARTIN, and N. M. GHONIEM, Experimental and Analytical Investigation of Single Event, Multiple Bit Upsets, in Poly-silicon Load, 64K x 1 NMOS SRAMs, *IEEE Trans. Nucl. Sci.*, **NS-35**, 1673 (1988).

SRINIVASAN 1994. G. R. SRINIVASAN, H. K. TANG, and P. C. MURLEY, Parameter-Free, Predictive Modeling of Single Event Upsets due to Protons, Neutrons, and Pions in Terrestrial Cosmic Rays, *IEEE Trans. Nucl. Sci.*, **41**, 2063 (1994).

SRINIVASAN 1996. Modeling the Cosmic-Ray-Induced Soft-Error Rate in Integrated Circuits: An Overview, *IBM J. Res. Dev.*, **40**(1), 77 (January 1996).

STAPOR 1990. W. J. STAPOR, J. P. MEYERS, J. B. LANGWORTHY, and E. L. PETERSEN, Two Parameter Bendel Model Calculations for Predicting Proton Induced Upset, *IEEE Trans. Nucl. Sci.*, **NS-37**, 1966 (1990).

STAPOR 1995. W. J. STAPOR, Single Event Effects (SEE) Qualification, IEEE NSREC Short Course Text (1995).

STRANATHAN 1942. J. D. STRANATHAN, *The "Particles" of Modern Physics*, Philadelphia: Blakiston Company, 1942; reprint 1948.

SWIFT 2001. G. M. SWIFT, F. F. FARMANESH, S. M. GUERTIN, F. IROM, and D. G. MILLWARD, Single-Event Upset in the PowerPC750 Microprocessor, *IEEE Trans. Nucl. Sci.*, **NS-48**, 1822–1827 (2001).

SWIFT 2008. G. M. SWIFT, G. R. ALLEN, C. W. TSENG, C. CARMICHAEL, G. MILLER, and J. S. GEORGE, Static Upset Characteristics of 90nm Virtex-4QV FPGAs, 2008 IEEE Radiation Effects Data Workshop Record, pp. 98–105 (2008).

TABOR 1993. A. TABOR and E. NORMAND, Single Event Upset in Avionics, *IEEE Trans. Nucl Sci.*, **NS-40**, 120 (April 1993).

TAKAGI 1993. S. TAKAGI, T. NAKAMURA, T. KOHNO, N. SHIONO, and F. MAKINO, Observation of Space Radiation Environment with EXOS-D, *IEEE Trans. Nucl. Sci.*, **NS-40**, 1491 (1993).

TAKAMI 1990. Y. TAKAMI, F. SHIRAISHI, T. GOD, Y. SHIMANO, M. SEKIGUICHI, K. SHIDA, N. KISHIDA, H. KADOTANI, T. KIKUCHI, N. HOSHINO, S. MURAKAMI, H. ANAYAMA, and A. MORIO, Investigation of Single Event Upset Subject to Protons of Intermediate Energy Range, *IEEE Trans. Nucl. Sci.*, **NS-37**, 1953 (1990).

TAKEDA 1989. E. TAKEDA, K. TAKEUCHI, D. HISAMOTO, T. TOYABE, K. OHSHIMA, and K. ITOH, A Cross Section of α-Particle Induced-Soft-Error Phenomena in VLSIs, *IEEE Trans. Electron Devices*, **ED-36**, 2567 (November 1989).

TANG 1996. H. H. K. TANG, Nuclear Physics of Cosmic Ray Interactions with Semiconductor Materials: Particle-Induced Soft Errors from a Physicist's Perspective, *IBM J. Res. Dev.*, **40**(1), 91 (January 1996).

TANG 2004. H. H. K. TANG and E. H. CANNON, SEMM-2: A Modeling System for Single Event Upset Analysis, *IEEE Trans. Nucl. Sci.*, **NS-51**, 3342 (2004).

TANG 2007. H. H. K. TANG, C. E. MURRAY, G. FIORENZA, K. P. RODBELL, and M. S. GORDON, Importance of BEOL Modeling in Single Event Effect Analysis, *IEEE Trans. Nucl. Sci.*, **NS-54**, 2162 (2007).

TAUSCH 2009. H. J. TAUSCH, Simplified Birthday Statistics and Hamming EDAC, *IEEE Trans. Nucl. Sci.*, **NS-56**, 474 (April 2009).

TIPTON 2006. A. D. TIPTON, J. A. PELLISH, R. A. REED, R. D. SCHRIMPF, R. A. WELLER, M. H. MENDENHALL, B. SIERAWSKI, A. K. SUTTON, R. M. DIESTELHORST, G. ESPINEL, J. D. CRESSLER, P. W. MARSHALL, and G. VIZKELETHY, Multiple-Bit Upset in 130 nm CMOs Technology, *IEEE Trans. Nucl. Sci.*, **NS-53**, 3259 (2006).

TRUSCOTT 2004. P. TRUSCOTT, F. LEI, C. S. DYER, A. FRYLAND, S. CLUCAS, B. TROUSSE, K. HUNTER, C. COMBER, A. CHUGG, and M. MOUTRIE, Assessment of Neutron- and Proton-Induced Nuclear Interaction and Ionization Models in GEANT4 for Simulating Single Event Effects, *IEEE Trans. Nucl. Sci.*, **NS-51**, 3369 (2004).

TSAO 1983. C. H. TSAO, R. SILBERBERG, J. H. ADAMS, Jr., and J. R. LETAW, Cosmic Ray Transport in the Atmosphere: Dose and LET-Distributions in Materials, *IEEE Trans. Nucl. Sci.*, **NS-30**, 4398–4402 (1983).

TSAO 1984. C. H. TSAO, R. SILBERBERG, and J. R. LETAW, Cosmic-Ray Heavy Ions at and above 40,000ft, *IEEE Trans. Nucl. Sci.*, **NS-31**, 1066–1068 (1984).

TYLKA 1994. A. J. TYLKA, P. R. BOBERG, and J. H. ADAMS, Jr., LET Spectra of Trapped Anomalous Cosmic Rays in Low-Earth Orbit, presented at COSPAR, Hamburg Germany, Summer 1994.

TYLKA 1996. A. J. TYLKA, W. F. DIETRICK, P. R. BOBERG, E. C. SMITH, and J. H. ADAMS, Jr., Single Event Upsets Caused by Solar Energetic Heavy Ions, *IEEE Trans. Nucl. Sci.*, **NS-43**, 2758 (1996).

TYLKA 1997. A. J. TYLKA, J. H. ADAMS, P. R. BOBERG, B. BROWNSTEIN, W. F. DIETRICH, E. O. FLUECKIGER, E. L. PETERSEN, M. A. SHEA, D. F. SMART, and

E. C. SMITH, CREME96: A Revision of the Cosmic Ray Effects on Microelectronics Code, *IEEE Trans. Nucl. Sci.*, **NS-44**, 2150 (1997).

UNDERWOOD 1992. C. I. UNDERWOOD, J. W. WARD, C. S. DYER, and A. J. SIMS, Observations of Single Event Upsets in Non-hardened High-Density SRAMs in Sun-Synchronous Orbit, *IEEE Trans. Nucl. Sci.*, **NS-39**, 1817 (1992).

UNDERWOOD 1993. C. I. UNDERWOOD, R. ECOFET, S. DUZELLIER, and D. FAGURERE, Observation of Single-Event Upset and Multiple-Bit Upset in Non-hardened High-Density SRAMs in the TOPEX/Poseidon Orbit, 1993 IEEE Radiation Effects Data Workshop Record, p. 85 (1993).

UNDERWOOD 1994. C. I. UNDERWOOD, D. J. BROCK, P. S. WILLIAMS, S. KIM, R. DILAO, P. R. SANTOS, M. C. BRITO, C. S. DYER, and A. J. SIMS, Radiation Environment Measurements with the Cosmic Ray Experiments On-board the KITSAT-1 and PoSAT-1 Micro-Satellites, *IEEE Trans. Nucl. Sci.*, **NS-41**, 2353 (1994).

VAMPOLA 1994. A. L. VAMPOLA, M. LAURIENTE, D. C. WILKINSON, J. ALLEN, and F. ALBIN, Single Event Upsets Correlated with Environment, *IEEE Trans. Nucl. Sci.*, **NS-41**, 2383 (1994).

VELACHERI 1994. S. VELACHERI, L. S. MASSENGILL, and S. E. KERNS, Single-Event-Induced Charge Collection and Direct Channel Conduction in Submicron MOS-FETs, *IEEE Trans. Nucl. Sci.*, **NS-41**, 2103 (1994).

WAGNER 1986. R. S. WAGNER, J. M. BRADLEY, C. J. MAGGIORE, J. G. BEERY, and R. B. HAMMOND, An Approach to Measure Ultrafast-Funneling-Current Transients, *IEEE Trans. Nucl. Sci.*, **NS-33**, 1651 (1986).

WAGNER 1988. R. S. WAGNER, N. BORDES, J. M. BRADLEY, C. J. MAGGIORE, A. R. KNUDSON, and A. B. CAMPBELL, Alpha-, Boron-, Silicon-, and Iron-Ion-Induced Current Transients in Low-Capacitance Silicon and GaAs Diodes, *IEEE Trans. Nucl. Sci.*, **NS-35**, 1578 (1988).

WALLMARK 1962. J. T. WALLMARK and S. M. MARCUS, Minimum Size and Maximum Packing Density of Nonredundant Semiconductor Devices, *Proc. IRE*, **50**, 286 (1962).

WARREN 1999. K. WARREN, L. MASSENGILL, R. SCHRIMPT, and H. BARNABY, Analysis of the Influence of MOS Device geometry on Predicted SEU Cross-Sections. *IEEE Trans. Nucl. Sci.*, **NS-46**, 1363–1369 (1999).

WARREN 2005. K. M. WARREN, R. A. WELLER, M. H. MENDENHALL, R. A. REED, D. R. BALL, C. L. HOWE, B. D. OLSON, M. L. ALLES, L. W. MASSSENGILL, R. D. SCHRIMPF, N. F. HADDAD, S. E. DOYLE, D. MCMORROW, J. S. MELINGER, and W. T. LOTSHAW, The Contribution of Nuclear Reactions to Heavy Ion Single Event Upset Cross-Section Measurements in a High-Density SEU Hardened SRAM, *IEEE Trans. Nucl. Sci.*, **NS-52**, 2125–2131 (2005).

WARREN 2007a. K. M. WARREN, B. D. SIERAWSKI, R. A. REED, R. A. WELLER, C. CARMICHAEL, A. LESEA, M. H. MENDENHALL, P. E. DODD, R. D. SCHRIMPF, L. W. MASSENGILL, T. HOANG, H. WAN, J. L. DE JONG, R. PADOVANI, and F. J. FABULA, Monte-Carlo Based On-Orbit Single Event Upset Rate Prediction for a Radiation Hardened by Design Latch, *IEEE Trans. Nucl. Sci.*, **NS-54**, 2419 (2007).

WARREN 2007b. K. M. WARREN, R. A. WELLER, B. D. SIERAWSKI, R. A. REED, M. H. MENDENHALL, R. D. SCHRIMPF, L. W. MASSENGILL, M. E. PORTER, J. D. WILKINSON, K. A. LABEL, and J. H. ADAMS, Application of RADSAFE to Model the Single Event Upset Response of a 0.25 μm CMOS SRAM, *IEEE Trans. Nucl. Sci.*, **NS-54**, 898 (August 2007).

WARREN 2008. K. M. WARREN, A. L. STERNBERG, R. A. WELLER, M. P. BAZE, L. W. MASSENGILL, R. A. REED, M. H. MENDENHALL, and R. D. SCHRIMPF, Integrating Circuit Level Simulation and Monte-Carlo Radiation Transport Code in Single Event Upset Analysis in SEU Hardened Circuitry, *IEEE Trans. Nucl. Sci.*, **NS-55**, 2886–2893 (2008).

WEATHERFORD 1986. T. R. WEATHERFORD, J. R. HAUSER, and S. E. DIEHL, Comparisons of Single Event Vulnerability of GaAs SRAMs, *IEEE Trans. Nucl. Sci.*, **NS-33**, 1590 (1986).

WEATHERFORD 2002. T. WEATHERFORD, From Carriers to Contacts, a Review of SEE Charge Collection Processes in Devices, IEEE NSREC Short Course Notes (July 2002).

WELLER 2003. R. A. WELLER, A. L. STERNBERG, L. W. MASSENGILL, R. D. SCHRIMPF, and D. M. FLEETWOOD, Evaluating Average and Atypical Response in Radiation Effects Simulations, *IEEE Trans. Nucl. Sci.*, **NS-50**, 2265 (2003).

WELLER 2009. R. A. WELLER, R. A. REED, K. M. WARREN, M. H. MENDENHALL, B. D. SIERAWSKI, R. D. SCHRIMPF, and L. W. MASSENGILL, General Framework for Single Event Effects Rate Prediction in Microelectronics, *IEEE Trans. Nuc. Sci.*, **NS-56**, 3098–3108 (2009).

WERT 2005. J. WERT, E. NORMAND, and C. HAFER, The Effects of Device Metal Interconnect Overlayers on SEE Testing, Radiation Effects Data Workshop, 2005 IEEE NSREC, pp. 20–25 (2005).

WEULERSSE 2006. C. WEULERSSE, G. HUBERT, G. FORGET, N. BUARD, T. CARRIERE, P. HEINS, J. M. PALAU, F. SAIGNE, and R. GAILLARD, DASIE Analytical Version: A Predictive Tool for Neutrons, Protons and Heavy Ion Induced SEU Cross Section, *IEEE Trans. Nucl. Sci.*, **NS-53**, 1876 (August 2006).

WILKINSON 1991. D. C. WILKINSON, S. C. DAUGHTRIDGE, J. L. STONE, H. H. SAUER, and P. DARLING, TDRS-1 Single Event Upsets and the Effect of the Space Environment, *IEEE Trans. Nucl. Sci.*, **NS-38**, 1708 (1991).

WILKINSON 2008. J. WILKINSON, Soft Error Case Study: Implantable Defibrillators and Pacemakers, 2008 IEEE NSREC Short Course Notebook (July 2008).

WOODRUFF 1993. R. L. WOODRUFF and P. J. RUDECK, Three-Dimensional Numerical Simulation of Single Event Upset of an SRAM Cell, *IEEE Trans. Nucl. Sci.*, **NS-40**, 1795 (1993).

WORLEY 1990. E. WORLEY, R. WILLIAMS, A. WASKIEWICZ, and J. GRONINGER, Experimental and Simulation Study of the Effects of Cosmic Particles on CMOS/SOS RAMs, *IEEE Trans. Nucl. Sci.*, **NS-37**, 1855 (1990).

WROBEL 2003. F. WROBEL, J. M. PALAU, M. C. CALVERT, and P. IACCONI, Contributions of SiO_2 in Neutron Induced SEU in SRAMs, *IEEE Trans. Nucl. Sci*., **NS-50**, 2055 (2003).

WROBEL 2006. F. WROBEL, G. HUBERT, and P. IACCONI, A Semi-empirical Approach for Heavy Ion SEU Cross Section Calculations, *IEEE Trans. Nucl. Sci*., **NS-53**, 3271 (2006).

WYATT 1979. R. C. WYATT, P. J. MCNULTY, P. TOUMBAS, P. L. ROTHWELL, and R. C. FILZ, Soft Errors Induced by Energetic Protons, *IEEE Trans. Nucl. Sci*., **NS-26**, 4905 (1979).

XAPSOS 1987. M. A. XAPSOS, L. W. MASSENGILL, W. J. STAPOR, P. SHAPIRO, A. B. CAMPBELL, S. E. KERNS, K. W. FERNALD, and A. R. KNUDSON, Single-Event, Enhanced Single-Event, and Dose Rate Effects with Pulsed Proton Beams, *IEEE Trans. Nucl. Sci*., **NS-34**, 1419 (1987).

XAPSOS 1989. M. A. XAPSOS, W. J. STAPOR, A. B. CAMPBELL, E. L. PETERSEN, and P SHAPIRO, Transient Upset Behavior in a Proton Beam Environment for a CMOS Static RAM, *J. Radiat. Effects Res. Eng*., **7**(1), 77 (January 1989).

XAPSOS 1992. M. A. XAPSES, Applicability of LET to Single Events in Microelectronics Structures, *IEEE Trans. Nucl. Sci*., **NS-39**, 1613 (1992).

XAPSOS 1993. M. A. XAPSOS, T. R. WEATHERFORD, and P. SHAPIRO, The Shape of Heavy Ion Upset Cross Section Curves, *IEEE Trans. Nucl. Sci*., **NS-40**, 1812 (1993).

XAPSOS 2000. M. A. XAPSOS, G. P. SUMMERS, J. L. BARTH, E. G. STASSINOPOULOS, and E. A. BURKE, Probability Model for Cumulative Solar Proton Event Fluences, *IEEE Trans. Nucl. Sci*., **NS-47**, 486 (June 2000).

XAPSOS 2004. M. A. XAPSOS, C. STAUFFER, G. B. GEE, J. L. BARTH, E. G. STASSINOPOULOS, and R. E. MCQUIRE, Model for Solar Proton Risk Assessment, *IEEE Trans. Nucl. Sci*., **NS-51**, 3394 (2004).

XAPSOS 2006a. M. A. XAPSOS, C. STAUFFER, J. L. BARTH, and E. A. BURKE, Solar Particle Events and Self-Organized Criticality: Are Deterministic Predictions of Event Possible? *IEEE Trans. Nucl. Sci*., **NS-53**, 1839 (August 2006).

XAPSOS 2006b. M. XAPSOS, Modeling the Space Radiation Environment, 2006 IEEE NSREC Short Course Notebook (July 2006).

XAPSOS 2007. M. A. XAPSOS, C. STAUFFER, T. JORDAN, J. L. BARTH, and R. A MEWALDT, Model for Cumulative Solar Heavy Ion Energy and Linear Energy Transfer Spectra, *IEEE Trans. Nucl. Sci*., **NS-54**, 1985 (2007).

ZEBREV 2005. G. I. ZEBREV, I. A. LADANOV, A. Y. NIKIFOROV, D. V. BOYCHENKO, V. N. ULIMOV, and V. V. EMELYANOV, PRIVIT—A Heavy Ion Induced Single Event Upset Rate Simulator in Space Environment, RADECS 2005 Proceedings, p. PC13-1 (2005).

ZIEGLER 1979. J. F. ZIEGLER and W. A. LANFORD, Effects of Cosmic Rays on Computor Memories, *Science*, **206**, 776 (1979).

ZIEGLER 1981. J. F. ZIEGLER and W. A. LANFORD, The Effect of Sea Level Cosmic Rays on Electronic Devices, *J. Appl. Phys*., **52**(6), 4305 (1981).

ZIEGLER 1996. J. F. ZIEGLER, *SRIM 96, The Stopping and Range of Ions in Matter*. http://www.srim.org/.

ZIEGLER 1996a. J. F. ZIEGLER et al., IBM Experiments in Soft Fails in Computer Electronics, *IBM J. Res. Dev.*, **40**(1), 3 (January 1996). (27 Authors, It was a big program.)

ZIEGLER 1996b. J. F. ZIEGLER, Terrestrial Cosmic Rays, *IBM J. Res. Dev.*, **40**(1), 19 (January 1996).

ZIEGLER 1996c. J. F. ZIEGLER, H. P. MUHFELD, C. J. MONTROSE, H. W. CURTIS, T. J. O'GORMAN, and J. M. ROSS, Accelerated Testing for Cosmic Soft-Error Rate, *IBM J. Res. Dev.*, **40**(1), 51 (January 1996).

ZIEGLER 1996d. J. F. ZIEGLER, P. A. SAUNDERS, and T. H. ZABEL, Portable Faraday Cup for Non-vacuum Proton Beams, *IBM J. Res. Dev.*, **40**(1), 73 (January 1996).

ZIEGLER 2004. J. F. ZIEGLER and H. PUCHNER, *SER—History, Trends, and Challenges*, Cypress Semiconductor, San Jose, CA 95134 (2004).

ZIEGLER 2008. http://www.srim.org/.

ZOUTENDYK 1984. J. A. ZOUTENDYK, C. J. MALONE, and L. S. SMITH, Experimental Determination of Single-Event (SEU) as a Function of Collected Charge in Bipolar Integrated Circuits, *IEEE Trans. Nucl. Sci.*, **NS-31**, 1167 (1984).

ZOUTENDYK 1985. J. A. ZOUTENDYK, L. S. SMITH, C. A. SOLI, P. THIEBERGER, and H. E. WAGNER, Single-Event Upset (SEU) Model Verification and Threshold Determination Using Heavy Ions in a Bipolar Static RAM, *IEEE Trans. Nucl. Sci.*, **NS-32**, 4264 (1985),

ZOUTENDYK 1986. J. A. ZOUTENDYK, L. S. SMITH, G. A. SOLI, S. L. SMITH, G. E. ATWOOD, and P. THIEBURGER, Empirical Modeling of Single Event-Upset (SEU) in NMOS Depletion-Mode-Load Static RAM (SRAM) Chips, *IEEE Trans. Nucl. Sci.*, **NS-33**, 1581 (1986).

ZOUTENDYK 1987. J. A. ZOUTENDYK, L. S. SMITH, G. A. SOLI, and R. Y. LO, Experimental Evidence for a New Single-Event (SEU) Mode in a CMOS SRAM Obtained from Model Verification, *IEEE Trans. Nucl. Sci.*, **NS-34**, 1292 (1987).

ZOUTENDYK 1987a. J. A. ZOUTENDYK, H. R. SCHWARTZ, R. K. WATSON, Z. HASNAIN, and L. R. NEVILL, Single-Event (SEU) in a DRAM with on-Chip Error Correction, *IEEE Trans. Nucl. Sci.*, **NS-34**, 1310 (1987).

ZOUTENDYK 1988. J. A. ZOUTENDYK, H. R. SCHWARTZ, and L. R. NEVILL, Lateral Charge Transport from Heavy-Ion Tracks in Integrated Circuit Chips, *IEEE Trans. Nucl. Sci.*, **NS-35**, 1644 (1988).

ZOUTENDYK 1989. J. A. ZOUTENDYK, L. D. EDMONDS, and L. S. SMITH, Characterization of Multiple-Bit Errors from Single-Ion Tracks in Integrated Circuits, *IEEE Trans. Nucl. Sci.*, **NS-36**, 2267 (1989).

ZOUTENDYK 1990. J. A. ZOUTENDYK, L. S. SMITH, and L. D. EDMONDS, Response of a DRAM to Single-Ion Tracks of Different Heavy-Ion Species and Stopping Powers, *IEEE Trans. Nucl. Sci.*, **NS-37**, 1844 (1990).

Author Index

Single Event Effects in Aerospace, First Edition. By Edward Petersen.
© 2011 the Institute of Electrical and Electronics Engineers, Inc.
Published 2011 by John Wiley & Sons, Inc.

Subject Index

Single Event Effects in Aerospace, First Edition. By Edward Petersen.
© 2011 the Institute of Electrical and Electronics Engineers, Inc.
Published 2011 by John Wiley & Sons, Inc.

Printed and bound by CPI Group (UK) Ltd, Croydon, CR0 4YY

16/09/2026

14963413-0006